Coatings Technology Handbook

Coatings Technology Handbook

Editor: Falicia Radcliff

NY RESEARCH PRESS

New York

Published by NY Research Press
118-35 Queens Blvd., Suite 400,
Forest Hills, NY 11375, USA
www.nyresearchpress.com

Coatings Technology Handbook
Edited by Falicia Radcliff

International Standard Book Number: 978-1-63238-628-1 (Hardback)

Cataloging-in-Publication Data

Coatings technology handbook / edited by Falicia Radcliff.
    p. cm.
Includes bibliographical references and index.
ISBN 978-1-63238-628-1
1. Coating processes. I. Radcliff, Falicia.
TP156.C57 C63 2019
667.9--dc23

Contents

Permissions

List of Contributors

Index

Preface

Coatings are generally applied to the surface of objects for decorative, protective and functional purposes. There are multiple techniques of coating that can be applied to alter properties of substrates, make surfaces anti-reflective or corrosion resistant, or conducive to the flow of electric current, etc. This book presents the science behind coating, its applications, conceptual developments and new techniques. It is a valuable compilation of topics, ranging from the basic to the most complex advancements in this area. It strives to provide a fair idea about this field and to help develop a better understanding of the latest advancements in the applications of coatings. Scientists and students actively engaged in this field will find this book full of crucial and unexplored concepts.

This book unites the global concepts and researches in an organized manner for a comprehensive understanding of the subject. It is a ripe text for all researchers, students, scientists or anyone else who is interested in acquiring a better knowledge of this dynamic field.

I extend my sincere thanks to the contributors for such eloquent research chapters. Finally, I thank my family for being a source of support and help.

Editor

Thickness Measurement Methods for Physical Vapor Deposited Aluminum Coatings in Packaging Applications

Martina Lindner [1,*] **and Markus Schmid** [1,2]

[1] Fraunhofer-Institute for Process Engineering and Packaging IVV, Giggenhauser Strasse 35, Freising 85354, Germany; markus.schmid@ivv.fraunhofer.de

[2] Chair for Food Packaging Technology, Technische Universität München, Weihenstephaner Steig 22, Freising 85354, Germany

* Correspondence: martina.lindner@ivv.fraunhofer.de

Academic Editor: Massimo Innocenti

Abstract: The production of barrier packaging materials, e.g., for food, by physical vapor deposition (PVD) of inorganic coatings such as aluminum on polymer substrates is an established and well understood functionalization technique today. In order to achieve a sufficient barrier against gases, a coating thickness of approximately 40 nm aluminum is necessary. This review provides a holistic overview of relevant methods commonly used in the packaging industry as well as in packaging research for determining the aluminum coating thickness. The theoretical background, explanation of methods, analysis and effects on measured values, limitations, and resolutions are provided. In industrial applications, quartz micro balances (QCM) and optical density (OD) are commonly used for monitoring thickness homogeneity. Additionally, AFM (atomic force microscopy), electrical conductivity, eddy current measurement, interference, and mass spectrometry (ICP-MS) are presented as more packaging research related methods. This work aims to be used as a guiding handbook regarding the thickness measurement of aluminum coatings for packaging technologists working in the field of metallization.

Keywords: PVD; aluminum; quartz micro balance; optical density; AFM; electrical conductivity; eddy current; interference; ICP-MS; metal coating; nano-scale coatings

1. Introduction

As early as 1994, researchers were looking for the absolute minimum of material usage for disposable packaging, pursuing the need for environmental protection. Additionally, mono materials are preferred due to easier recyclability [1]. However, pure polymeric materials often fail to fulfill the barrier requirements needed to sufficiently protect packed goods, e.g., food against light, moisture, oxygen, and other gases. In contrast to that, aluminum foil with a thickness of about 6 to 40 μm has extremely high barrier properties. Therefore it is predominantly used for the packaging of highly sensitive pharmaceutical products in flexible packaging or blister packs [2,3]. Yet, aluminum has a rather negative environmental impact, which is why we must strive to further reduce its amount [4]. In order to maintain the high barrier properties of aluminum while simultaneously minimizing material usage, nanometer-thin aluminum coatings are applied on polymers via PVD (vacuum evaporation). In this process, aluminum is heated until it evaporates in a vacuum chamber. The polymer substrate is moved across the aluminum gas cloud so that the metal condenses on the polymer surface. In this way, thicknesses of only a few nanometers can be realized. However, in order to reach suitable barrier properties, an approximate thickness of 40 nm of aluminum is commonly necessary. Apart from pure aluminum, aluminum oxide and silicon oxide are common inorganic coating materials (ceramics). In the following review, we refer to aluminum only. The aluminum coating will be referred to as aluminum or coating. When writing about a single atomistic or molecular layer in the aluminum coating, this is referred to as a layer. The substrate polymer is named a substrate or polymer.

1.1. Deposition Techniques in Packaging Applications

In general, according to Seshan [5], all deposition techniques can be subdivided into evaporative methods (such as vacuum evaporation), glow-discharge processes combined with either sputtering or plasma processes, gas-phase chemical processes with either chemical vapor deposition or thermal forming processes, and liquid phase chemical techniques combined with either electro processes or mechanical techniques. Thin film application techniques are widely used in industries like microelectronics, photovoltaic devices, and optics. Apart from that, especially vacuum evaporation is used in the packaging industry. However, other methods find their way into the packaging market, such as CVD (chemical vapor deposition), PECVD (plasma-enhanced chemical vapor deposition), ALD (atomic layer deposition), magnetron sputtering and sol-gel coating. Apart from financial and production speed considerations, the decisive factor for choosing one of the methods will be the required characteristics of the coating. The most important task of the coating in the packaging industry is to achieve a suitable barrier against gases. Thus, the coating must be pure and not contain cracks or pores [5]. In the following, vacuum deposition and other coating methods that appear in the packaging industry are described.

During vacuum deposition, commonly only one inorganic material can be deposited. This happens by heating the coating material and the subsequent condensation on the substrate surface. However, if alloys are used, the heating of the target material might lead to a disintegration of the compound. Moreover, undercuts and roughness are difficult to coat. Furthermore, the process requires high-performance cooling and vacuum systems. The vacuum deposition process has the advantage that due to the vacuum the energy to melt the material is reduced, and furthermore, the incorporation of gaseous atoms into the coating is drastically reduced. The deposition rate is highly dependent on the gas pressure in the chamber. The flux distribution can be altered by the source geometry. In comparison to other methods, high deposition rates and extremely pure coatings can be obtained [6].

Sputtering is a process sometimes used in packaging applications, in which atoms from the target are attacked by accelerated ions, which transfer their momentum to the targeted atoms. Thus the targeted atoms are knocked out from the bulk, leave it in a cosine distribution, and then hit the substrate surface and adhere there. The magnetron sputtering technique has the advantage, that secondary electrons generated in the target are trapped so that they cannot hit the substrate surface. Sputtering has a high amount of materials usable for deposition and typically the coating has the same composition as the target material. However, if reactive gases are present in the chamber, they can react with the target and alter the composition. Moreover, the particles have a higher kinetic energy compared to thermally vaporized atoms. It is possible that temperature-sensitive substrates can be coated without excessive temperature impact [6].

In contrast to that, CVD offers the opportunity to introduce precursors into the process chamber, which then react with the metal to form, e.g., metal oxides, nitrides, carbides, borides and others. However, in this process, volatile and partially toxic gases may be produced, which have to be handled separately. At the same time, powerful vacuum pumps are not necessary like in vacuum deposition. As the impacting molecules or atoms have a high kinetic energy, this process might lead to an increase in the substrate temperature. Moreover, the interaction with precursor atoms may lead to a scattering of the evaporated atoms. This in turn can lead to a roughening of the surface, the penetration of atoms or molecules into the material, pinholes and chemical reactions with the rest gases. However, the achieved thicknesses can be up to centimeter scales. As the coating is not limited to line-of-sight areas, the process has the advantage that three dimensional structures, voids, and peaks can also be covered evenly [7].

ALD is a layer-by-layer process in which only one monomolecular layer is applied in each production cycle. It is based on alternate pulsing of the precursor gases onto the substrate surface, followed by the chemisorption or surface reaction [8]. The advantage is that the produced layers are perfectly dense and free of pinholes. Therefore, they offer extremely low gas permeabilities. Moreover, temperature and pressure conditions are less intense than in vacuum deposition [9]. Although lately a continuous operation mode was developed, the production speed is rather low and therefore costly.

One example for liquid phase chemical techniques in packaging applications is Ormocer®, i.e., organically modified ceramics [10]. Those are applied via a sol-gel process in order to achieve a nanometer thin coating [11,12]. The very basic form of a sol-gel process is the draining and evaporation of the solvent, followed by condensation reactions [13]. This coating leads to smoothening of the surface of polymers. Like this, the negative impact of surface inhomogeneities on barrier performance can be reduced [14]. The advantage in comparison to other processes is the lower amount of required equipment and lower costs. Moreover similar to CVD it provides the possibility to tailor the microstructure of the coating [13].

1.2. Application of Aluminum via Vacuum Evaporation and Layer Growth

Physical vapor deposition is not only used for packaging materials, but also e.g., for capacitor films, holographic coatings, transparent conducting oxides, energy conservation windows, solar cells and absorbers, flexible circuits, or thin film batteries [15]. The basic construction for a vacuum evaporation deposition chamber is illustrated in Figure 1. The chamber is divided into two parts, where chamber (A) includes the unwinding (C) and rewinding (D) of the substrate web and holds a low pressure of about 1×10^{-3} mbar. The lower part (B) is set under a vacuum of about 1×10^{-5} mbar. There are lots of data available about aluminum vapor pressure curves, onset, offset and melting temperature as values change a lot depending on the exact metal composition [16]. However, it can be said that the vacuum reduces the evaporation temperature of aluminum from approximately 2742 °C at 1013 mbar to 813 °C at 1×10^{-5} mbar [17,18]. Moreover the vacuum avoids the scattering of aluminum atoms and their reaction with other gas atoms or molecules (e.g., with oxygen to aluminum oxide). The chilled process roll (I) together with the conductance rolls (E) separate the chamber into (A) and (B), leaving a small opening for the substrate to pass from one zone to the other. The process roll (I) is positioned above the evaporator (F). Two main principles are available for evaporation. The aluminum could either be fed as a wire onto a resistance heated boat, from where it evaporates. Otherwise the aluminum could be fed as granulate in a target and then be heated via an electron beam. In each case the aluminum evaporates and condenses on the surface of the substrate web, which is moved across the process roll (I). The thickness of the aluminum can be adjusted by the web speed or the evaporation rate. The evaporation rate is regulated via the energy input in the evaporator in combination and/or the speed of aluminum feed. The thickness is commonly monitored either by a quartz micro balance (QMB) right at the place of evaporation (G) or by the measurement of optical density (H) before rewinding [15].

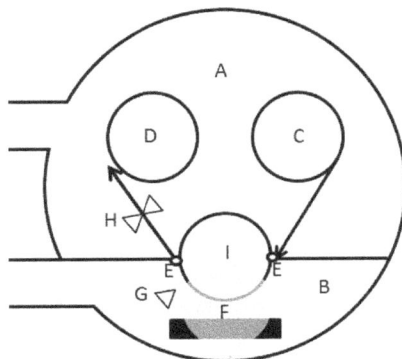

Figure 1. Basic construction of a vacuum deposition chamber (adapted from [1]).

The growing of condensed atoms to a closed coating and the developing micro structure are dependent on various parameters. The kind of substrate as well as its orientation, evaporation temperature and rate play an important role. Coating thickness, angle of deposition and the energy of condensed atoms or molecules also affect the material structure [19]. Apart from that, the substrate has a major effect on the formation of the coating. In case dust is present on the polymer surface, these particles might be coated with aluminum but then fall off after deposition. In this case, the particle leaves behind a non-coated pinhole, which reduces the gas barrier. One other issue is the substrate roughness. According to [15] and others, valleys in the surface will lead to preferential nucleation and epitaxy as these valleys are energetically favorable for aluminum atoms. Consequently, layer growth is not evenly distributed but rather inhomogeneous. One technique to overcome this problem is the biaxial orientation of polymeric substrates, which leads to the reduction of surface roughness due to the stretching and smoothening of polymeric chains for PET [20]. Generally it is assumed that by stretching the polymer chains align, so that the crystallinity of the material increases and the surface is smoothened [21]. Conversely, a roughening effect due to stretching was reported in [22]. The negative effect of substrate roughness on the gas barrier of inorganic coatings is described in [23,24]. Moreover, the attachment of atoms on the polymer surface depends on the surface tension of the substrate. Surface tension can be increased by plasma treatment, by which e.g., carbonyl, carboxyl, hydroxyl, peroxide, and other groups are introduced onto the polymer surface. There is a tendency that with higher surface tension, more atoms attach on the surface and thus increase the barrier [24]. Additionally, it was reported that plasma treatments can simultaneously alter the surface geometry, as low weight molecules migrate to the charged areas on the surface and consequently built peaks [25].

Layer growth starts with the first nuclei of condensed atoms, which define the subsequent layer structure. The arriving atoms can either deflect from the substrate or lose enough kinetic energy that they are loosely bound as adatoms on the surface. Those adatoms predominantly bind to existing material clusters or to other energetically favorable places. Those could be steps, edges or cavities in the surface topography. If the residual energy is high enough, the atoms further diffuse on the substrate surface. Like this, single adatoms can form stable or metal stable clusters [26,27].

Generally, three models are described, which cover a variety of possible interactions between substrate and adatoms of which model in Figure 2a,c are the two extremes and Figure 2b is a combination of both [26–29]. In the model of Frank van der Merwe (Figure 2a), the cohesion between the adatoms is weaker than the binding to the substrate. Then, a monolayer grows, which is first fully closed, before the next layers start to grow. In comparison to that, the Stranski-Krastanov model (Figure 2b) introduces the idea that first single layers grow like in (Figure 2a), on which islands then start to appear. The Volmer-Weber model (Figure 2c) goes to the other extreme and assumes that the interactions between adatoms are stronger then the interaction with the substrate. After the nucleation of small clusters, these grow to small three-dimensional islands. Only at high layer thicknesses can closed coating be achieved [29–32].

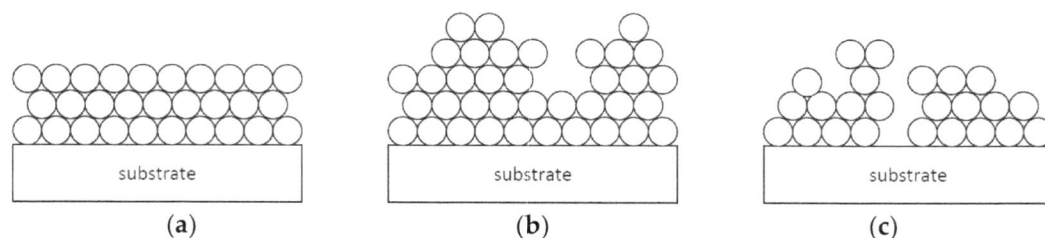

Figure 2. Models for layer growth: Frank van der Merwe (**a**); Stranski-Krastanov (**b**); Volmer-Weber (**c**) (adapted from [1,15]).

1.3. Pores and Defects

Normally, metals have a crystalline structure, i.e., a strictly periodical arrangement of atoms. Therefore, Bravais [33] introduced the concept of a space lattice, which equals a three-dimensional mathematical point pattern. Herein, each point can be imagined as the center of an atom. Metallic elements mostly crystallize in three lattice types [34]: cubic body centered, cubic face centered, and hexagonal. Aluminum typically crystallizes in a cubic face centered manner [35]. However, crystalline structures are never free of defects. Defects are subdivided into (i) blank spaces and interstitials; (ii) displacements; (iii) grain and phase limits [34,36] and can affect physical properties like barrier towards gases, electrical conductivity, and optical density.

Apart from the defects due to the lattice, additional macroscopic defects might appear due to contamination with dust prior to the deposition process. These dust particles might stick on the substrate surface, are then covered with aluminum and fall off the substrate afterwards as they are only loosely bound. In such a case, this leaves behind a defect in the coating. Additionally some films contain so-called anti-block particles on one side of the film. When the material is winded, the backside of the film touches the front side of the film. Consequently, the anti-block particles can injure the inorganic nano-scale aluminum coating [37].

1.4. Formation of Aluminum Oxide

Aluminum is a reactive metal that easily oxidizes when it is taken out of the vacuum chamber and set under atmospheric conditions. Hence it builds up a stable oxide layer that has an amorphous structure and protects the underlying metal from further corrosion and oxidation. Then the oxide layer is hydrated to aluminum oxide hydroxide or aluminum hydroxide [38,39]. The oxide layer has a thickness of 3 to 10 nm [39–42]. In [42] it was observed, that the oxide layer grows not only on the outer surface but also between the aluminum and the substrate web.

1.5. Permeation through Organic Substrates and Inorganic Coatings

A barrier is defined as the resistance against permeation, i.e., against the mass transfer of gaseous substances through a solid body. The process of permeation involves four main stages: adsorption, absorption, diffusion and desorption [43]. In the first step the molecules are adsorbed onto the substrate surface of the solid material and build up a thin molecule layer by adhesive power. Then, permeating molecules are absorbed and transported through the polymer. The permeation coefficient (P) of molecules is determined by two factors, (S) and (D) (Equation (3)). The solubility coefficient (S) describes the concentration (c) of dissolved molecules in the polymer in dependence of the partial pressure (p) (Equation (1)). The diffusion coefficient (D) describes how fast molecules permeate along the concentration gradient (Δc) using intermolecular and intramolecular spaces until a solution equilibrium is established. The diffusion flux (J) of molecules permeating through a homogenous polymer over a certain distance (Δd) along the concentration gradient is explained by Fick's first law of diffusion (Equation (2)). Finally molecules are desorbed from the polymer surface and evaporated or removed by other mechanisms. The overall permeability of a polymer (Q_{poly}) is then described by the permeation coefficient (P) and the thickness of the material (d) (Equation (4)).

$$c = S \times p \tag{1}$$

$$J = -D \times \frac{\Delta c}{\Delta d} \tag{2}$$

$$P = D \times S \tag{3}$$

$$Q_{poly} = \frac{P}{d} \tag{4}$$

The barrier of inorganic coatings is subjected to its chemical composition, the microstructure and homogeneity [19,26]. Although a thin evaporated inorganic coating clearly improves the barrier properties of polymer substrates, it still shows a permeability that is several magnitudes higher than it would be for a perfectly crystalline material. This can be partially explained by the layer growth and the connected degree of imperfection. Moreover, contamination during the evaporation by other gases or dust, anti-block particles in the substrate, as well as mechanical tensions between the substrate and the inorganic coating play an important role. All these factors may lead to cracks or pores in the coating and thus increase the gas transmission of the polymer and coating ($Q_{poly+coat}$) [44,45]. In relation to the pore size, different mechanisms for permeation are presented in [46]. Nevertheless, the diffusion through macroscopic defects (100 nm) plays a major role [47]. Given such macroscopic defects, permeation continues as if there was no inorganic coating. According to the current state of the art, a completely defect free inorganic coating cannot be produced by vacuum deposition [48]. However, the determination of the amount and size of pores is rather complex and time-consuming. This is why the barrier improvement factor (*BIF*) is commonly used to describe the quality of the inorganic coating. This value sets into relation the permeabilities of the polymer with ($Q_{poly+coat}$) and without (Q_{poly}) the inorganic coating. It is affected by the amount and size of defects as well as the thickness of the underlying polymer substrate:

$$BIF = \frac{Q_{poly}}{Q_{poly+coat}} \tag{5}$$

Herein, the permeability of the polymer with inorganic coating ($Q_{poly+coat}$), actually consists of the transmission through all defects, assuming that the rest of the surface does not let pass any gas. Trying to describe the permeation through these defects, various models were developed. The first models were created by Prins and Hermans [44], based on which further ideas were presented by e.g., [49–53]. For technically relevant substrate thicknesses and defect sizes, the models can be rewritten approximately as in Equation (6). Herein it is visible, that the transmission $Q_{poly+coat}$ does not depend on the substrate thickness (within the range of validity), but only on the permeability coefficient (*P*) of the material, the amount of the defects per area (n_d) and the effective average of defect area (*a*) [37]:

$$Q_{poly+coat} \approx 2 \times P \times n_d \times \sqrt{a} \tag{6}$$

1.6. Importance of Coating Thickness for Barrier Properties

The thickness and consistency of nanodeposited layers (nanocoatings) have a high impact on their performance [54]. Several studies [37,50,55] showed that the permeability $Q_{poly+coat}$ negatively correlates with the increasing thickness of the inorganic coating until a certain point. Then, even for higher thicknesses, $Q_{poly+coat}$ stays almost constant. In [37] it was shown that only for thicknesses that are one to three magnitudes higher the permeability further decreases. It was concluded, that the idea of Volmer-Weber growth helps us to understand the steady decrease of permeability for low coating thicknesses. Nonetheless, the few existing investigations that have been made to analyze the coating structure rather indicate a Frank-van-der-Merwe growth.

Figure 2 raised the question of how the thickness of an inorganic coating should be defined. Does "thickness" take blank spaces or defects into account or not? Is "thickness" an average value over a broader surface or is it measured at a certain distinct point? Does "thickness" include aluminum oxide or only pure aluminum? Accordingly, when having a look at various publications [37,39,42,56–60], the thickness is measured by numerous different methods. Because the thickness of these aluminum coatings is much shorter than the wavelength of visible light, traditional microscopy is not usable for this application.

As stated by Mattox [61], these methods can be subdivided into mass, geometrical, and property thicknesses. Mass thickness is measured in $\mu g/cm^2$ (e.g., by mass spectroscopy) but does not take into account the density, micro structure, composition, or surface morphology. In contrast, the geometrical thickness in measured in μm, nm, or Å (e.g., AFM, profilometry). This value is affected by the surface morphology (e.g., roughness). Just like the mass thickness, geometrical thickness does not consider composition, thickness, or microstructure. By measuring property thicknesses, a physical value is obtained (e.g., electrical resistance, optical density, interference) which is then mathematically related to the coating thickness (see Figure 3).

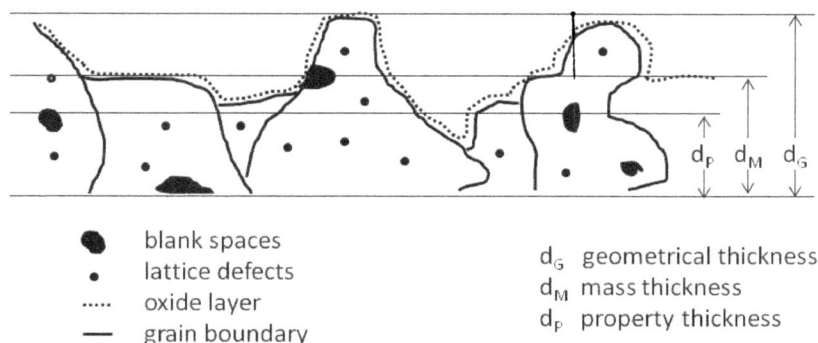

blank spaces
lattice defects
..... oxide layer
— grain boundary

d_G geometrical thickness
d_M mass thickness
d_P property thickness

Figure 3. Mass, geometrical, and property thickness.

In the following sections, the quartz micro balance (QMB), mass spectroscopy (ICP-MS), atomic force microscopy (AFM), optical density (OD), interference, electrical surface resistance and eddy current measurements are presented as examples for the methods explained above.

For each technique the theoretical background, the method, and the analyses are explained. Limitations and especially parameters affecting the measured values are highlighted. Yet, especially concerning limitation values, equipment specifications will give more specialized information to the user. This work aims to be used as a guiding handbook for packaging technologists working in the field of metallization. For more detailed information, literature hints are provided for the interested reader.

2. Characterization Techniques

2.1. Mass Thickness

There are various measurement techniques available today, for determining mass thickness, such as X-ray fluorescence, ion probe, radioactivation analysis, chemical balance, micro balance, or torsional balance [62]. However, the most common method in the field of PVD is the quartz crystal micro balance (QCM), with which the evaporation rate is usually monitored [63]. Taking into account the coating speed and geometrical constraints under which the QCM, the evaporation source and the substrate are arranged, the coating thickness can be calculated in g/m^2. The second method presented here is the analyses via ICP-MS, which is a certain kind of mass spectrometry. This, of course, can only be used when the sample is inserted as a liquid. Therefore, the evaporated aluminum needs to be dissolved from the substrate of distinct area before the amount of aluminum in the sample can be determined.

2.1.1. QCM

QCM: Theory

When alternating current (AC) is applied on to the gold electrodes, which are evaporated on a piezoelectric quartz crystal, the crystal starts to oscillate in its resonance frequency (see Figure 4). A quartz with a thickness of 3.317×10^{-4} has a resonance frequency of 5 MHz [64]. For doing so, the crystal needs ideally to be cut from a mono crystal at an angle of $35.1°$ toward the optical axis, so that it oscillates in thickness shear mode. The frequency of this oscillation depends on the geometry of the crystal and therefore drops when the thickness increases [65,66]. By monitoring the change in frequency and relating it to the deposited mass, this effect is used to gain information about the amount of aluminum that is evaporated on the QCM [67].

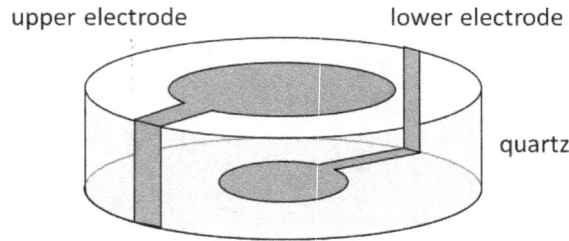

Figure 4. Typical QCM geometry, diameter ca. 10–30 mm, thickness ca. 0.3 mm (adapted from [64]).

The relation between frequency and mass is based on physical principles, as explained in the following. Commonly, the software will use mathematics to export data about the thickness.

Under the assumption, that the evaporated material (aluminum) behaves similar to the quartz and that the evaporated mass is small in comparison to the quartz (about 2%), the resonance frequency changes linearly with the mass increase. Resonance appears, when the thickness of the quartz (d) is half of the transversal wavelength (λ) (Equation (7)). The speed of sound (v_s) in quartz is a constant and dictates possible wavelengths (λ) and frequencies (f) (Equation (8)). Consequently, (d) and (f) always have the same proportion, as (v_s) is constant (Equation (9)). When the evaporated material arrives on the surface, the thickness and mass increase. Therefore, Sauerbrey [68] introduced the mass density (M_E) for the evaporated material and for the quartz material (M_Q) which is defined by the thickness (d) and density (δ) as in Equations (10) and (11). Accordingly, when the thickness (d) increases, also the mass density (M_E) increases. Consequently, the change in frequency (Δf) is related to the mass density (M_E) as in Equation (12). By combining Equations (8), (11) and (12), Equation (13) can be derived and it becomes obvious, that (Δf) directly relates to the mass density of aluminum (M_E), as (v_s), (δ_Q) and (f) are constant. This correlation was first mentioned by Sauerbrey [68] and is therefore denoted as the Sauerbrey-equation [66]. For further improvement of the mathematical model, the impedance values for each plane can be introduced when modeling them as coupled resonators from two planes. A precise overview is given in [65].

$$d = \frac{\lambda}{2} \tag{7}$$

$$v_s = \lambda \times f \tag{8}$$

$$\frac{\Delta d}{d} = -\frac{\Delta f}{f} \tag{9}$$

$$M_E = d_E \times \delta_E \tag{10}$$

$$M_Q = d_Q \times \delta_Q \tag{11}$$

$$\Delta f = -f \times \frac{\Delta M_E}{M_Q} \tag{12}$$

$$\Delta f = -2f^2 \times \frac{M_E}{\delta_Q \times v_s} \tag{13}$$

QCM: Method

The QCM is commonly integrated in the machine at some point close to the evaporation source and the process roll. It has to be taken into account, that at the place where the aluminum is deposited on the QCM, there is a shadowing effect so the aluminum will not reach the substrate surface. Additionally, the angle (α) and distance towards the evaporation source should be taken into account, as this might alter the measured results (see Figure 5).

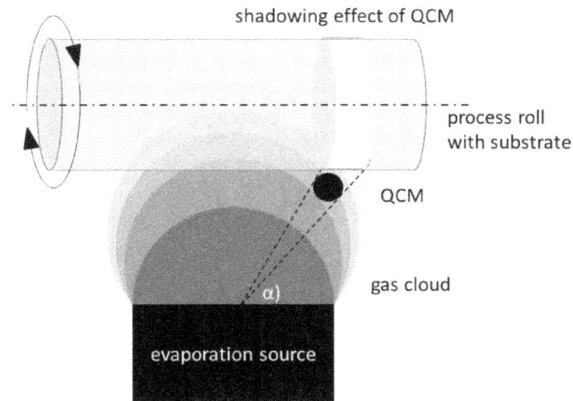

Figure 5. Arrangement of a QCM in the vacuum deposition unit.

QCM: Practical Aspects and Analyses

The software commonly needs the material density as input and its output will commonly be the deposition rate in Å/s. The accuracy is about 2%. The accuracy decreases with increasing thickness, which is why there are a few methods to reduce this effect. One option is to use a shutter, which only lets aluminum pass on the QCM in a certain frequency. Or a filter could be used, which only lets a certain number of atoms pass through to the QCM. Like this, the actually arriving amount of aluminum can be extrapolated, while at the same time the layer thickness on the QCM is reduced. Of course, this will reduce the accuracy of the QCM [15,67,69]. It has to be taken into account that the value obtained is the deposition rate and has the unit Å/s. The faster the material moves, the lower the coating thickness will be. The coating thickness (t) can be approximated by the following Equation (14), taking into account the deposition rate (x), the web speed (w) and the shutter width (y):

$$t = x \times \frac{w}{v} \tag{14}$$

2.1.2. ICP-MS

ICP-MS: Theory

ICP-MS (inductively coupled plasma mass spectrometry) is based on the chemical dissolution of the aluminum and the subsequent measurement of the concentration of aluminum in the dissolution. There are quite a few different ICP methods available for detecting unknown substances. The major concurring method that needs to be mentioned is ICP-OES (inductively coupled plasma optical emission spectrometry), in which electrons are excited, and the photons that are emitted when the electrons fall back to their ground state, are detected. In comparison, in ICP-MS, the plasma is used to ionize atoms and the ion charge is used to detect them. However, even for ICP-MS there are various equipment designs available today, of which three are outlined. An extensive overview is given in [70].

The ICP-MS basically consists of three main parts: the probe- and plasma unit, the cones, and the mass spectrometer. Firstly, the liquid probe is conveyed over the pump and the vaporizer into the spray chamber. The small droplets are then transferred into the plasma unit. The plasma unit involves an induction coil that produces a high-frequency electromagnetic field. Argon is inserted in this field and becomes plasma, which is "seeded" with electrons [70]. When the small gas droplets reach the plasma, they are dried, decomposed, atomized and ionized. Afterwards, the positively charged ions pass through two concentric quartz cylinders (sampling cone, skimmer cone) where non-ionized atoms are excluded. The cones also lead to a focusing of the ion beam (Figure 6).

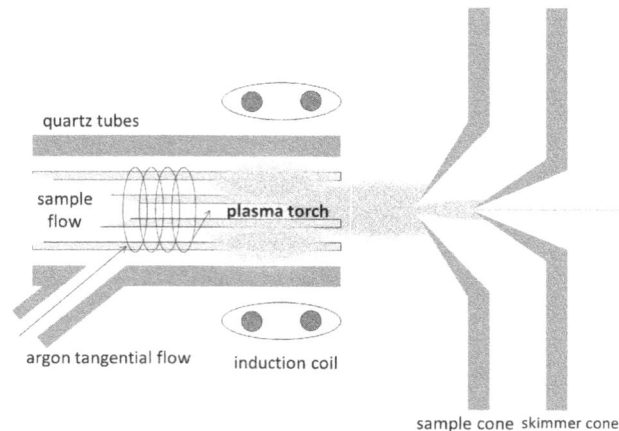

Figure 6. Plasma unit and torch in ICP-MS (adapted from [70–72]).

In a mass spectrometer, three basic principles are used, namely quadrupole filters, magnetic analyzers, and time-of-flight analyzers. In the first case, the ions pass a quadrupole or even octopole reaction system. This element consists of four or eight dipoles with opposite directions. On the opposing dipoles, a combination of DC and AC potentials is applied. A positively charged ion will be torn towards the negatively charged rod and discharged, unless the voltage changes before the collision. When the voltage changes before collision, the ion changes its direction. Like this, depending on their mass to charge ratio (m/z) ions start to oscillate and can only pass the quadrupole if they have a distinct m/z-value for a given frequency (Figure 7) [65,71]. All other ions will impact on the surface of the quadrupole or the sled wall and are discharged. Within one measurement, typically one mass is detected after another. Additionally, a helium collision mode is available. In this mode, helium is inserted in the quadrupole and leads to collisions with the sample ions. As molecular ions have bigger dimensions, they tend to collide more often with helium atoms and consequently lose kinetic energy (KED, kinetic energy discrimination). Like this, they are rather easily torn towards the quadrupole rods and can be excluded more easily while passing the magnetic field.

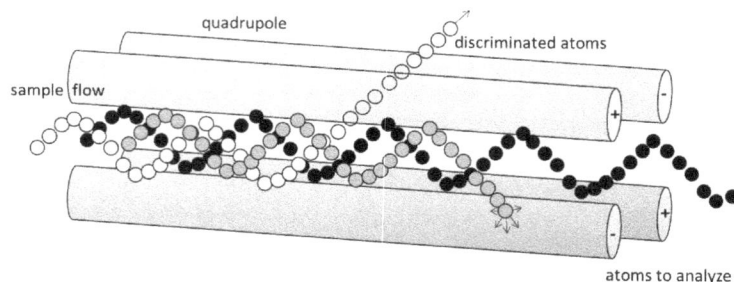

Figure 7. Quadrupole filter (adapted from [70–72]).

In a magnetic analyzer, the ions pass an electric field. When a particle, charged with the load (q) (Equation (15)) is accelerated in an electric field, its kinetic energy (E_k) and velocity (v) depend on the voltage (U) (Equation (16)). When this accelerated ion passes a magnetic field that is perpendicular to the trajectory, it is deflected and follows a circular path (Figure 8), on which the centrifugal force equals the magnetic force (Equation (17)). The magnetic force is defined by the magnetic field strength (B). By combining Equations (16) and (17), Equation (18) is deducted. It becomes obvious, that the radius (r) is related to the m/z-ratio and that by a local dissolution of impinging ions on the detector, the abundance of each m/z-specie can be observed [73].

$$q = z \times e \qquad\qquad (15)$$

$$E_k = \frac{m \times v^2}{2} = q \times U \tag{16}$$

$$q \times v \times B = m \times \frac{v^2}{r} \tag{17}$$

$$r = \frac{\sqrt{2 \times m \times E_k}}{q \times B} \tag{18}$$

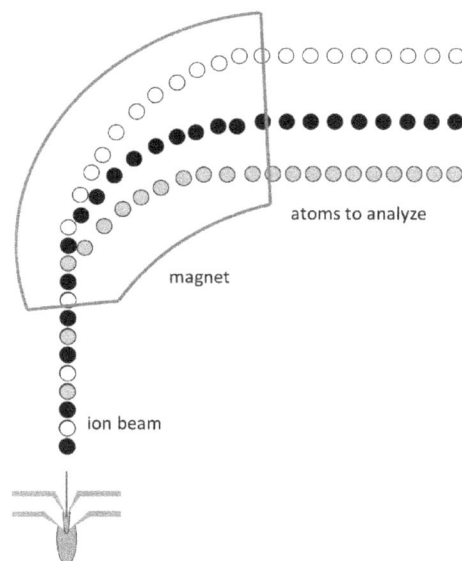

Figure 8. Magnetic analyzer (adapted from [70]).

In a time-of-flight measurement the ions are first accelerated in an electric field and then enter a field free region (Figure 9). As already revealed in Equation (16), the speed at the end of the electric field depends on the mass (m) and charge (q). As all ions have the same charge, the speed (v) is defined by the mass (m). For then passing a certain distance (L), the ions will need the time (t) at a given velocity (v) (Equation (19)). By combining Equations (15), (16) and (19), the time of flight (t) can be used to calculate the m/z-ration of ions, when (L) and (U) are kept constant (Equation (20)) [71]. An overview of interferences between atoms with the same m/z-ratio is given in [74].

$$L = v \times t \tag{19}$$

$$t^2 = \frac{m \times L^2}{z \times e \times 2 \times U} \tag{20}$$

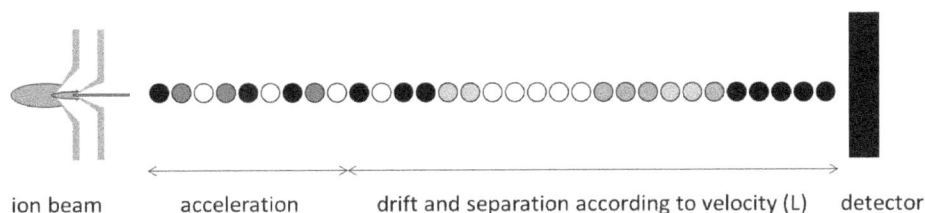

ion beam   acceleration   drift and separation according to velocity (L)   detector

Figure 9. Time of flight measurement (adapted from [70]).

After the mass spectrometer, the ions need to be detected and quantified. As detection systems, three main working functions are available. Either the charge is directly measured (Faraday cup), or the kinetic energy transfer (which depends on the mass and velocity of the ions) leads to secondary electrons which are detected. In each case, the impacting ions lead to an electric current that is proportional to the amount of ions that are counted per second. Correspondingly, the resulting value is CPS (count per second) [71].

A quantitative identification of atoms in a probe is possible by standard dissolutions, as there is a linear relationship between the signal intensities of the ions (counts per second, CPS) and the concentration of the element [72,73].

ICP-MS: Method

For calibration, an aluminum standard dissolution is used. This should be diluted with double distilled water to concentrations of, e.g., 0.25, 0.50, 1.00, 1.50, 2.00, and 3.00 mg/L aluminum. The software automatically calculates the counts per second (CPS), which is associated with the relating dilution and draws a standard graph (Figure 10).

For the analysis of the aluminum content of the samples, those should be cut to a defined surface (A), e.g., 10 cm^2. In the next step, the aluminum of each sample is dissolved by a defined factor (f_1) in 1 molar NaOH, e.g., 50 mL. After dissolving the aluminum for a certain time, the sample can again be diluted with double-distilled water by a factor (f_2). The concentration (c) in µg/L of this double diluted sample is then analyzed in the ICP-MS as counts per second (CPS) and then related to the concentration by the standard graph.

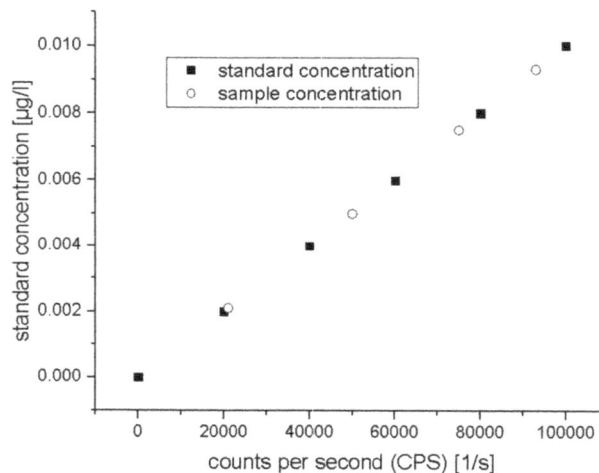

Figure 10. Example of a standard graph for correlation of CPS and standard concentrations.

In order to reach trustworthy results, the following needs are formulated in [70]:

- Complete dissolution;
- Highly pure reagents;
- No chemical interaction between equipment and reagents;
- No loss of analyte.

When it comes to the interpretation of the measured values, it needs to be considered that this value is related to a certain area of the tested material. Whereas the thickness determined via QCM is only valid for a certain location in relation to the evaporation source, the thickness determined via ICP-MS is more of an average value for a whole area. When the measurement area is taken from the same place, where the QCM is usually measuring, the values should be comparable. However, it cannot be compared to, e.g., the area right above the evaporation source, as the thickness will be higher there due to the cosine distribution.

ICP-MS: Practical Aspects and Analyses

Under consideration of the dilution factor (b_1) and (b_2), the sample surface (a_s) and the theoretical density (δ) of aluminum, the thickness (d) of the evaporated coating can be calculated according to Equation (21).

$$d = \frac{c \times b_1 \times b_2}{a_s \times \delta} \tag{21}$$

When using this value, it should be considered that this method determines all the aluminum in the sample: aluminum existing as oxide as well as the pure metal, which might explain the deviations from other methods [42]. Additionally, interferences between atoms with the same m/z-ratio might alter the results. An overview of those is given in [74]. Limits of detectable concentrations (c) for diverse elements are given in [71] and range from 0.001 to >10 µg/L. For aluminum, the range is 0.001–0.1 µg/L. This range can further be exploited by increasing or decreasing the dilution factors (b_1) and (b_2) or sample surface (a_s), as given in Equation (21).

2.2. Geometrical Thickness: AFM

AFM is a contact profilometry method. In contrast to contact profilometry, non-contact profilometry uses electrons or photons for scanning the surface and gathering information about the depth profile. Contact profilometry methods are mechanical stylus profilometers, AFM and STM (scanning tunneling microcopy). In the latter, a cantilever is brought in such a short distance to the surface, that electrons start to tunnel and the current is measurable. As the current is then proportional to the distance, a surface topography image can be obtained. The AFM is based on a mechanical scanning of surfaces with the help of a cantilever with a sharp probe on top. This probe is moved across the surface line by line. In consonance with the surface topography, the cantilever is deflected and the extent of bending can be measured with capacitive or optical sensors. This information is translated into a surface topography image.

2.2.1. AFM

AFM: Theory

When the probe approaches the surface, depending on the distance, different interaction forces superimpose. Van-der-Waals forces which are indirect proportional to $-(r^6)$, appear due to charge transfer and act attractive. At smaller distances, orbital overlaps produce repulsive forces. They are indirect proportional to r^{12}. When superimposing both potentials, the resulting Lenard-Jones-potential shows the dependency of attractive and repulsive forces of the distance between probe and surface (Figure 11). As the probe approaches the surface, the cantilever ideally shows no bending. Then a "snap in" occurs, when attractive forces start to dominate. At this point, the probe starts to touch the surface. When the cantilever is further pressed against the surface, it is deflected and the repulsive forces dominate. Due to this effect, not only the imaging of topography but even the resolution of single atom orbitals is possible. For operating the AFM two different procedures are possible, one of which is in the field of "attractive forces" and the other one in the area of "repulsive" forces [75].

AFM: Method

Basically, three different modes are available for AFM measurements: contact mode, non-contact mode, and intermitting mode. In the contact mode with constant height, the probe is moved so close to the surface that it is bent. According to the extent of bending (repulsive forces), the surface topography is imaged. However, in this mode, the surface and probe might be injured and hence the nature of the surface might affect the results.

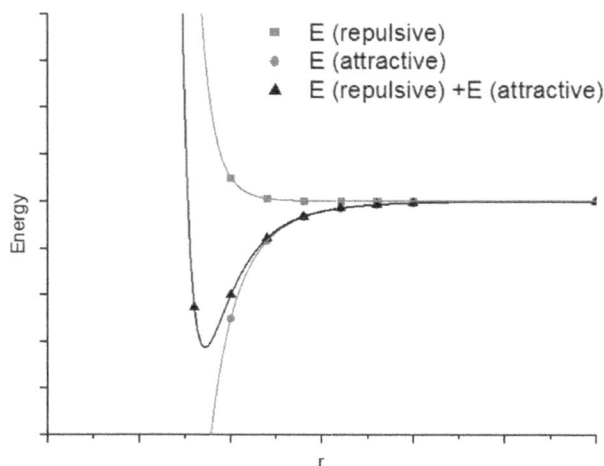

Figure 11. Lennard Jones potential (adapted from [76]).

In the contact mode with constant force, the deflection of the cantilever is adjusted permanently to a set point that is especially useful for soft matter. In order to optimize the outcome, the set point needs to be adjusted in dependence of the cantilever stiffness and the surface nature. Information about the topography can then be extracted from the adjusted height of the cantilever. The advantages of this mode are the high resolution and that the probe and surface are protected.

In the non-contact mode, the cantilever oscillates at its resonant frequency. As the cantilever comes near the surface, it interacts with the force field (attractive forces), which quenches the oscillation and affects frequency and amplitude. Based on the phase and amplitude shift, information about the surface topography is deduced.

In the intermitting mode, the cantilever oscillates just like in the non-contact mode. Still, the distance between probe and surface is continuously adjusted to a constant level by keeping the interaction steady [75]. The advantage is that, apart from topography, stiffness and adhesion can also be determined by separately capturing attractive and repulsive forces [77]. As stated by Eaton and West [75], more than 20 different modes of AFM are available today. Therefore only the basic principles are outlined in this review.

All modes use the same basic function, as explained in the following. The surface of the sample is scanned line by line. Therefore, either the sample or the cantilever can be moved. As described before, based on the interaction between cantilever and sample surface the cantilever is deflected vertically. While the probe moves across the surface, the bending of the cantilever is measured by optical sensors (Figure 12). In this optical sensor a laser beam impinges on the backside of the reflective cantilever and the movement of the reflected light spot is measured by a photodetector. Thus when the probe touches the surface, the light spot moves and the feedback control reacts by increasing the voltage output. Following the increase of voltage, a piezoelectric device will expand and lets the probe move away from the surface (approximately 0.1 nm per applied volt). The voltage used to move the piezo element in the z-direction is monitored and the height topography can be imaged accordingly [75,77,78].

The resolution power of the AFM is limited by the geometry of the cantilever. Due to the respective geometries there is a real and an imaginary point of contact between probe and surface during scanning. Mathematically speaking, the measurement is a convolution between probe geometry and the surface (Figure 13). Consequently, elevations are depicted larger (Figure 13c,d) and indentations are represented smaller (Figure 13a,b) than they really are. Thus the geometry of the probe is critical to the quality of the images measured with an AFM [75].

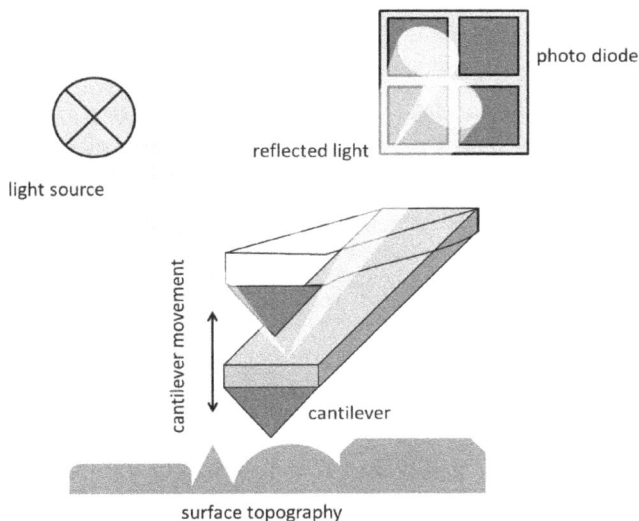

Figure 12. AFM equipment (adapted from [77]).

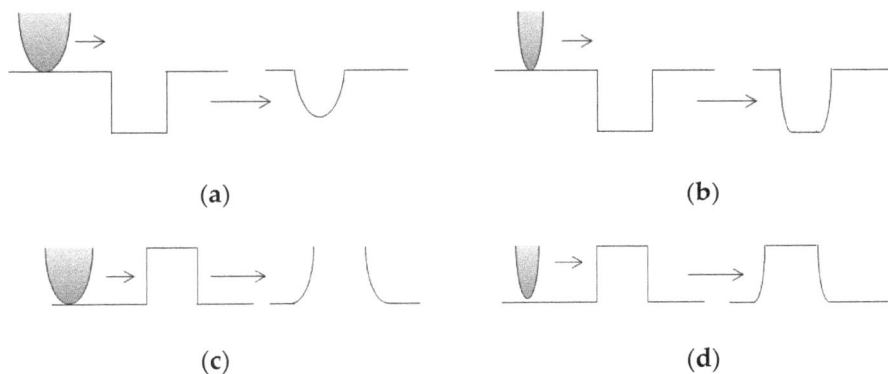

Figure 13. AFM operation (adapted from [75]) (**a**,**b**) smaller representation of indentations and (**c**,**d**) larger representation of elevations.

The cantilever with the nm-scale probe on top is usually produced by MEMS (Micro-Electro-Mechanical Systems) technology. The materials used are commonly Si_3N_4 or Si which show a diamond-like structure. Whereas SiN_4 cantilevers tend to bend due to residual stress, Si probes have a tendency to chip while contacting the surface [75].

For measuring the thickness of evaporated aluminum, an adhesion tape needs to be applied on the surface before evaporation (Figure 14). Afterwards, the adhesion tape is removed and a clear edge between the aluminum coated surface and the area, that was covered by the adhesive tape, appears. In this area, the measurements can be done by AFM. The measurement area can be adjusted between a few µm to up to 100 µm × 100 µm.

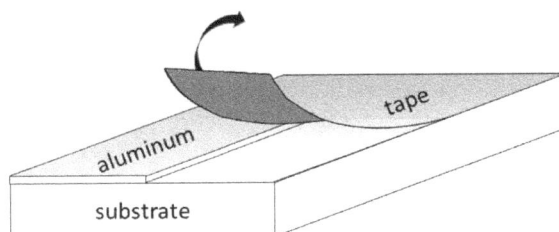

Figure 14. AFM measurement surface.

AFM: Practical Aspects and Analyses

Based on the measured topography (Figure 15) cross sections perpendicular to the edge are drawn and height profiles are extracted (Figure 16). The coating thickness can now be calculated by the height difference between the surface with and without aluminum. The peak in the middle of the graph in Figure 16 is a result of the tearing of the adhesive tape. By tearing, the aluminum is slightly lifted on the edges. Therefore, the peak area must not be considered. Moreover, the base line might not be as ideally horizontal as in Figure 16. In this case, the extracted height profile must be transcribed into an Excel file. Here, the data points from the base line of the substrate can be used to approximate a linear curve. Then the intercept of the curve is moved up so that the base line for the substrate plus coating can be fitted. By using a cosine function, a perpendicular can be dropped between the two baselines in order to calculate the distance between them.

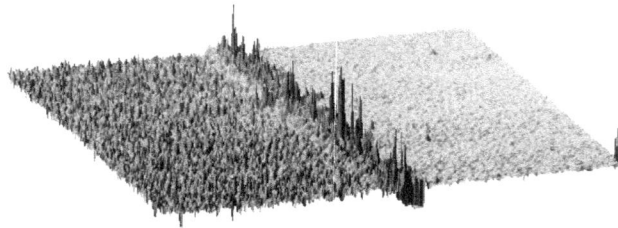

Figure 15. Example of a surface topography.

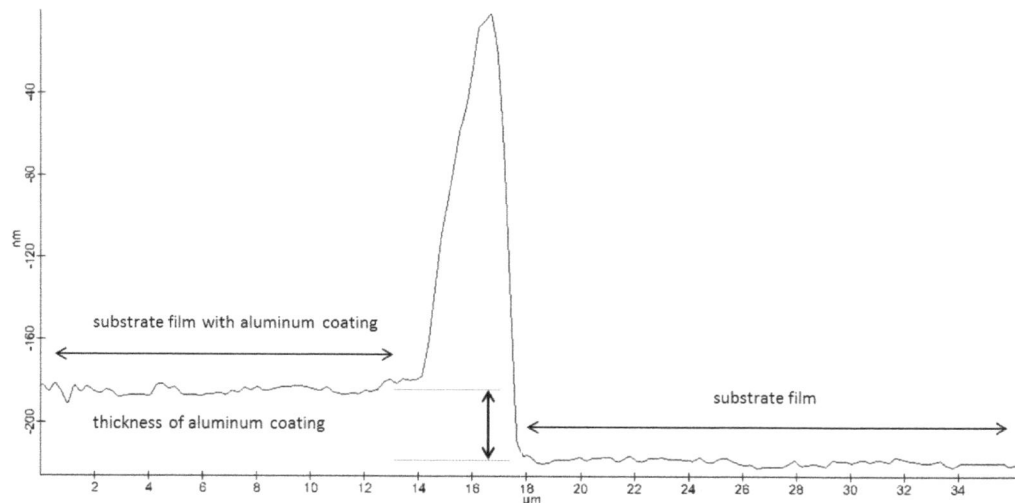

Figure 16. Example of an extracted profile.

What has to be taken into account when using this method is that that aluminum oxide is not captured separately but is included in the measured value. Additionally, inclusions, blank spaces, voids, etc. are not recorded either. The measurement is quite punctual, because it is not practical to make measurements on areas greater than about 100 μm × 100 μm. Scanning larger areas would lead to long scanning times in the range of a few hours. However, the resolution is extremely high and apart from height information, knowledge about the single molecule interactions between surface and cantilever (e.g., surface tension) can be gathered [75]. As displayed in Figure 13, the image that is acquired is strongly affected by the cantilever that is used. This effect can be diminished, when profile information is not only gathered in one scan direction but in backward and forward measurement modes. That means that each unevenness is captured from both sides. The resolution of AFM is commonly greater than 100 nm, reaching up to atomistic scales. Commercial specimens for Z calibrations are even available to a size of 2 nm [75].

2.3. Property Thickness

Whereas the last two chapters described the geometrical and mass thickness of thin coatings, this chapter focuses on the so-called property thickness. This denomination originates from the fact, that not the material itself, but the interaction with another physical phenomenon, e.g., ions, neutral particles, electrical field, or thermal energy, is monitored. In the present case, the interaction with photons (optical density, interference) and electrons (electrical surface resistance, eddy current measurement) is used to further describe thin aluminum coatings [65]. Other methods that can be counted as property thickness measurements are the Hall voltage measurement, interference spectra, polarization analysis, beta backscattering, X-ray fluorescence, X-ray emission, energy dispersive X-ray spectroscopy (EDX), and others [65].

2.3.1. Optical Density

Any optical phenomenon is connected with the reaction of light with matter. Light can generally be considered as a superposition of electromagnetic waves, which interact with the substrate so that amplitude, wave length, angular frequency, intensity, polarization, and propagation direction changes. For all optical methods, one or more of these characteristics is observed and then related to the material features [79]. Examples are: total reflection X-ray fluorescence analysis (TXRF), energy-dispersive X-ray spectroscopy (EDXS), grazing incidence X-ray, glow discharge optical emission spectroscopy (GD-OES), reflection absorption IR spectroscopy (RAIRS), surface-enhanced Raman scattering (SERS), and UV-Vis-IR ellipsometry (ELL) [65,80]. When the correspondent signals are analyzed, one will usually achieve information about the surface of the probe and/or about the material itself. However, especially when thin coatings are analyzed, both effects superimpose. This problem is often solved by building mathematical models that incorporate both phenomena. Then the model is fitted to the measured values. Like this, information about both (geometry and material) can be derived. Though for doing so, a minimum knowledge about the materials (e.g., refractive indices, absorption coefficients, approximate thicknesses, etc.) in the probe is necessary before fitting the measurement with the mathematical model [79]. The method of measuring the optical density uses the intensity of light as a characteristic feature. The material specific property that has to be known for this kind of analyses is the absorption coefficient (α), as explained in the following.

Optical Density: Theory

When light impacts a surface, it will partially be reflected (R), absorbed (A), scattered (S), and transmitted (T). Because of the conservation of energy, the sum of all parts is always 1 (Equation (22)). By knowing three of them, the fourth one can easily be calculated. However, reflection (R) and transmission (T) are rather easy to capture. Consequently, (A) and (S) are often summarized to the optical loss (L), as they are rather hard to define (Equation (22)). Commonly, the parts of reflected, absorbed, scattered, and transmitted light are defined by their relation to the initial intensity of the light beam (Equation (23)) [79,81].

$$R + A + S + T = 1 = R + L + T \tag{22}$$

$$R \equiv \frac{I_R}{I_0} \quad S \equiv \frac{I_S}{I_0} \quad A \equiv \frac{I_A}{I_0} \quad T \equiv \frac{I_T}{I_0} \tag{23}$$

Apart from that, Lambert and Beer introduced a mathematical law concerning absorption, which is commonly used in biology and chemistry to find the amount of dispersed particles in a liquid. They found that the intensity of transmitted light correlates with the absorption coefficient of the particles (α) and the distance (l) that light travels through the dispersion according to Equation (24).

The absorption coefficient (α) in turn is linked to the frequency (ω), particle concentration (c) and extinction coefficient (K), as illustrated in Equation (27) [79].

$$I_T = I_0 \times e^{-\alpha l} \tag{24}$$

$$\alpha = 2 \times \frac{\omega}{c} \times K(\omega) \tag{25}$$

This means that according to the Lambert Beer law, absorbance (A') is defined as the part of the light that is not transmitted (Equation (26)). This definition of "absorbance" does not equal the definition given at the beginning of the paragraph but is rather defined by absorption (A), scattering (S), and reflectance (R), as becomes obvious in Equation (27). Nevertheless, the optical density (OD) (Equation (27)) is derived from Equation (24) and is used to characterize the thickness of thin coatings. That means a coating having an optical density of 1, 2 or 3 lets pass 10%, 1% and 0.1% of light, respectively. Consequently, when using the concept of optical density, one needs to be aware that scattering, reflectance and absorption might alter the measured value although the originally defined absorbance might have kept constant. Additionally all four values are subjected to the wavelength, material, and the material's structure [79,81,82].

$$T = 1 - A' \tag{26}$$

$$\alpha l = -\log \frac{I_T}{I_0} = -\log(1 - A - S - R) = OD \tag{27}$$

Optical Density: Method

Because of the above-mentioned simplification, only the transmittance is measured for optical density [39]. Some simpler or rather sophisticated methods are commercially available. They all have in common that, for measuring the transmittance, the object is placed under a focused light source and transmittance is measured by a photo diode on the backside of the material. Depending on the equipment, one single wave length or a whole range can be measured (e.g., with FTIR). As evaporated coatings are typically applied on polyethylenetherephthalate or polypropylene substrates, the optical density of the pure substrate should also be captured and subtracted from the measured value.

Optical Density: Practical Aspects and Analyses

Weiss [83] showed in his work a linear correlation between optical density and thickness in the range of optical density of 0.3 to 3.5 (this means a transmission of 50% to 0.03%, respectively). However, the thickness suddenly increases at optical thicknesses <3.5. Similarly Hertlein [81] declares that OD is not useful for values >3.2. Apart from that, Copeland and Astbury [39] showed, that the optical density decreases over time, as the light-absorbing aluminum reacts to the transparent aluminum oxide. Therefore, the acquired value only takes into account the metallic part of the coating.

Apart from that, the challenge while calculating the thickness of the coating based on the optical density is the definition of the absorption coefficient (α), and the extinction coefficient (K). As mentioned before, these parameters are highly susceptible on material, wavelength, and structure. It was calculated for various other evaporated substances in [84–88]. The interrelation of optical density or transmittance with process conditions, coating thickness and wavelength was evaluated by [42,89,90]. According to Schulz [91], the absorption coefficient for aluminum has a value of approximately 4–9 for wavelengths of 0.4–0.9 μm. As stated by Lehmuskero and Kuittinen [89] the coefficient had a value of 3–21 for a wavelength of 0.3–2 μm. Heavens [92] gave an overview of coefficients determined by various sources. Here, the extinction coefficient attained values of 0–12 in the region of 0.1–100 μm wavelength. Moreover, Lehmuskero and Kuittinen [89] found that the values for atomic layer deposited aluminum were higher than for physical vapor deposited aluminum. This was explained by the higher grain size for evaporated coatings, which leads to a reduced scattering of electrons, which in turn increases the absorption coefficient. Apart from that, the coating thickness showed an effect over the whole range of wave lengths without showing a clear trend. One way to explain this effect was to take into account the aluminum oxide. Depending on the overall thickness, the oxide layer might occupy different portions of this thickness and might therefore have different impacts on the optical density. However, this was not evaluated further. As a conclusion, Lehmuskero and Kuittinen [89] recommended, not to use literature values for extinction and absorption coefficients, as these might lead to deviations of approximately 20% but rather identify them for each new process and process equipment. Consequently, the method can be used as a fast way of obtaining approximate thickness evaluations, especially when the process conditions are kept constant. However, the coefficients need to be clearly determined when very accurate values shall be obtained. In this case, the identification of these values might be extremely time consuming.

2.3.2. Interference (Tolansky Method)

In 1827 the physicist Jacques Babinet proposed the idea of using the wavelength of light as a measure of length, which is basically done in interferometry [93]. As elucidated previously, the interaction of light with matter alters the amplitude, wave-length, angular frequency, intensity, polarization, and/or propagation direction [79]. The interference method uses the reflectance (R) of surfaces as well as the intensity (I) of light.

Interference is the superimposition of rays, which can be non-destructive or destructive in the case of coherent waves, and a phase shift of a half wavelength. If rays are reflected on a surface and superimpose with itself at a phase shift of a half wavelength, then the intensity becomes zero. This effect is visible as bright and dark lines (interference pattern) and can therefore be used to measure coating thicknesses, as is demonstrated in the following Section [93].

For thin coating analyses, a huge number of measurements based on interference is available, which are principally divided into single and multiple beam interferometry. The latter has the advantage that the intensity at each interference band is an accumulated intensity of each reflection and therefore the bands are rather sharp and easy to identify. Further examples are the Michelson interferometry, Fourier spectroscopy, and Fabry-Perot Interferometry.

Interference (Tolansky Method): Theory

The thickness of a deposited aluminum coating can be determined by the light interference method according to Tolansky [94], which is based on Newtonian interference bands. Knowledge about material density, electrical conductivity, etc. is not necessary [95]. The prerequisite for this method is access to an edge on the substrate, as illustrated in Figure 14. Additionally, both surfaces (aluminum and underlying polymer) should have a high (ideally identical) reflection. If this is not the case, an additional nano meter thin coating of e.g., gold should be sputtered on both surfaces. For creating an interferable system, a semitransparent reference glass is positioned on the sample surface under a small angle (α). Rays will pass through the glass, impact on the probe surface, and be reflected by it with a phase shift of 180°. Incoming and reflected rays will then interfere. For doing so, these rays need to have a minimum distance. This is achieved by a small distance between the glass and the surface as well as a small angle between both. When the distance between the substrate surface and the reference glass is a multiple of 0.5λ, waves superimpose in a destructive way and interference bands appear in a defined distance (a). Due to the small angle and low distance, interference bands are highly contrasting with a step loss in intensity. They will appear very thin compared to their distances (a) and are therefore usable for an evaluation [96]. Because of the step due to the aluminum coating, there is an offset (a') between the interference bands caused by the polymer surface and the bands caused by the aluminum surface. The higher the aluminum thickness (d), the bigger the offset (a') (Figure 17) [94,95,97]. A mathematical description of the dependencies based on simple trigonometric relations allows for deriving Equations (28) and (29). By combining both, the thickness (d) is then calculated like in Equation (30) [98,99].

$$\tan \alpha = \frac{d}{a'} \tag{28}$$

$$\tan \alpha = \frac{\lambda/2}{a} \tag{29}$$

$$d = \frac{\lambda \times a'}{2 \times a} \tag{30}$$

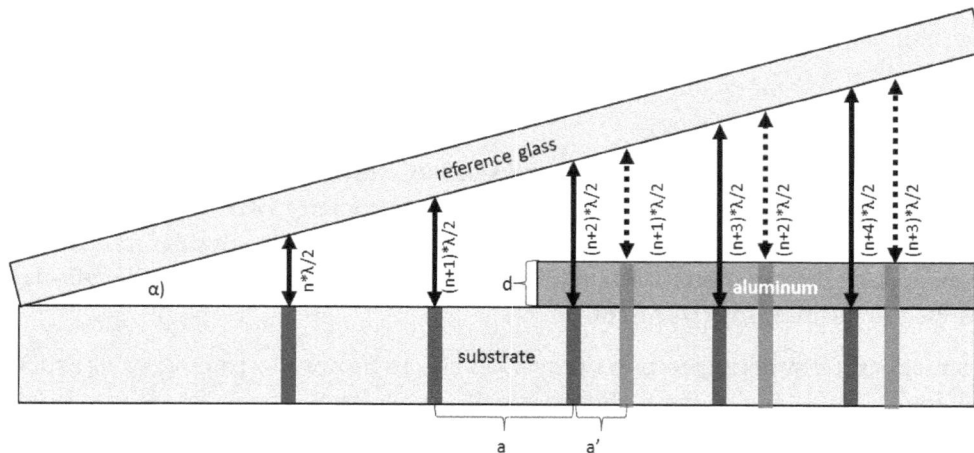

Figure 17. Appearance of interference bands on the semitransparent reflecting coating.

Interference (Tolansky Method): Method

According to the theory described before, only the wavelength (λ) as well as the distance (a) and offset (a') need to be known/measured, for calculating the coating thickness (d). Ideally, both surfaces should have the same refractive index. This is why an additional sputtered coating is often applied on both the polymer and the aluminum. For generating the interference bands, a Tolansky objective is applied on the light microscope and a filter is added for adjusting the light to monochromatic rays of 550 nm. The microscopic objectives that are necessary for interference measurements are commercially available. They are working on the basic principle depicted in Figure 18. Light is emitted and condensed by a lens, then filtered by a monochrome filter to a wave-length of typically 550 nm. Condensed by a second lens and mirrored towards the reference glass, the step between the aluminum and the substrate triggers interference bands, which are then visible and ready to be evaluated by microscope software. The interference bands should be perpendicular to the direction of the coating edge. Once the resolution is adjusted, (a) and (a') is measured and (d) is calculated as in Equation (30).

Interference (Tolansky Method): Practical Aspects and Analyses

Figure 19 shows typical interference bands of an aluminum coating on a PET surface. Care needs to be taken to correctly identify (a) and (a'). One possibility to facilitate this was presented by Hanszen [95], who explained that, due to the different phase shifts of monochromatic light on the substrate and coating, both should be covered with an additional, highly reflective 10 nm coating (e.g., gold or silver). This is especially important, when the thickness of the transparent aluminum oxide layer should be included in the measured value. However, roughness, blank spaces between aluminum coating, and sputtered coating, as well as gaps in the sputtered coating, still have an effect. Thus, if the surface to measure is rough, the actually measured coating thickness is not the average height (d), but a certain factor higher than (d). Only if the roughness of the substrate and the coating are the same is this effect negligible. Gaps in the sputtered coating appear due to island growth (see Figure 19).

When this happens, light is partially reflected on the sputtered island and partially on the coating. This might lead to deviations in the measured value of the coating thickness. Moreover, even this sputtered coating might show different growth behavior, which might again manipulate the results. Piegari and Masetti [59] followed that, in order to avoid deviations, the sputtered coating should be applied in a high vacuum at low condensation temperatures. However, the accuracy is reported to be approximately 1%.

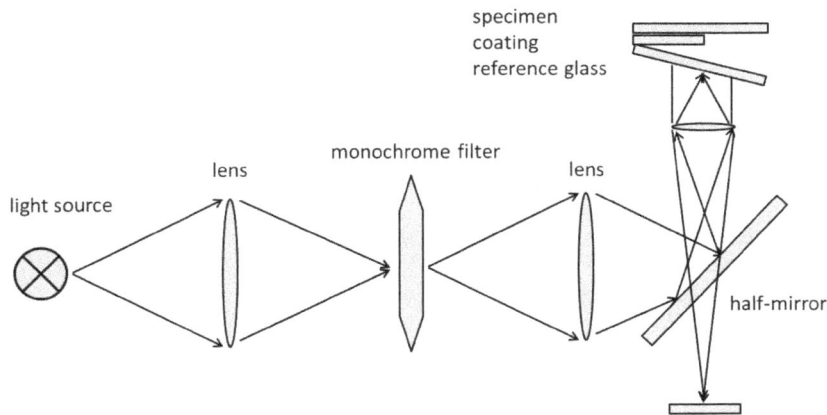

Figure 18. Set up for interference measurement.

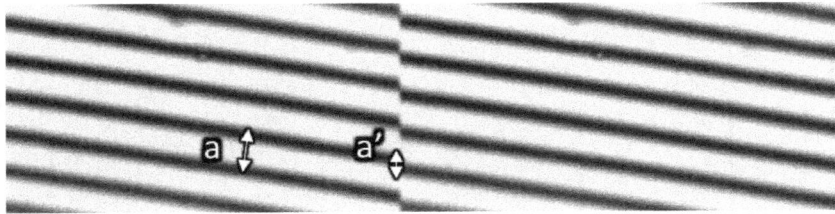

Figure 19. Microscopic image of interference bands.

2.3.3. Electrical Surface Resistance

Electrical Surface Resistance: Theory

Electrical resistivity (R) is the property of a current carrying conductor to inhibit the current flow. It is inhibited by collisions with lattice defects and impurities as well as oscillations of the crystal lattice atoms [100]. It is defined as the relation of voltage (U) to current (I). The inverse value represents the conductivity (G) (Equation (31)) [75]. The electrical resistivity depends on the geometry and material characteristics of the current carrying conductor. With the help of the specific electrical resistivity (ρ), length (l), width (b) and thickness (d) of the conductor, the expected resistivity (R) can be calculated (Equation (32)). If length (l) and width (b) are equal, both can be deducted from the formula which leads to the value of the so called surface resistivity ($R_\blacksquare$) (Equation (33)). As this resistivity is measured in a squarish setup, it is often indexed with a small square ($R_\blacksquare$). The advantage is now, that by using a defined construction (with $l = b$) for measuring the resistance, the coating thickness (d) can easily be calculated from Equation (33).

$$R = \frac{U}{I} \quad G = \frac{1}{R} = \frac{I}{U} \tag{31}$$

$$R = \rho \times \frac{l}{A} = \frac{\rho \times l}{d \times b} \tag{32}$$

$$R_\blacksquare = \frac{\rho \times l}{d \times b} = \frac{\rho}{d} \ (\text{with } l = b) \tag{33}$$

Pursuant to the Drude-Lorentz-Sommerfeld theory, the conductivity of metals is subjected to the density of the free electrons in the metal (n), the electron charge (e), the mean free path of the conducting electrons (l), the effective mass of an electron (m), and the average speed of free electrons (v_F), like in Equation (34). However, the mean free path (a) is related to the amount of imperfections and defects as well as the structure of the conductor as electrons are scattered on these lattice imperfections [100]. Kinds of imperfections are revealed in [34]. For small amounts of impurities, the effective resistivity is a sum of the bulk resistivity ($\rho_b(T)$) in dependence of the temperature (T) and the resistivity of the defects ($\rho_d(c)$) in dependence of the defect concentration (c) [100]. According to Mattheissen's rule for thin metal coatings, an additional term for the scattering at the boundary surfaces ($\rho_h(h)$) can be added, which correlates with the thickness (d) of the thin coating [100]:

$$\rho_0 = \frac{m \times v_F}{n \times e^2 \times a} \tag{34}$$

$$\rho_0 = \rho_b(T) + \rho_d(c) + \rho_d(d) \tag{35}$$

Electrical Surface Resistance: Method

In order to eliminate (b) and (l) from Equation (33), a defined geometry of the measured surface needs to be ensured where $b = l$. For accuracy reasons, usually 4 point set ups are used which are depicted in Figure 20. Therefore, electrodes might either be arranged in a linear or in a squarish way (van-der-Pauw method). In each case, the current is introduced between point (A) and (B) and the decrease in voltage is determined via (C) and (D).

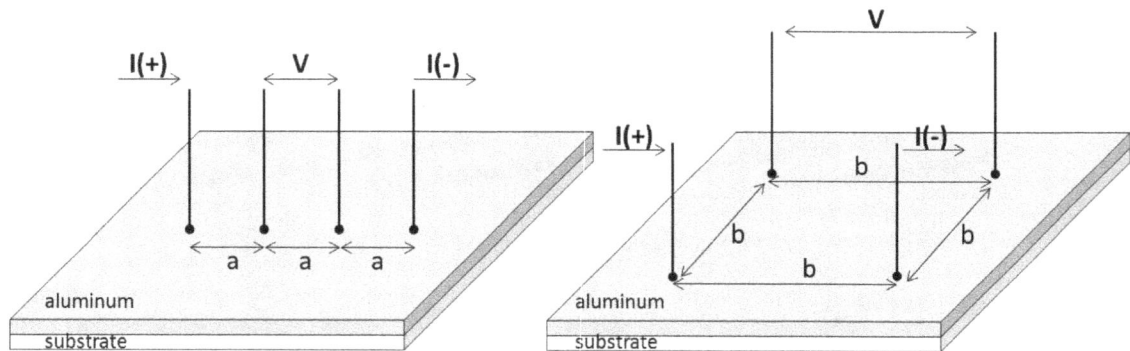

Figure 20. Common arrangements for measurement of surface resistance.

Electrical Surface Resistance: Practical Aspects and Analyses

Especially for nanometer thin coatings, the gravitational pressure of the tool might already cause cracks in the material. To reduce the influence of the pressure of the measurement setup on the thin coating, the electrodes are sometimes spring loaded. Moreover, oxide layers (which can have a resistivity 20-fold higher than the pure metal) can increase the determined value. Once the surface resistivity ($R_\blacksquare$) is measured, the coating thickness (h) can easily be calculated. However, some more effects should be taken into account, when interpreting this value.

Aluminum oxide has a resistivity of 10^{18} ($\Omega \cdot mm^2$)/m, which is 20-fold higher than aluminum. Consequently, the thickness of aluminum oxide is not captured in the measured value.

Another factor is the effect of electron scattering on surfaces and grain boundaries, of surface roughness and of island growth on electrical conductivity for different other metals such as copper, silver or gold were evaluated by [101–119] and fitted to models of Fuchs [120], Sondheimer [121], Soffer [122], Namba [107], Mayadas, and Shatzkes [123,124].

Rider and Foxon [125] quantified the dislocation density in cold-worked and partially annealed aluminum. They found that the dislocation resistivity was independent of dislocation density and arrangement. Additionally, Mayadas and Feder [126] measured the resistivity of thin aluminum coatings in the range of 700 to 10,000 Å and modeled the effect of electron scattering on surfaces as well as on grain boundaries. However, the fitting of the curves to the Fuchs theory was not successful. However, Mayadas and Shatzkes [123] showed that if the grain size increases with coating thickness, then a distinct effect on the resistivity exists. In [124] the resistivity of a thin aluminum coating is modeled, taking into account the background scattering of phonons and point defects, grain boundaries, and scattering on external surfaces. The conclusions were that the effect of thickness on resistivity is due to the grain-boundary scattering and the Fuchs size effect. The grain-boundary reflection coefficient in aluminum was found to be ≈0.15. Apart from that, in [127] results are presented, where the effect of pore size, its volume fraction, and direction on the electrical resistivity was measured, modeled, and simulated.

The methods limitation is mainly fixed by the set ups measurement range. Especially when aluminum coatings are thin and the distance between single atoms or clusters is large, no current can flow. Consequently, the resistivity of the aluminum coating is high and the measured value will principally be that of the substrate and therefore probably not detectable. This effect has been experimentally and theoretically investigated by [103,106,128,129].

2.3.4. Eddy Current Measurement

Just like with the setup for the electrical surface resistance, the eddy current technology is used to measure the resistivity—or in this case the conductivity—of the aluminum surface. The eddy current measurement is extendable to the impedance spectroscopy by varying the inserted frequency of the current (I), which allows for extracting some more information about Ohmic and capacitive resistivities (as e.g., in [130]).

Eddy Current Measurement: Theory

Just like surface resistance measurement, the eddy current measurement is an electrical method. However, it can be applied for non-destructive material testing, which is especially used for defining quality characteristics like coating thickness, resistivity, material homogeneity, and other physical changes in the material. By applying an alternating voltage (100 kHz to some MHz) on the induction coil, an electromagnet field (primary field) is generated. If a conducting sample is placed in this electromagnetic field, eddy currents are triggered in the sample. The notation "eddy current" is based on the movement of the current carriers on circular paths. This eddy current then leads to a secondary electromagnetic field, which impinges on the primary field. This impingement can be measured and related to the thickness of the coating [65]. The basic physical principles are outlined in the following. However, the exact mathematical description of the interaction of coil and a flat metal sheet can be found in [131].

The aluminum coating is characterized by the thickness (d), the electrical conductivity ($1/R$), and the magnetic permeability (μ_r) [65]. When an alternating current passes a coil, this leads to the development of an electromagnetic field (Figure 21), defined by the magnetic flux density ($B1$) and electric flux density (E), which leads to the movement of current carriers with the velocity (v) in a flat metal sheet supposed to the field. The three vectors (B), (E), and (v) are perpendicular to each other (Lorentz rule) (Equation (38)) and lead to the circular movement of the current carriers in the aluminum (eddy current). The magnetic flux density (B_1) is triggered depending on the electrical current (I) passing through the coil, the length of the coil (l), its coil number (N), and the magnetic constant (μ_0) (Equation (36)). These eddy currents trigger a secondary magnetic field (B_2), which is opposite directed and alters the primary field (B_1). Pursuant to the rule of Lenz, the secondary field superimposes the first one and reduces it to the value of (B_1'). The relation of (B_1') to (B_1) is denoted as the permeability (μ_r). In a nutshell, the eddy currents will be higher when the aluminum coating is thicker, and the inductance in the coil will be lower [65,132,133]. This energy loss consists of three parts: the hysteresis loss (W_h), the classical loss (W_{cl}), and the excess loss (W_{exc}) (see Equation (42)).

Herein the classical loss (W_{cl}) depends on the thickness as in Equation (43). Here, (σ) is the conductivity, (d) is the material thickness, (J_p) is the peak polarization, (f) is the frequency, and (δ) is the material density [134].

$$B = \mu_0 \times I \times \frac{N}{l} \tag{36}$$

$$F_M = Q \times v \times B \tag{37}$$

$$F_L = Q \times v \times B + Q \times E \tag{38}$$

$$\mu_r = \frac{B}{B_0} \tag{39}$$

$$\mu = \mu_0 \times \mu_r \tag{40}$$

$$Z = \frac{U_{eff}}{I_{eff}} \tag{41}$$

$$W = W_h + W_{cl} + W_{exc} \tag{42}$$

$$W_{cl} = \frac{\pi^2}{6} \times \frac{\sigma d^2 J_p^2 f}{\delta} \tag{43}$$

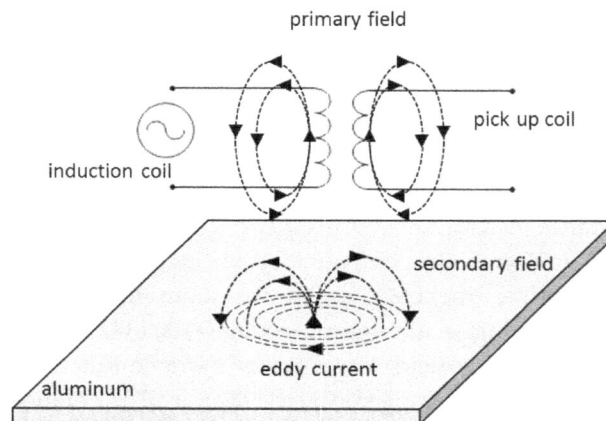

Figure 21. Schematic representation of eddy current (adapted from [133]).

Eddy Current Measurement: Method

Basically, two different methods are used for eddy current measurements: the surface and transmission technique. In the first case, the sending coil is also the receiving coil and the specimen is placed at a defined distance to it. In the second case, the specimen is placed between two coils: the first one triggers the electromagnetic field; in the second coil the electromagnetic field leads to a current in the coil. The second method has the advantage that misalignments as well as the distance between coil and specimen only have a minor effect on test results [135]. Various different methods and setups are reviewed in [136].

In order to reveal the shift of inductance, a circuit can be used, as in Figure 22. This consists of a power supply (U_0), the coil with inductance (L), the coils' resistivity (r), a capacitor (C), and an external resistor (R). Because of the shift of energy between the coil and the capacitor, the circuit oscillates and shows a typical resonance frequency, which can be monitored via the LC circuit voltage (U_{LC}). When the sample is introduced to the system, the coils inductivity (L) changes and the resonance voltage (U_{LC}) shifts away. The change of (U_{LC}) is amplified and related to the samples' surface resistivity ($R_\blacksquare$). Further variations in the eddy current measurements are described in [130,131,135,137–141].

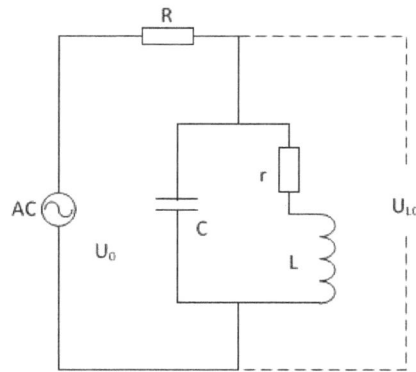

Figure 22. Resonant circuit used of eddy current measurements (adapted from [142]).

Eddy Current Measurement: Practical Aspects and Analyses

Once the surface resistance ($R_\blacksquare$) is measured, the material thickness (d) can then be calculated from Equation (33). New systems exist, that automatically or semi-automatically measure the surface resistance in a raster of e.g., 1 cm squares. Although eddy current measurements area a rather punctual method, this offers the opportunity to gain an average value which characterizes a whole material area. In [143] simulations have been carried out that show a clear relationship between bulk resistivity of the specimens and the eddy current induced in the thin conductor metals. Measurement setups are available e.g., in the range of 0.001–3000 Ω/sq, that equals of an ideal aluminum coating thickness of approximately 26 μm–0.01 nm [133]. Just like in the electrical surface resistance measurement, aluminum oxide is not captured in the measured value because of the extremely high resistivity.

However, several factors might affect the results. Heuer and Hillmann [139] stated, that subjected to the frequency, the depth where eddy currents are triggered might vary. Therefore, for thicker materials, a lower frequency is necessary to ensure the full penetration of the material.

Qu, Zhao [142] varied the coil resistance, its capacity, and induction in the LC circuit and found that, by reducing the resistance by using multi strand of wires for the coil the sensitivity increases. The maximum sensitivity reached was 2 mV/nm. The importance of sensitivity was also emphasized by Angani, Ramos [141]. Moulder, Uzal [140] determined the effect of the coil size to specimen thickness ratio on the sensitivity of the instrument. They found that for thin specimens only one feature (thickness or conductivity) can be measured when the other one is known. This effect is also comprehensively explained in [135], where a solution is presented as to avoid this problem by choosing the right frequency. Similar to Moulder, Uzal [140], Rajotte [138] also recommended, that the specimen should be at least 1.5 times larger than the outside diameter of the spiral coil.

Hillmann, Klein [135] evaluated the effect of sensor-to-sensor distances and material thickness on the deviation from expected values and found a non-linear behavior. The deviation was small (approximately 0.25 to 1.5 nm) and showed a further decrease for thicknesses below 20 nm. However, the deviation was much higher (approximately 0.5 to 4.0 nm) but still showed a decrease for thicknesses of $\geq$50 nm. A smaller distance between the two sensors (6 cm, 4 cm and 2 cm were tested) seemed to reduce the deviation.

Heuer, Hillmann [139] and Hillmann, Klein [135] even stated that the kind of material in mono and multilayers, the depth profile, thicknesses, and hardnesses as well as microstructure properties can be revealed. However, they were partially working in a thickness range of several micrometers and it is not mentioned which method was used to apply the coating. Concerning the differentiation of materials, Hillmann, Klein [135] emphasized the importance of choosing the right frequency, in order to properly distinguish between different materials with different conductivities and thicknesses. Angani, Ramos [141] proposed using this method to evaluate the corrosion of metals.

3. Methods Overview and Conclusions

As illustrated before, each method has its challenges when it comes to the interpretation of the measured values. No measurement is the "wrong" or the "right" one, but they need to be interpreted based on the characteristic that is actually acquired. Only in accordance with the scope of application, can the "most suitable" system be chosen. Table 1 gives an overview that provides indications for choosing the most suitable method for different purposes. The indications are denoted as follows.

Table 1. Overview over methods.

Characteristic	Mass Thickness		Geometrical Thickness	Property Thickness				
	QCM	ICP-MS	AFM	Eddy Current Measurement	Electrical Resistivity Linear	Electrical Resistivity Squarish	Optical Density	Interference
Measurement range	++	+++	++	+++	+++	+++	+	++
Time needed for one measurement	+	+++	++	+	+	+	+	++
Non destructive	✓	✗	✗	✓	✗	✗	✓	✗
Punctual measurement	(✓)	✗	✓	✓	✗	✗	✓	✓
Measurement within multilayer is possible	✗	✓	(✓)	✓	✗	✗	✗	✗
Impact of pores and defects	✗	✗	++	+++	+++	+++	++	++
Is only metallic aluminum detected?	✓	✗	✗	✓	✓	✓	✓	✓
Usable as inline measurement	✓	✗	✗	✓	✗	✗	✓	✗
Financial invest	+	+++	+++	++	+	+	+	++

+++/++/+: big/intermediate/small; ✓/✗: yes/no; ( ): with restrictions.

From this overview it can be concluded, that a higher financial investment does not necessarily lead to shorter measuring times or higher measurement ranges. Whereas ICP-MS, AFM, and interference seem to be more interesting methods for science related questions, especially QCM, eddy current, electrical resistivity, and optical density are commonly used in packaging material producing industries. QCM and optical density are widely used as inline measurement method, as they are also non-destructive.

When it comes to the correlation of thickness values with barrier effects against water vapor and oxygen, the awareness about the subdivision into mass, geometrical, and property thickness is useful. As permeation appears mainly through areas where the aluminum coating is not yet closed or is defective, the interpretation of measured and derived thickness values is critical.

For the calculation of mass thickness measurements, it is assumed that all the aluminum in the sample is arranged in a perfectly crystalline manner on the substrate surface. It is not taken into account that there might be defects or irregularities in the atomic lattice or aluminum that reacted to aluminum oxide and might influence the gas barrier.

In contrast to that, the geometrical measurement based on AFM is a very punctual measurement. The measured thickness consists of both the pure aluminum and also of aluminum oxide. Here one needs to be aware that the molar volume of aluminum oxide is higher (values depending on the exact composition) than that of pure aluminum. Additionally, the measurement should be repeated at different areas of the sample as it cannot be excluded that one measurement is done on a defect or an area with extraordinarily high or thin aluminum coverage.

As the name suggests, property thickness measurements measure a certain property of the aluminum, the thickness of which is then calculated based on the assumption that the nanometer thin coating behaves like an ideal bulk material. However, it is known, that nanometer materials behave differently from bulk material. Imagine a thick coating, full of pores and defects. These pores and defects might lead to increased electrical resistivity. Thus, a thickness would be calculated, that is thinner than it actually is (compare Equation (33)). For the case of optical density, an increasing relative amount of aluminum oxide decreases the measured value for the optical density. The amount of aluminum oxide is in turn affected by the residual oxygen in the recipient.

From this overview it becomes obvious, that the thickness can be measured rather quickly; however, a full characterization of the coating can only be done by the combination of the above mentioned methods.

Acknowledgments: This work was supported by the German Research Foundation (DFG) and the Technical University of Munich (TUM) in the framework of the Open Access Publishing Program. Therefore the authors thank the DFG and TUM for their support. Moreover, the authors thank Horst-Christian Langowski and Florian Höflsauer for their support.

Author Contributions: Martina Lindner wrote the manuscript and was in charge of the overall outline and editing of the manuscript. She was involved in the revision and completion of the work. Markus Schmid contributed to the outline as well as to the revision, completion, and editing of the manuscript.

Conflicts of Interest: The authors declare no conflict of interest.

References

1. Bichler, C.; Langowski, H.C.; Moosheimer, U.; Bischoff, M. Transparente Aufdampfschichten aus Oxiden von Si, Al und Mg für Barrierepackstoffe. Available online: https://www.mysciencework.com/publication/show/430561f397d6f9ad469136be1369ee3b (accessed on 29 December 2016).
2. Pilchik, R. Pharmaceutical blister packaging, Part I. *Pharm. Technol.* **2000**, *24*, 68–78.
3. Dean, D.A.; Evans, E.R.; Hall, I.H. *Pharmaceutical Packaging Technology*; Taylor & Francis: London, UK, 2005.
4. Huang, C.C.; Ma, H.W. A multidimensional environmental evaluation of packaging materials. *Sci. Total Environ.* **2004**, *324*, 161–172. [CrossRef] [PubMed]
5. Seshan, K. *Handbook of Thin Film Deposition Processes and Techniques*; William Andrew: Norwich, NY, USA, 2002.
6. Mattox, D.M. Physical vapor deposition (PVD) processes. *Metal Finish.* **2001**, *99*, 409–423. [CrossRef]
7. Pierson, H.O. *Handbook of Chemical Vapor Deposition: Principles, Technology and Applications*; William Andrew: Norwich, NY, USA, 1999.
8. Leskelä, M.; Ritala, M. Atomic layer deposition (ALD): From precursors to thin film structures. *Thin Solid Films* **2002**, *409*, 138–146. [CrossRef]
9. Hirvikorpi, T.; Vähä-Nissi, M.; Mustonen, T.; Harlin, A.; Iiskola, E.; Karppinen, M. Thin inorganic barrier coatings for packaging materials. In Proceedings of the PLACE 2010 Conference, Albuquerque, NM, USA, 18–21 April 2010.
10. Mackenzie, J.D.; Bescher, E.P. Physical properties of sol-gel coatings. *J. Sol-Gel Sci. Technol.* **2000**, *19*, 23–29. [CrossRef]
11. Logothetidis, S.; Laskarakis, A.; Georgiou, D.; Amberg-Schwab, S.; Weber, U.; Noller, K.; Schmidt, M.; Kuecuekpinar-Niarchos, E.; Lohwasser, W. Ultra high barrier materials for encapsulation of flexible organic electronics. *Eur. Phys. J. Appl. Phys.* **2010**, *51*, 33203. [CrossRef]
12. Noller, K.; Mikula, M.; Amberg-Schwab, S.; Weber, U. Multilayer coatings for flexible high-barrier materials. *Open Phys.* **2009**, *7*, 371–378.
13. Brinker, C.J.; Frye, G.C.; Hurd, A.J.; Ashley, C.S. Fundamentals of sol-gel dip coating. *Thin Solid Films* **1991**, *201*, 97–108. [CrossRef]
14. Schultrich, B. Physikalische dampfphasenabscheidung: Bedampfen. In Proceedings of the Surface Engineering und Nanotechnologie SENT, Dresden, Germany, 5–7 December 2006.
15. Bishop, C.A. *Vacuum Deposition onto Webs, Films and Foils*, 2nd ed.; Elsevier: Amsterdam, The Netherlands, 2011.
16. Mondolfo, L.F. *Aluminum Alloys: Structure and Properties*; Elsevier: Amsterdam, The Netherlands, 2013.
17. Ans, J.; Lax, E.; Synowietz, C. *Taschenbuch für Chemiker und Physiker*; Springer: Berlin/Heidelberg, Germany, 1967.

18. Hatch, J.E.; Association, A.; Metals, A.S. *Aluminum: Properties and Physical Metallurgy*; American Society for Metals: Geauga County, OH, USA, 1984.

19. Kaßmann, M. *Grundlagen der Verpackung: Leitfaden für die Fächerübergreifende Verpackungsausbildung*; Beuth Verlag: Berlin, Germany, 2014.

20. Iwakura, K.; Wang, Y.D.; Cakmak, M. Effect of Biaxial Stretching on Thickness Uniformity and Surface Roughness of PET and PPS Films. *Int. Polym. Process.* **1992**, *7*, 327–333. [CrossRef]

21. Cakmak, M.; Wang, Y.; Simhambhatla, M. Processing characteristics, structure development, and properties of uni and biaxially stretched poly (ethylene 2,6 naphthalate)(PEN) films. *Polym. Eng. Sci.* **1990**, *30*, 721–733. [CrossRef]

22. Lin, Y.J.; Dias, P.; Chum, S.; Hiltner, A.; Baer, E. Surface roughness and light transmission of biaxially oriented polypropylene films. *Polym. Eng. Sci.* **2007**, *47*, 1658–1665. [CrossRef]

23. Müller, K.; Schönweitz, C.; Langowski, H.C. Thin Laminate Films for Barrier Packaging Application—Influence of Down Gauging and Substrate Surface Properties on the Permeation Properties. *Packag. Technol. Sci.* **2012**, *25*, 137–148. [CrossRef]

24. Utz, H. Barriereeigenschaften Aluminiumbedampfter Kunststofffolien. Ph.D. Thesis, Technische Universität, Fakultät für Brauwesen, Lebensmitteltechnologie und Milchwissenschaft, Berlin, Germany, 1995. (In German)

25. Kim, C.; Goring, D. Surface morphology of polyethylene after treatment in a corona discharge. *J. Appl. Polym. Sci.* **1971**, *15*, 1357–1364. [CrossRef]

26. Neugebauer, A. Condensation, nucleation, and growth of thin films. In *Handbook of Thin Film Technology*; Maissel, L.I., Glang, R., Eds.; McGraw-Hill: New York, NY, USA, 1970.

27. Haefer, R.A. *Oberflächen- und Dünnschicht-Technologie*; Springer: Berlin, Germany, 1987.

28. Jacobs, K. H. Frey, G. Kienel. Dünnschichttechnologie. VDI-Verlag GmbH, Düsseldorf 1987. 691 + XVIII pages, numerous figures and tables, 395.00 DM, ISBN 3-18-400670-0. *Cryst. Res. Technol.* **1989**, *24*, 1232. [CrossRef]

29. Vook, R.W. Structure and growth of thin films. *Int. Met. Rev.* **1982**, *27*, 209–245. [CrossRef]

30. Reichelt, K.; Jiang, X. The preparation of thin films by physical vapour deposition methods. *Thin Solid Films* **1990**, *191*, 91–126. [CrossRef]

31. Kern, R.; Metois, G.L. Basic mechanisms in the early stage of epitaxy. *Curr. Top. Mater. Sci.* **1979**, *3*, 135–419.

32. Stoyanov, S. Nucleation theory for high and low supersaturations. *Curr. Top. Mater. Sci.* **1979**, *3*, 421–462.

33. Bravais, A. *Abhandlung über Die Systeme von Regelmässig auf Einer Ebene Oder Raum Vertheilten Punkten*; Wilhelm Engelmann: Leipzig, Germany, 1897. (In German)

34. Gottstein, G. *Materialwissenschaft und Werkstofftechnik: Physikalische Grundlagen*; Springer: Berlin, Germany, 2014.

35. Weitze, M.D.; Berger, C. Strukturen und Eigenschaften. In *Werkstoffe: Unsichtbar, Aber Unverzichtbar*; Springer: Berlin, Germany, 2013; pp. 9–66.

36. Bollmann, W. *Crystal Defects and Crystalline Interfaces*; Springer Science & Business Media: Berlin, Germany, 2012.

37. Miesbauer, O.; Schmidt, M.; Langowski, H.C. Stofftransport durch Schichtsysteme aus Polymeren und dünnen anorganischen Schichten. *Vak. Forsch. Prax.* **2008**, *20*, 32–40. [CrossRef]

38. Barker, C.P.; Kochem, K.-H.; Revell, K.M.; Kelly, R.S.A.; Badyal, J.P.S. The Interfacial Chemistry of Metal Oxide Coated and Nanocomposite Coated Polymer Films. *Thin Solid Films* **1995**, *257*, 77–82. [CrossRef]

39. Copeland, N.J.; Astbury, R. Evaporated aluminium on polyester: Optical, Electrical, and Barrier Properties as a Function of Thickness and Time (Part I). In Proceedings of the AIMCAL Technical Conference, Myrtle Beach, SC, USA, 14 October 2010.

40. Hass, G.; Scott, N.W. On the structure and properties of some metal and metal oxide films. *J. Phys. Radium* **1950**, *11*, 394–402. [CrossRef]

41. McClure, D.; Struller, C.; Langowski, H.C. Evaporated Aluminium on Polypropylene: Oxide-Layer Thickness as a Function of Oxygen Plasma-Treatment Level. Available online: http://www.aimcal.org/uploads/4/6/6/9/46695933/mcclure_abs.pdf (accessed on 29 December 2016).

42. McClure, D.J.; Copeland, N. Evaporated Aluminium on Polyester: Optical, Electrical, and Barrier Properties as a Function of Thickness and Time (Part II). Available online: http://dnn.convertingquarterly.com/magazine/matteucci-awards/id/2420/evaporated-aluminum-on-polyester-optical-electrical-and-barrier-properties-as-a-function-of-thickness-and-time-part-1.aspx (accessed on 29 December 2016).

43. Menges, G. *Werkstoffkunde der Kunststoffe*; Walter de Gruyter: Berlin, Germany, 1971; Volume 2620.

44. Prins, W.; Hermans, J.J. Theory of Permeation through Metal Coated Polymer Films. *J. Phys. Chem.* **1959**, *63*, 716–720. [CrossRef]

45. Langowski, H.C. Stofftransport durch polymere und anorganische Schichten Transport of Substances through Polymeric and Inorganic Layers. *Vak. Forsch. Prax.* **2005**, *17*, 6–13. [CrossRef]

46. Langowski, H.C.; Utz, H. Dünne anorganische Schichten für Barrierepackstoffe. *Int. Z. Lebensm. Technol. Mark. Verpack. Anal.* **2002**, *9*, 522.

47. Roberts, A.P.; Henry, B.M.; Sutton, A.P.; Grovenor, C.R.; Briggs, G.A.; Miyamoto, T.; Kano, M.; Tsukahara, Y.; Yanaka, M. Gas permeation in silicon oxide/polymer (SiOx/PET) barrier films: Role of oxide lattice, nano-defects and macrodefects. *J. Membr. Sci.* **2002**, *208*, 75–88. [CrossRef]

48. Lohwasser, W. Not only for packaging. In Proceedings of the 43rd Annual Technical Conference of the Society of Vacuum Coaters, Denver, CO, USA, 23–28 April 2000.

49. Hanika, M.; Langowski, H.C.; Moosheimer, U.; Peukert, W. Inorganic layers on polymeric films—Influence of defects and morphology on barrier properties. *Chem. Eng. Technol.* **2003**, *26*, 605–614. [CrossRef]

50. Hanika, M. Zur Permeation Durch Aluminiumbedampfte Polypropylen-und Polyethylenterephtalatfolien. Ph.D. Thesis, Technical University of Munich, Munich, Germany, 2004.

51. Mueller, K.; Weisser, H. Numerical simulation of permeation through vacuum-coated laminate films. *Packag. Technol. Sci.* **2002**, *15*, 29–36. [CrossRef]

52. Rossi, G.; Nulman, M. Effect of local flaws in polymeric permeation reducing barriers. *J. Appl. Phys.* **1993**, *74*, 5471–5475. [CrossRef]

53. Jamieson, E.H.H.; Windle, A.H. Structure and oxygen-barrier properties of metallized polymer film. *J. Mater. Sci.* **1983**, *18*, 64–80. [CrossRef]

54. Bugnicourt, E.; Kehoe, T.; Latorre, M.; Serrano, C.; Philippe, S.; Schmid, M. Recent Prospects in the Inline Monitoring of Nanocomposites and Nanocoatings by Optical Technologies. *Nanomaterials* **2016**, *6*, 150. [CrossRef]

55. Utz, H. Barriereeigenschaften Aluminiumbedampfter Kunststofffolien. Ph.D. Thesis, Technical University of Munich, Munich, Germany, 1995.

56. Kääriäinen, T.O.; Maydannik, P.; Cameron, D.C.; Lahtinen, K.; Johansson, P.; Kuusipalo, J. Atomic layer deposition on polymer based flexible packaging materials: Growth characteristics and diffusion barrier properties. *Thin Solid Films* **2011**, *519*, 3146–3154. [CrossRef]

57. McCrackin, F.L.; Passaglia, E.; Stromberg, R.R.; Steinberg, H.L. Measurement of the thickness and refractive index of very thin films and the optical properties of surfaces by ellipsometry. *J. Res. Natl. Bur. Stand. Phys. Chem. A* **1963**, *67*, 363–377. [CrossRef]

58. Chatham, H. Oxygen diffusion barrier properties of transparent oxide coatings on polymeric substrates. *Surf. Coat. Technol.* **1996**, *78*, 1–9. [CrossRef]

59. Piegari, A.; Masetti, E. Thin film thickness measurement: A comparison of various techniques. *Thin Solid Films* **1985**, *124*, 249–257. [CrossRef]

60. Pulker, H.K. Thickness measurement, rate control and automation in thin film coating technology. In Proceedings of the 1983 International Techincal Conference, Geneva, Switzerland, 18 April 1983.

61. Mattox, D.M. Film characterization and some basic film properties. In *Handbook of Physical Vapor Deposition (PVD) Processing*; Mattox, D.M., Ed.; William Andrew Publishing: Westwood, NJ, USA, 1998; pp. 569–615.

62. Martin, P.M. *Handbook of Deposition Technologies for Films and Coatings: Science, Applications and Technology*; Elsevier: Amsterdam, The Netherlands, 2009.

63. Juzeliūnas, E. Quartz crystal microgravimetry-fifty years of application and new challenges. *Chemija* **2009**, *20*, 218–225.

64. Zeitvogl, J. Quarzkristallmikrowaage-QCM. Ph.D. Thesis, University of Erlangen-Nuremberg, Erlangen, Germany, 2009.

65. Frey, H.; Khan, H.R. *Handbook of Thin Film Technology*; Springer: Berlin, Germany, 2010.

66. Höpfner, M. Untersuchungen zur Anwendbarkeit der Quarzmikrowaage für Pharmazeutisch Analytische Fragestellungen. Ph.D. Thesis, Martin Luther University of Halle-Wittenberg, Halle, Germany, 2005.

67. MacLeod, H.A. *Thin-Film Optical Filters*, 3rd ed.; CRC Press: Cleveland, OH, USA, 2001.

68. Sauerbrey, G. Verwendung von Schwingquarzen zur Wägung dünner Schichten und zur Mikrowägung. *Z. Phys.* **1959**, *155*, 206–222. [CrossRef]

69. Lu, C.; Czanderna, A.W. *Applications of Piezoelectric Quartz Crystal Microbalances*; Elsevier: Amsterdam, The Netherlands, 2012.

70. Thomas, R. *Practical Guide to ICP-MS: A Tutorial for Beginners*, 2nd ed.; Taylor & Francis: London, UK, 2008.

71. De Hoffmann, E.; Stroobant, V. *Mass Spectrometry: Principles and Applications*; Wiley: New York, NY, USA, 2007.

72. Zoorob, G.K.; McKiernan, J.W.; Caruso, J.A. ICP-MS for elemental speciation studies. *Microchim. Acta* **1998**, *128*, 145–168. [CrossRef]

73. Broekaert, J.A.C. ICP-Massenspektrometrie. In *Analytiker-Taschenbuch*; Günzler, H., Bahadir, A.M., Danzer, K., Engewald, W., Fresenius, W., Galensa, R., Huber, W., Linscheid, M., Schwedt, G., Tölg, G., Eds.; Springer: Berlin/Heidelberg, Germany, 1988; pp. 127–163.

74. May, T.W.; Wiedmeyer, R.H. A table of polyatomic interferences in ICP-MS. *At. Spectrosc.* **1998**, *19*, 150–155.

75. Eaton, P.; West, P. *Atomic Force Microscopy*; Oxford University Press: Oxford, UK, 2010.

76. Voigtlaender, B. *Scanning Probe Microscopy: Atomic Force Microscopy and Scanning Tunneling Microscopy*; Springer: Berlin/Heidelberg, Germany, 2015.

77. Schieferdecker, H.G. *Bestimmung Mechanischer Eigenschaften von Polymeren Mittels Rasterkraftmikroskopie*; Fakultät für Naturwissenschaften, Universität Ulm: Ulm, Germany, 2005.

78. Meyer, G.; Amer, N.M. Simultaneous measurement of lateral and normal forces with an optical-beam-deflection atomic force microscope. *Appl. Phys. Lett.* **1990**, *57*, 2089–2091. [CrossRef]

79. Stenzel, O. *The Physics of Thin Film Optical Spectra*; Springer: Berlin, Germany, 2005.

80. Bubert, H.; Rivière, J.C.; Arlinghaus, H.F.; Hutter, H.; Jenett, H.; Bauer, P.; Palmetshofer, L.; Fabry, L.; Pahlke, S.; Quentmeier, A.; et al. *Surface and Thin-Film Analysis*; Wiley Online Library: New York, NY, USA, 2002.

81. Hertlein, J. *Untersuchungen über Veränderungen der Barriereeigenschaften Metallisierter Kunststoffolien Beim Maschinellen Verarbeiten*; Utz, Wiss: München, Germany, 1998.

82. Miller, D.A. *Optical Properties of Solid Thin Films by Spectroscopic Reflectometry and Spectroscopic Ellipsometry*; ProQuest: Ann Arbor, MI, USA, 2008.

83. Weiss, J. Einflussfaktoren auf die Barriereeigenschaften metallisierter Folien. *Verpak. Rundsch.* **1993**, *44*, 23–28.

84. Anna, C.; Cosslett, V.E. The optical density and thickness of evaporated carbon films. *Br. J. Appl. Phys.* **1957**, *8*, 374–376.

85. Deb, S.K. Optical and photoelectric properties and colour centres in thin films of tungsten oxide. *Philos. Mag.* **1973**, *27*, 801–822. [CrossRef]

86. Johnson, P.B.; Christy, R.W. Optical Constants of the Noble Metals. *Phys. Rev. B* **1972**, *6*, 4370–4379. [CrossRef]

87. Agar, A.W. The measurement of the thickness of thin carbon films. *Br. J. Appl. Phys.* **1957**, *8*, 35–36. [CrossRef]

88. Moss, T.S. Optical Properties of Tellurium in the Infra-Red. *Proc. Phys. Soc. Sec. B* **1952**, *65*, 62–66. [CrossRef]

89. Lehmuskero, A.; Kuittinen, M.; Vahimaa, P. Refractive index and extinction coefficient dependence of thin Al and Ir films on deposition technique and thickness. *Opt. Express* **2007**, *15*, 10744–10752. [CrossRef] [PubMed]

90. Hass, G.; Waylonis, J.E. Optical constants and reflectance and transmittance of evaporated aluminum in the visible and ultraviolet. *JOSA* **1961**, *51*, 719–722. [CrossRef]

91. Schulz, L.G. The Optical Constants of Silver, Gold, Copper, and Aluminum. I. The Absorption Coefficient *k*. *J. Opt. Soc. Am.* **1954**, *44*, 357–362. [CrossRef]

92. Heavens, O.S. Optical properties of thin films. *Rep. Prog. Phys.* **1960**, *23*, 1–65. [CrossRef]

93. McMillan, G.K.; Considine, D. *Process/Industrial Instruments and Controls Handbook*, 5th ed.; McGraw-Hill: New York, NY, USA, 1999.

94. Pulker, H.K. Einfaches Interferenz-Wechselobjektiv für Mikroskope zur Dickenmessung nach Fizeau-Tolansky. *Naturwissenschaften* **1966**, *53*, 224. [CrossRef] [PubMed]

95. Hanszen, K.J. Der Einfluss von Strukturunregelmässigkeiten beim Zusammenwachsen zweier Aufdampfschichten auf das Schichtdickenmessverfahren mit Hilfe von Vielstrahl-Interferenzen. *Thin Solid Films* **1968**, *2*, 509–528. [CrossRef]

96. Großes Interferenzmikroskop. In *Vertriebsabteilung Feinmessgeräte*; Carl Zeiss Jena, Ed.; Druckerei Fortschritt: Jena, Germany, 1965.

97. Tippmann, H.; Schawohl, J.; Kups, T. *Schichtdickenmessung*; TU Ilmenau—Fakultät für Elektrotechnik und Informationstechnik Institut für Werkstofftechnik: Ilmenau, Germany, 2013.

98. Hammer, A.; Hammer, H.; Hammer, K. *Physikalische Formeln und Tabellen*; Lindauer: Munich, Germany, 1994.

99. Pitka, R. *Physik: Der Grundkurs*; Harri Deutsch Verlag: Frankfurt am Main, Germany, 1999.

100. Zhigal'skii, G.P.; Jones, B.K. *The Physical Properties of Thin Metal Films*; CRC Press: Cleveland, OH, USA, 2003; Volume 13.

101. Liu, H.D.; Zhao, Y.P.; Ramanath, G.; Murarka, S.P.; Wang, G.C. Thickness dependent electrical resistivity of ultrathin (<40 nm) Cu films. *Thin Solid Films* **2001**, *384*, 151–156.

102. Philipp, M. Electrical Transport and Scattering Mechanisms in Thin Silver Films for Thermally Insulating Glazing. Available online: http://www.qucosa.de/fileadmin/data/qucosa/documents/7092/Dissertation_Martin_Philipp.pdf (accessed on 29 December 2016).

103. Hoffmann, H.; Vancea, J. Critical-Assessment of Thickness-Dependent Conductivity of Thin Metal-Films. *Thin Solid Films* **1981**, *85*, 147–167. [CrossRef]

104. Leung, K.M. Electrical resistivity of metallic thin films with rough surfaces. *Phys. Rev. B* **1984**, *30*, 647–658. [CrossRef]

105. Ke, Y.; Zahid, F.; Timoshevskii, V.; Xia, K.; Gall, D.; Guo, H. Resistivity of thin Cu films with surface roughness. *Phys. Rev. B* **2009**, *79*, 155406. [CrossRef]

106. Borziak, P.G.; Kulyupin, Y.A.; Nepijko, S.A.; Shamonya, V.G. Electrical conductivity and electron emission from discontinuous metal films of homogeneous structure. *Thin Solid Films* **1980**, *76*, 359–378. [CrossRef]

107. Namba, Y. Resistivity and Temperature Coefficient of Thin Metal Films with Rough Surface. *Jpn. J. Appl. Phys.* **1970**, *9*, 1326–1329. [CrossRef]

108. Darevskii, A.S.; Zhdan, A.G. Real structure and electrical conductivity of island films of metals. *Sov. Microelectron.* **1978**, *7*, 356–359.

109. Bassewitz, A.V. Der Einfluß der Unterlage auf die Struktur und Leitfähigkeit metallischer Aufdampfschichten. *Z. Phys.* **1967**, *201*, 350–367. [CrossRef]

110. Jannesar, M.; Jafari, G.R.; Farahani, S.V.; Moradi, S. Thin film thickness measurement by the conductivity theory in the framework of born approximation. *Thin Solid Films* **2014**, *562*, 372–376. [CrossRef]

111. Palasantzas, G.; Zhao, Y.P.; Wang, G.C.; Lu, T.M.; Barnas, J.; De Hosson, J.T. Electrical conductivity and thin-film growth dynamics. *Phys. Rev. B* **2000**, *61*, 11109. [CrossRef]

112. Munoz, R.C.; Finger, R.; Arenas, C.; Kremer, G.; Moraga, L. Surface-induced resistivity of thin metallic films bounded by a rough fractal surface. *Phys. Rev. B* **2002**, *66*, 205401. [CrossRef]

113. Timalsina, Y.P.; Horning, A.; Spivey, R.F.; Lewis, K.M.; Kuan, T.S.; Wang, G.C.; Lu, T.M. Effects of nanoscale surface roughness on the resistivity of ultrathin epitaxial copper films. *Nanotechnology* **2015**, *26*, 075704. [CrossRef] [PubMed]

114. Ketenoğlu, D.; Ünal, B. Influence of surface roughness on the electrical conductivity of semiconducting thin films. *Phys. A Stat. Mech. Appl.* **2013**, *392*, 3008–3017. [CrossRef]

115. Arenas, C.; Henriquez, R.; Moraga, L.; Muñoz, E.; Munoz, R.C. The effect of electron scattering from disordered grain boundaries on the resistivity of metallic nanostructures. *Appl. Surf. Sci.* **2015**, *329*, 184–196. [CrossRef]

116. Lim, J.W.; Mimura, K.; Isshiki, M. Thickness dependence of resistivity for Cu films deposited by ion beam deposition. *Appl. Surf. Sci.* **2003**, *217*, 95–99. [CrossRef]

117. Zhang, W.; Brongersma, S.H.; Richard, O.; Brijs, B.; Palmans, R.; Froyen, L.; Maex, K. Influence of the electron mean free path on the resistivity of thin metal films. *Microelectron. Eng.* **2004**, *76*, 146–152. [CrossRef]

118. Camacho, J.M.; Oliva, A.I. Surface and grain boundary contributions in the electrical resistivity of metallic nanofilms. *Thin Solid Films* **2006**, *515*, 1881–1885. [CrossRef]

119. Camacho, J.M.; Oliva, A.I. Morphology and electrical resistivity of metallic nanostructures. *Microelectron. J.* **2005**, *36*, 555–558. [CrossRef]

120. Fuchs, K. The conductivity of thin metallic films according to the electron theory of metals. *Math. Proc. Camb. Philos. Soc.* **1938**, *34*, 100–108. [CrossRef]

121. Sondheimer, E.H. The mean free path of electrons in metals. *Adv. Phys.* **1952**, *1*, 1–42. [CrossRef]

122. Soffer, S.B. Statistical Model for the Size Effect in Electrical Conduction. *J. Appl. Phys.* **1967**, *38*, 1710–1715. [CrossRef]

123. Mayadas, A.F.; Shatzkes, M.; Janak, J.F. Electrical resistivity model for polycrystalline films: The case of specular reflection at external surfaces. *Appl. Phys. Lett.* **1969**, *14*, 345–347. [CrossRef]

124. Mayadas, A.F.; Shatzkes, M. Electrical-Resistivity Model for Polycrystalline Films: The Case of Arbitrary Reflection at External Surfaces. *Phys. Rev. B* **1970**, *1*, 1382–1389. [CrossRef]

125. Rider, J.G.; Foxon, C.T.B. An experimental determination of electrical resistivity of dislocations in aluminium. *Philos. Mag.* **1966**, *13*, 289–303. [CrossRef]

126. Mayadas, A.F.; Feder, R.; Rosenberg, R. Resistivity and structure of evaporated aluminum films. *J. Vac. Sci. Technol.* **1969**, *6*, 690–693. [CrossRef]

127. Nakajima, H. *Porous Metals with Directional Pores*; Springer: Berlin, Germany, 2013.

128. Lux, F. Models proposed to explain the electrical conductivity of mixtures made of conductive and insulating materials. *J. Mater. Sci.* **1993**, *28*, 285–301. [CrossRef]

129. Siegel, A.C.; Phillips, S.T.; Dickey, M.D.; Lu, N.; Suo, Z.; Whitesides, G.M. Foldable printed circuit boards on paper substrates. *Adv. Funct. Mater.* **2010**, *20*, 28–35. [CrossRef]

130. Parfenov, E.V.; Yerokhin, A.L.; Matthews, A. Impedance spectroscopy characterisation of PEO process and coatings on aluminium. *Thin Solid Films* **2007**, *516*, 428–432. [CrossRef]

131. Dodd, C.V.; Deeds, W.E. Analytical Solutions to Eddy-Current Probe-Coil Problems. *J. Appl. Phys.* **1968**, *39*, 2829–2838. [CrossRef]

132. Hillmann, S.; Heuer, H.; Klein, M. Schichtdicken-Charakterisierung dünner, leitfähiger Schichtsysteme mittels Wirbelstromtechnik. In Proceedings of the DGZfP-Jahrestagung, Erfurt, Germany, 10–12 May 2010.

133. Suragus GmbH. EddyCus® TF Lab 4040. Available online: https://www.suragus.com/en/products/thin-film-characterization/sheet-resistance/eddycus-tf-lab-4040/ (accessed on 29 December 2016).

134. Fiorillo, F. *Measurement and Characterization of Magnetic Materials*; Elsevier: Amsterdam, The Netherlands, 2004.

135. Hillmann, S.; Klein, M.; Heuer, H. In-line thin film characterization using eddy current techniques. *Stud. Appl. Electromagn. Mech.* **2011**, *35*, 330–338.

136. García-Martín, J.; Gómez-Gil, J.; Vázquez-Sánchez, E. Non-destructive techniques based on eddy current testing. *Sensors* **2011**, *11*, 2525–2565. [CrossRef] [PubMed]

137. Singh, S.K. *Industrial Instrumentation & Control*, 2nd ed.; McGraw-Hill Education: Noida, India, 2003.

138. Rajotte, R.J. Eddy-current method for measuring the electrical conductivity of metals. *Rev. Sci. Instrum.* **1975**, *46*, 743–745. [CrossRef]

139. Heuer, H.; Hillmann, S.; Roellig, M.; Schulze, M.H.; Wolter, K.J. Thin film characterization using high frequency eddy current spectroscopy. In Proceedings of the 9th IEEE Conference on Nanotechnology (2009 IEEE-NANO), Genoa, Italy, 26–30 July 2009.

140. Moulder, J.C.; Uzal, E.; Rose, J.H. Thickness and conductivity of metallic layers from eddy current measurements. *Rev. Sci. Instrum.* **1992**, *63*, 3455–3465. [CrossRef]

141. Angani, C.S.; Ramos, H.G.; Ribeiro, A.L.; Rocha, T.J.; Prashanth, B. Transient eddy current oscillations method for the inspection of thickness change in stainless steel. *Sens. Actuators A Phys.* **2015**, *233*, 217–223. [CrossRef]

142. Qu, Z.; Zhao, Q.; Meng, Y. Improvement of sensitivity of eddy current sensors for nano-scale thickness measurement of Cu films. *NDT E Int.* **2014**, *61*, 53–57. [CrossRef]

143. Mehrabad, M.J.; Ehsani, M.H. An Investigation of Eddy Current, Solid Loss, Induced Voltage and Magnetic Torque in Highly Pure Thin Conductors, Using Finite Element Method. *Procedia Mater. Sci.* **2015**, *11*, 412–417. [CrossRef]

2

State of the Art of Antimicrobial Edible Coatings for Food Packaging Applications

Arantzazu Valdés, Marina Ramos, Ana Beltrán, Alfonso Jiménez and María Carmen Garrigós *

Analytical Chemistry, Nutrition & Food Sciences Department, University of Alicante,
03690 San Vicente del Raspeig (Alicante), Spain; arancha.valdes@ua.es (A.V.);
marina.ramos@ua.es (M.R.); ana.beltran@ua.es (A.B.); alfjimenez@ua.es (A.J.)
* Correspondence: mc.garrigos@ua.es

Academic Editor: Stefano Farris

Abstract: The interest for the development of new active packaging materials has rapidly increased in the last few years. Antimicrobial active packaging is a potential alternative to protect perishable products during their preparation, storage and distribution to increase their shelf-life by reducing bacterial and fungal growth. This review underlines the most recent trends in the use of new edible coatings enriched with antimicrobial agents to reduce the growth of different microorganisms, such as Gram-negative and Gram-positive bacteria, molds and yeasts. The application of edible biopolymers directly extracted from biomass (proteins, lipids and polysaccharides) or their combinations, by themselves or enriched with natural extracts, essential oils, bacteriocins, metals or enzyme systems, such as lactoperoxidase, have shown interesting properties to reduce the contamination and decomposition of perishable food products, mainly fish, meat, fruits and vegetables. These formulations can be also applied to food products to control gas exchange, moisture permeation and oxidation processes.

Keywords: antimicrobial; edible coatings; food packaging

1. Introduction

The search for more natural and healthy food, based on minimally-processed, easily-prepared and ready-to-eat fresh products, has resulted in an increase in consumer requirements for safe and high-quality food [1]. These new social trends joined with changes in the usual procedures to make food processing faster and more efficient have caused a rising interest to obtain food products with a longer shelf-life. These properties are directly related to the development of new improved packaging materials, including active, intelligent and edible systems [2,3].

Food products are perishable by nature, as they can be subjected to degradation by many different environmental effects, including contamination by bacteria and fungi. Therefore, processed food requires protection from spoilage during preparation, storage and distribution to improve shelf-life and quality. Many of these microorganisms, in particular pathogens, could cause severe health problems to consumers, especially if food is handled and distributed under inappropriate conditions. In addition, undesirable reactions can occur to modify odor, flavor, color and textural properties in fresh food [4]. Traditional preservation techniques, such as heat treatment, salting or acidification, have been applied for a long time by the food industry to prevent the growth of spoilage and pathogenic microorganisms in food, but they often result in unacceptable losses in their nutritional value [1]. However, new strategies related to preservation techniques have been demonstrated to be more effective in protecting food without hampering organoleptic and nutritional properties. This is particularly the case for edible antimicrobial films and coatings that have attracted the interest of researchers and the food industry in the last decade, since they can improve the safety, quality and functionality of food products while inhibiting the growth of undesirable microorganisms during storage, transportation and handling [2].

Different biopolymers, such as proteins, lipids, polysaccharides or their combinations, have been used as carriers to produce edible coatings with antimicrobial properties (Figure 1) [5–8]. They can be directly extracted from biomass and easily processed to get films to be used as coatings to control gas exchange, moisture permeation or oxidation processes in food while also reducing the microorganism's growth. These biopolymer films can be also used to host additives and nutrients to be released at a controlled rate to food, forming the basis of active packaging systems [9]. Other biopolymers, such as chitosan, pectin and pullulan, have been used with sodium benzoate and potassium sorbate to obtain edible coatings to improve the quality and shelf-life of fresh food [10]. The functionality of edible coatings can be expanded by incorporating antimicrobial additives to protect food products from microbial spoilage, extending their shelf-life and safety. Some reports have been recently published with the aim to evaluate and demonstrate the effectiveness against several pathogens of antimicrobial substances incorporated as active additives into edible coatings [5,11–13]. The incorporation of antimicrobial agents into edible coatings and their effectiveness could be improved by using nano-emulsification, which would also reduce essential oils losses by volatilization [14].

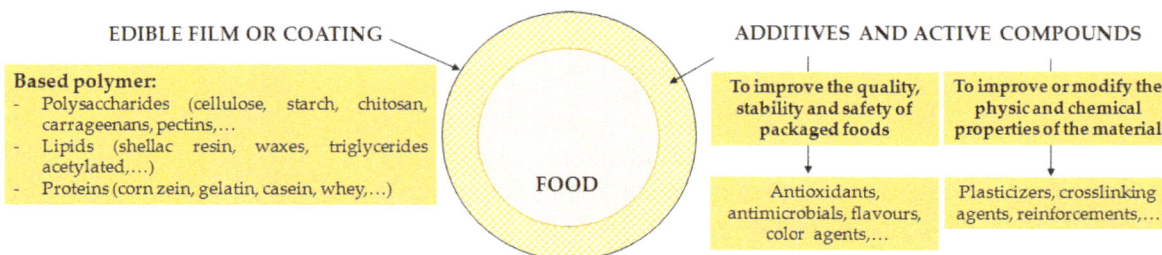

Figure 1. Edible films' and coatings' compositions.

Essential oils derived from plants [14,15], organic acids [16], nisin or natamycin from microbial sources [17], enzymes obtained from animal sources [18], like lysozyme and lactoferrin, and polymers, such as chitosan [19], have been also proposed as active agents against microorganisms in food.

Bioactive edible coatings can be applied to food surfaces by dipping or spraying, and they have been tested in meat, fish, dairy products or minimally-processed fruit and vegetables. Krašniewska et al. reported that pullulan with 1.0% of oregano essential oil was an effective active material to maintain the safety and quality of fresh Brussels sprouts stored at 16 °C for 14 days [20]. Eggs coated with chitosan and lysozyme-chitosan combinations maintained their internal quality for a long-term storage since the albumen properties (including pH, dry matter, viscosity and relative whipping capacity) were kept as in their original state [21]. Shakila et al. evaluated the antimicrobial activity against *Staphylococcus aureus* (*S. Aureus*), *Aeromonas hydrophila* and *Listeria monocytogenes* (*L. monocytogenes*) of gelatin coatings with different additives (chitosan, clove and pepper) in vacuum-packed fish steaks. Satisfactory results were obtained for coatings with chitosan and clove, which extended shelf-life from four to eight days at 4 °C [6]. Other applications of these antimicrobial coatings in meat products can be also found in the recent literature. For example, Matiacevich et al. studied the effect of alginate-based edible coatings with propionic acid and thyme essential oil on the microbiological growth in chicken breast fillets stored under refrigerated conditions, demonstrating the inhibition of *Salmonella* after seven days [22].

This review focuses on the use of antimicrobial agents in edible coatings, including incorporation methods and the main characteristics of these innovative combinations. An overview of the state of the art of antimicrobial coatings and their use in different food products is presented in the following sections.

2. Antimicrobial Agents

A wide variety of antimicrobial agents has been described for their use in coating formulations for packaged fresh food [9]. Some of them are derived from natural sources, and they have been traditionally used as food additives and been awarded with the Generally Recognized as Safe (GRAS) label. Organic acids, enzymes, bacteriocins and plant-derived compounds and by-products, including essential oils, such as thymol or carvacrol, have been proposed for their antimicrobial performance [4]. The selection of the most appropriate antimicrobial agent is an important issue, and some relevant aspects should be considered, particularly the good interaction between the polymer matrix and the active agent, the presence of other functional additives in the formulation, the type and properties of the packaged food and their effectiveness against the target microorganisms. These aspects can result in major modifications of the coating final properties and the antimicrobial activity against the target microorganisms.

Organic acids and their salts are those compounds most commonly used as antimicrobial agents due to their well-known effectiveness and low cost. In addition, many of them are also labelled with the GRAS status, and they are accepted as food preservatives in the current European legislation [16]. Some of them have been traditionally incorporated into coatings, such as lauric, acetic, sorbic, citric, benzoic or propionic acids. Their antimicrobial effect has been related to the increase in the proton concentration thereby decreasing the external pH. In this way, the integrity and permeability of the microbial cell membranes can be altered, as well as some disturbance in the nutrients transport can be observed, causing cell inactivation and death [1]. Jin et al. [23] reported the effect of chitosan and the combination of three organic acids (lactic, levulinic and acetic) on fresh American ginseng roots. The antimicrobial activity of this coating was evaluated in terms of microbial stability by using total aerobic bacteria, yeasts and molds, showing that the microbial loads on coated samples were relatively stable during the 38-week storage period, with populations ranging between 2.2 and 2.9 log $CFU \cdot g^{-1}$. They also concluded that the use of a mixture of these antimicrobials can be more effective than the separate addition of the individual compounds due to synergistic or additive effects, permitting the inhibition or even killing certain microorganisms [24].

The antimicrobial activity of different spices and herbs has been known from ancient times, and they have been traditionally added to food as seasoning additives due to their aromatic properties [25]. Several studies have reported the preservation abilities of plant-derived compounds in food applications, as well as factors influencing their effectiveness. Among them, essential oils and their main components, such as thymol, carvacrol, p-cymene and γ-terpinene in *Thymus* [26], are gaining interest due to the presence of phenolic compounds or other hydrophobic components [27]. These phenolic groups are responsible for damage to the cell wall, interaction with and disruption of the cytoplasmic membrane, damage of membrane proteins, leakage of cellular components, coagulation of cytoplasm and depletion of the proton motive force [28]. All of these effects produce death to microorganisms by modification of the structure and composition of the bacteria cell walls [27]. However, their structural diversity and variations in their chemical composition can produce significant differences in their effectiveness against pathogens. For example, the antimicrobial effect of *Mosla chinensis* methanolic extract was evaluated against eight bacterial and nine fungal strains [29]. Results showed that this essential oil, whose main components are carvacrol (57%), p-cymene (14%), thymol acetate (13%), thymol (7%) and c-terpinene (2%), showed great potential against two Gram-positive bacteria commonly found in many food products, *S. aureus* and *L. monocytogenes*.

In general terms, essential oils are rich in monoterpenes, sesquiterpenes, esters, aldehydes, ketones, acids, flavonoids and polyphenols, which are well known for their antimicrobial performance [30]. Their mechanism of action against bacteria is not clear, since each compound in each essential oil exhibits a unique antimicrobial mechanism, which is specific to a particular range of food and microorganisms [31]. Table 1 shows some antimicrobial agents with antimicrobial activity on edible coatings.

Table 1. Antimicrobial activity of some edible films.

Antimicrobial Agent	Matrix	Microorganisms Tested	Reference
Citral and eugenol	Alginate and pectin	Aerobic mesophilic microorganisms, yeast and molds	[32]
Oregano essential oil	Basil seed gum	Aerobic mesophilic microorganisms, yeast and molds	[33]
Oregano essential oil	Mucilage	*L. monocytogenes, Salmonella typhimurium, Bacillus cereus, Yersinia enterocolitica, P. aeruginosa, S. aureus, E. coli, E. coli* O157:H7	[34]
Lemongrass	Alginate	*E.coli*, psychrophilic bacteria, molds and yeast	[35]
Clove	Gelatin	Total bacterial counts, *pseudomonas, Enterobacteriaceae*, lactic acid bacteria	[36]
Orange essential oil	Gelatin	Total viable counts, psychrotrophic bacteria and *Enterobacteriaceae*	[37]
Oregano, thyme essential oils	Soy protein	*E. coli* O157:H7, *S. aureus, P. aeruginosa* and *Lactobacillus plantarum*	[38]
Carvacrol and cinnamaldehyde	High methoxyl pectin and apple, carrot or hibiscus puree films	*L. monocytogenes*	[39]

Alparslan et al. added different amounts of essential oils obtained from orange (*Citrus sinensis* (*L.*) Osbeck) leaves into an edible gelatin coating solution with noticeable effects on the quality and shelf-life of shrimp covered by this edible film coating. The total viable counts, psychrotrophic bacteria counts and *Enterobacteriaceae* were determined [37]. Zhang et al. [40] used cinnamon bark and soybean oils in antimicrobial sodium alginate coatings, showing complete inhibition of *L. monocytogenes*, *Salmonella enterica* (*S. enterica*) and *Escherichia coli* O157:H7 (*E. coli* O157:H7) inoculated on cantaloupes after 15 days of storage. Other essential oils coming from different plants have been also used in edible coatings to improve the shelf-life of food products: garlic [41], basil [42], clove and pepper [6], oregano [20], thyme [22,43] and mandarin [15], among others.

Among natural antimicrobials from animal origin, chitosan has captured the attention of researchers and the food industry for commercial applications. This material offers the possibility of obtaining coatings to cover fresh or processed foods to extend their shelf-life, being an excellent film-forming material showing antifungal and antimicrobial activity due to its polycationic nature [44]. Regarding the mechanism of the antimicrobial activity, different ideas have been reported, although the precise mechanism has not been yet determined. Some authors have drawn theories based on: (i) interactions by electrostatic forces between chitosan amine groups and microbial cell membranes; (ii) the action of chitosan as a chelating agent; (iii) the penetration of low molecular weight chitosan molecules through the cell membrane; or (iv) modifications on cell surfaces that may affect the integrity of the microbial cell membrane interfering with energy metabolism and nutrient transport in bacteria cells [19,45]. Carrión-Granda et al. reported the combined effect of chitosan coatings with oregano and thyme essential oils (0.5%) in modified atmosphere packaging conditions onto peeled shrimp tails. The antimicrobial effect of this coating agent was evaluated for 12 days under chilling, and they further determined the total viable counts, lactic acid bacteria, *Enterobacteriaceae* and total psychrotrophic bacteria. Results showed that antimicrobial activity of the combination of chitosan with thyme was significantly increased in relation to samples treated just with chitosan, since the reduction was about 1 log CFU·g^{-1} at the end of the storage period [46].

Other antimicrobials obtained from bacteria, such as nisin, pediocin, natamycin or reuterin, have been also used against target microorganisms. Particularly, bacteriocins are proteinaceous compounds with antimicrobial activity produced by lactic acid bacteria [18]. Ribosomally-synthesized peptides with antimicrobial activity, such as nisin, have shown their ability to inhibit the growth of Gram-positive and spore-forming bacteria associated with food [1]. Nisin is able to penetrate the cytoplasmatic membrane of bacteria causing the leakage of cytoplasmic contents and dissipation of the membrane potential [17]. Nisin also inhibits Gram-negative bacteria when chelating agents, such as EDTA or lysozyme, are present [9]. Nisin has been tested as an antimicrobial additive in coated surfaces in different products, such as dairy foods [47], fruits and vegetables [48,49], meats and fishes [50–53].

3. Application of Antimicrobial Edible Coatings

An edible coating material can be defined as a thin layer of the selected formulation, which is applied directly over food in liquid form by using different techniques, such as immersion, spraying, etc. [54]. In addition, an edible film is defined as a packaging material, which is a thin layer placed on or between food components, used as wraps or separation layers [55]. The edibility of films and coatings is only achieved when all components including biopolymers, plasticizers and other additives are food-grade ingredients, while all of the involved processes and equipment should be also acceptable for food processing [56]. The main coating techniques used in these materials are described below.

3.1. Spraying

The interest in the industrial application of sprays in packaging is greatly increasing not only due to the potential cost reduction that the spraying technique may imply, but also by the high quality of the final product. This could be achieved when compared to those obtained by using conventional techniques [57], which mainly involve high temperatures, leading to important losses in volatile antimicrobial agents. The spraying technique offers uniform coating, thickness control and the possibility of multilayer applications [58,59]. Spraying systems do not contaminate the coating solution, allow temperature control and can facilitate automation of continuous production. In this sense, temperature directly affects the permanence of the antimicrobial agents due to their high volatility. In addition, the thickness control is very important to establish the amount of antimicrobial agent necessary to be released [60].

On the other hand, sprayed coatings combine hydrophobic and hydrophilic substances. Indeed, the spray can generate a coating with two solutions, by applying the emulsion solution directly, which is formed before atomization, or by forming a bilayer after two spray pulverizations. The application of a bilayer has the disadvantage of requiring four steps (two spray applications and two drying processes) [61].

By using this technique low viscosity coating solutions can be easily sprayed at high pressures [62]. The drop-size distribution of a sprayed coating solution can be up to 20 μm, whereas electrospraying can produce uniform particles lower than 100 nm from polymer and biopolymer solutions. Furthermore, the formation of polymer coatings by spraying can be also affected by other factors, such as the drying time and temperature [63].

A spray is a collection of moving droplets as the result of atomization processes to break up bulk liquids into droplets [64], including the following considerations:

- Increase in the liquid surface area, which is an important issue in processes where rapid vaporization is required. In fact, in antimicrobial applications, it is important to obtain homogeneous coatings where the additive is available to release quickly to the surrounding environment.
- The formation of an even surface, since the droplets dispersion generates coatings with homogeneous spatial patterns and controlled thicknesses. This is essential to evaluate the kinetics release of the antimicrobial additive.

- Cost reduction, since spraying techniques are usually fast and efficient processes in terms of solvent and material consumption.

Different spraying techniques have been proposed [65]:

- Air-spray atomization: In this case, the fluid emerging from a nozzle at low speed is surrounded by a high-speed stream of compressed air (up to 8 bar). The friction between the liquid and air molecules accelerates and disrupts the fluid stream and causes atomization.
- Pressure (airless) atomization: High pressures (34–340 bar) force the fluid through a small nozzle (spray tip) to emerge as a sheet. The friction between the fluid and the air molecules disrupts the stream, breaking it initially into fragments and ultimately into droplets. The fast-moving, high-pressure liquid stream provides energy enough to overcome the fluid's viscosity and surface tension by forming small droplets.
- Air-assisted airless atomization: This technique combines the features of air spraying and airless techniques. It is based on the principle of the airless atomization with the addition of a concentrated airflow to obtain droplets in a more controlled way.

A variety of factors can affect the droplet size and the liquid stream [66], corresponding to the fluid (surface tension, viscosity, density and temperature) and technique (spray tip size and shape, fluid and air pressure). Heat-sensitive antimicrobial agents are volatile compounds, and they should be preferably incorporated onto food matrices by non-heating methods, such as spray coating. In this sense, Peretto et al. [67] used spraying as an innovative and efficient technique for the application of an edible alginate coating enriched with carvacrol and methyl cinnamate (natural antimicrobials) onto fresh strawberries demonstrating superior performance on firmness, color retention and weight loss reduction in comparison with uncoated samples. Other examples of spayed-antimicrobial edible coating films include those based on a tapioca starch with green tea extracts that have reduced the growth of aerobic microorganisms and yeasts when applied to fruit-based salads, romaine hearts and pork slices [68].

3.2. Dipping

The dipping method has been used to form coatings onto fruits, vegetables and meat, among other food products [69]. Properties such as density, viscosity and surface tension of the coating solution are important to determine the film thickness [5], since dipping techniques are able to form thick coating layers [62]. In this method, a membranous film is formed over the product surface by directly dipping the product into the aqueous coating formulation and further air-drying. This process may be separated into three stages [70]:

- Immersion and dwelling: The substrate is immersed into the precursor solution at a constant speed followed by dwelling to ensure that interaction of the substrate with the coating solution is enough for complete wetting.
- Deposition: A thin layer of the precursor solution is formed on the food surface by deposition. The liquid excess drains from the surface and is removed.
- Evaporation: The solvent excess evaporates from the fluid, forming the thin film.

It is important to highlight that, when using this method, the coating solution must be diluted, and significant residual coating material is produced. The optimal amounts of coating solution cannot be easily controlled by dipping, and a further processing step to dry off surplus solution is needed, requiring extra time and hindering industrial applications of this technique [57].

Different examples of the use of the dipping technique for edible coating processing have been reported. A significant shelf-life extension for oysters was achieved by dipping them into a sodium acetate (10 g/L), solution resulting in a coating with sodium alginate (40 g/L) and further use of modified atmosphere packaging (MAP) conditions (0:75 O_2:CO_2) [71]. In addition, other authors evaluated the dipping method by coating papaya fruit with k-carrageenan [72] and carrots with sodium alginate [73] as packaging strategies to extend the shelf-life of fresh foods.

3.3. Spreading

This method, also known as brushing, consists of the controlled spreading of a suspension onto the material surface to be further dried. This method is considered a valid alternative for the preparation of films with dimensions larger than those prepared by casting procedures. The thickness of the coating suspension is controlled by a blade attached to the lower part of the spreading device, and the film drying is held on the support itself, by circulation of hot air. This method can be applied to the production of polysaccharides and protein-based films [74]. Two parameters can be used to characterize the spreading of liquid droplets: the wetting degree and the spreading rate [75]. In this sense, contact angle measurements are commonly used to evaluate the degree of spreading/wettability of a surface by a particular liquid. Spreading is affected by several factors, such as the substrate properties (surface roughness and geometry), system conditions, such as temperature and relative humidity, and liquid properties (viscosity, surface tension and density) [76]. Viscosity was found to have a major effect as it defines the resistance of a liquid to spreading on solid surfaces. Therefore, spreading of highly viscous liquids is more difficult than liquids with low viscosity [77].

As an example of antimicrobial spread surfaces, polyethylene (PE) was coated with chitosan, where the polymer surface was previously corona-treated to enhance the chitosan adhesion. The antimicrobial activity against Gram-positive (*L. monocytogenes*) or Gram-negative (*E. coli, Salmonella*) bacteria was proven in uncoated and chitosan-coated PE films [78]. Active packaging films containing partially purified antibacterial peptide solutions (ppABP) produced by *Bacillus licheniformis* Me1 were developed by using low density polyethylene (LDPE) and cellulose films by spread coating, showing a remarkable antibacterial activity. The release study of ppABP from the coated films showed that ppABP was released from LDPE films as soon as they are in contact with water, while a more gradual release of the coated ppABP was observed in the case of cellulose films [79].

In summary, although coating application methods are very diverse, their selection depends on the desired product, coating thickness, solution rheology and the drying technique.

4. Properties of Antimicrobial Coatings for Food Packaging Applications

Antimicrobial edible films and coatings have demonstrated their ability to protect foodstuff against spoilage and to decrease the risk of pathogens growth by controlling the diffusion and gradual release of embedded antimicrobial agents onto the food surface [80]. The selection of the most adequate antimicrobial agents is based on the consideration of their effectiveness against the target microorganisms, as well as their possible interactions with the film-forming polymers and the packaged food. These interactions can modify the antimicrobial activity and their own film characteristics, constituting key factors for the development of antimicrobial films and coatings [9]. The most common functions of edible films and coatings in food packaging are described in Figure 2.

Different physical tests to determine the mechanical and barrier properties of edible coatings and films have been reported. Quasi-static tension or puncture tests are applied to edible films to determine mechanical parameters, such as the elastic modulus, tensile strength and strain at break [81]. Water vapor permeability (WVP) of these films is determined in accordance with the ASTM E-96 static method. The resistance of films to water is critical for the potential application of these films since much food is stored in aqueous solutions. Sometimes, high solubility in water is desired, particularly when the film or coating will be consumed simultaneously with food. On the other hand, as a consequence of the generally poor barrier to water vapor and low mechanical strength of biopolymers, some edible films and coatings have still limited applications in food packaging [44]. In this sense, the use of polysaccharides is limited by their water solubility and high WVP. Blending with different biopolymers or addition of hydrophobic materials such as oils or waxes can be useful to overcome this shortcoming [82]. Chemical modification of the biopolymer structure, in particular by crosslinking [83], has been also proposed for such purposes [84,85]. Gutierrez et al. used this method to modify starch by cross-linking with sodium trimetaphosphate [86]. Edible films showed hydrophilic characteristics and some increase in the degradation temperature. In addition, the films developed in their studies were actually edible and, consequently, considered safe by the U.S. Food and Drug administration (FDA). Schmid et al. carried out swelling studies to demonstrate the improvement of the structural stability of whey protein isolate-based coatings. Results demonstrated that the denaturation degree had a significant influence on the cross-linking density, consequently a direct proportion of the number of disulfide bonds in the WPI-based network [87].

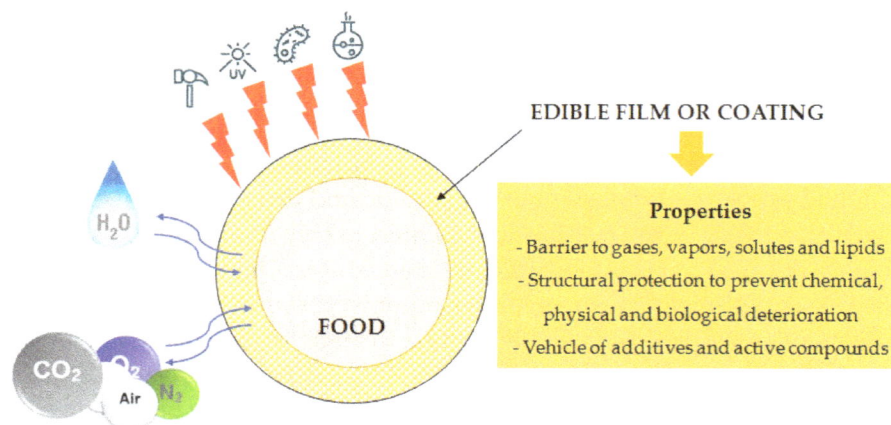

Figure 2. Main functions of edible films and coatings in food packaging applications.

Physico-chemical properties (thickness, WVP, puncture strength, tensile strength and elongation at break) of chitosan films enriched with oregano essential oil were evaluated [88]. The results of these tests showed that this combination resulted in increased thickness, higher elasticity, reduced puncture and tensile strength and lower WVP compared to pure chitosan films and chitosan films with surfactant.

Antimicrobial properties of edible coatings can be evaluated by using different types of microorganisms and different testing methods, including the film disk agar diffusion assay, the enumeration by plate count of microbial population and the film surface inoculation test [9]. These methods have been applied for the in vitro evaluation of the antimicrobial films performance. However, when coatings are applied onto food, their effectiveness is evaluated through the enumeration of indigenous or inoculated microbial population during food storage. For example, bologna-ham slices covered with chitosan films with 1% oregano essential oil under storage for five days at 10 °C absorbed 60 ppm of the essential oil and were effective against *L. monocytogenes* and *E. coli* O157:H7. The application of pure chitosan films reduced the pathogen counts on meat products from 1–3 logs, and chitosan films enriched with 1 and 2% oregano essential oil were sufficient for a four log reduction of *L. monocytogenes* and *E. coli* O157:H7.

5. Food Packaging Applications

This section is focused on the main studies reported in the scientific literature of antimicrobial coatings for food packaging purposes. In general terms, due to their chemical characteristics and their high degradation rates, antimicrobial coatings have been widely studied for fish and meat products, while some studies have been also reported with fruits and vegetables. Table 2 summarizes recent studies reported in the literature regarding the applicability of new antimicrobial coatings for food products.

Table 2. Antimicrobial coatings developed for fish and meat products.

Food	Product	Matrix	Antimicrobial Agent	Ref
Fish products	Sliced fresh *Channa argus*	Chitosan and polyethyleneimine	Thyme essential oil	[89]
	Fish sausages	Chitosan and warm-water fish gelatin	Shrimp concentrate from *Litopenaeus vannamei* cooking juice	[90]
	Trout fillets	Chitosan	Lactoperoxidase enzyme	[91]
		Whey protein	Lactoperoxidase enzyme	[92]
	Fresh Indian salmon	Gelatin from waste of *Nemipterus japonicas*	Garlic (*Allium sativum*) Lime (*Citrus aurantifolia*)	[93]
	Silvery pomfret	Chitosan	Gallic acid	[94]
	Rainbow trout	Soy, whey, egg, wheat gluten, corn, collagen and fish proteins	–	[95]
	Gilthead seabream fillets	Methylcellulose	*Satureja thymbra* (L.) essential oil	[96]
	Fresh silver carp fillets	Methylcellulose	*Pimpinella affinis* essential oil	[97]
	Surimi	Zein	Iron chelator	[98]
Meat products	Cooked cured chicken breasts	k-carrageenan and chitosan	Mustard extract	[99]
	Fresh chicken breasts	k-carrageenan and chitosan	Mustard extract	[100]
	Pork meat	Sodium Alginate	Thyme and propionic acid	[101]
		Oleic acid as part of starch	Lactic acid, nisin and lauric arginate	[22]
		Chitosan	Clove oil and/or ethylenediaminetetraacetate	[50]
	Dry-cured ham	Propylene glycol, xanthan gum and carrageenan with propylene glycol alginate	–	[102]
	Ham	Soybean meal and xanthan	Lactoperoxidase	[103]
	Frankfurters and ham	Sodium alginate	Ethanol	[104]
	Roast beef	Chitosan	Lauric arginate ester, lactic and levulinic acids	[105]

5.1. Fish Products

In general terms, fish products are susceptible to fast degradation mainly due to their high content in lipids, moisture losses and deterioration of the sensory and chemical quality of their muscles. As a result, coatings based on proteins, polysaccharides and lipid materials have been recently applied as antimicrobial active packaging systems to extend the shelf-life of seafood by the reduction of pathogenic microorganisms at their surface [106]. In this context, important results have been reported to control the proliferation of *L. monocytogenes* on cold smoked salmon [107]. Neetoo et al. [108] also found that alginate coatings supplemented with 2.4% sodium lactate and 0.25% sodium diacetate solutions significantly delayed the growth of *L. monocytogenes* during a 30-day storage at 4 °C on cold-smoked salmon slices and fillets. As is seen in Table 2, changes in the quality of *Channa argus*, treated with thyme essential oil (1% v/v) added to chitosan (2% w/v) and poly(ethylene imine) (PEI) (1% v/v) edible compounds, were used as film-forming substrates, and they were studied for 10 days of storage at 4 °C [90]. Results underlined the beneficial effect of the addition of PEI or thyme essential oil in combination with the chitosan biopolymer matrix, reducing the total viable bacterial growth and increasing by 4–5 days the shelf-life of the packaged product. In a different study, *S. aureus* or lactic acid bacteria were not detected in batches of fish sausages during chilled storage for at least 15 days when they were coated with chitosan (1% w/v) and warm-water fish gelatin (1% w/v). Shrimp (*Penaeus* spp.) cooking juice was also used as an antimicrobial agent at two different concentrations (1% and 2% w/v) [90]. In this study, both formulations showed antimicrobial activity against some fish spoilers and pathogenic organisms during chilled storage of Alaska pollock sausages. Enterobacteria counts in sausages wrapped in films of chitosan-gelatin-shrimp concentrate remained below the detection limit during storage (45 days), whereas lactic acid bacteria and *S. aureus* were not detected.

Chitosan was used as an edible antimicrobial coating of rainbow trout for storage at $4 \pm 1\ °C$ for 16 days [91]. In this case, lactoperoxidase (LPO) is one of the most important enzymes used as a natural antimicrobial agent, since it has a broad antimicrobial spectrum, showing bactericidal effect on Gram-negative bacteria, a bacteriostatic effect on Gram-positive bacteria, antifungal and antiviral activities [107]. LPO (5% v/v) was coated onto a chitosan matrix (1.5% w/v), showing significant reduction in *Shewanella putrefaciens*, *Pseudomonas fluorescens* and psychrotrophic and mesophilic bacteria. Other combinations of LPO with glucose oxidase, D-(+)-glucose and potassium thiocyanate (1.00, 0.35, 108.70 and 1.09 weight ratio) were suggested, while chitosan concentration was always 1.5% (w/v). Reductions in growth rate for *S. putrefaciens*, *P. fluorescens*, psychrotrophic and mesophilic bacteria at 4 °C were observed, in particular for formulations of chitosan with LPO, since samples did not reach the control values (6–7 log $CFU \cdot g^{-1}$) in psychrotrophic and mesophilic bacteria after 12 and 16 storage days. Similar studies incorporated LPO at concentrations up to 7.5% (v/v) in whey protein coatings in rainbow trout preservation under refrigeration temperatures for 16 days [92]. Results suggested the bactericidal capacity of LPO, which is related to its high content in glucose-oxidase that produces oxidizing products and, consequently, decelerating the growth of certain microorganisms in fish products. Other studies used chitosan as supporting material for the development of antimicrobial coatings for fresh Indian salmon (*Eleutheronema tetradactylum*) fillets [93]. In this case, combinations formed by gelatin extracted from the processing waste of *Nemipterus japonicus* collected from a commercial surimi processing plant (10 g in 60 mL of distilled water) with 1.5 mL of chitosan solution and concentrations of 30%, 40% and 50% v/v of lime and garlic natural extracts were used as the coating material. These combinations resulted in acceptable protection to salmon up to 12 and 16 days, respectively. The garlic extract showed better antimicrobial activity than lime since lower total plate counts were reported in this study. Finally, gallic acid (0.206%, w/v) was added to the chitosan matrix (2%, w/v) as an edible coating for silvery pomfret (*Pampus argenteus*) stored at 4 °C for 15 days with a reduction of the microbial growth with lower total psychrotrophic count and total plate count values [94].

Rainbow trout is one of the most economically important freshwater-cultured fish species, and some authors have proposed different antimicrobial coatings to preserve their quality and nutrition facts against the microbial growth. Table 2 shows different coatings obtained from protein sources for such purpose. Thus, smoked rainbow trout (*Oncorhynchus mykiss*) was protected with different protein-based edible coatings, in particular soy protein isolates, whey protein isolates, egg white powder proteins, wheat gluten, corn proteins, gelatin, collagen and proteins from two different fish species (rainbow trout and Atlantic mackerel) in combination with vacuum packaging and refrigeration for a period up to six weeks [95]. It was reported that the formation of acetic acid solutions reduced pH to 3–4, and the formed ethyl alcohol showed some antimicrobial effect. The highest reductions in microbiological activity were obtained for samples coated with soy protein isolates, and this result was related to its high content in isoflavones, which reduced the growth of mesophilic aerobic bacteria. In addition, coatings of wheat gluten, gelatin, collagen and proteins obtained from rainbow trout and Atlantic mackerel fish also inhibited to some degree the bacterial growth. Zein has been also studied as edible coating with the incorporation of a polymeric iron chelator molecule, based on hexadentate 3-hydroxypyridinone, to preserve commercially-manufactured fish balls [98]. This product is a hot-water bath cooked surimi product, which is popular in many countries, but it is highly perishable because of its high moisture content and abundance of fish muscle. The addition of 1 mg/mL of the iron chelator has been described as a potential antimicrobial coating since the inhibition zone of the four tested strains (two Gram-positive bacteria, *B. subtilis* and *S. aureus*, and two Gram-negative bacteria, *E. coli* and *Salmonella* spp.) significantly increased by using these edible coating films.

Methylcellulose is another polysaccharide widely used as an antimicrobial edible coating matrix. In this context, the effectiveness of the *Satureja thymbra L.* essential oil was evaluated as an antimicrobial coating of gilthead seabream fillets stored at 0 °C with a shelf life extension between 25% and 35% when fillets were coated with a combination of the essential oil (2%, v/v) and methylcellulose (1.5%, w/v), which was attributed to its high content in carvacrol [96]. In a different study, methylcellulose matrix (3%) was enriched with the *Pimpinella affinis* essential oil (1.5%, v/v) to be used as coating of fresh silver carp fillets for refrigerated storage (4 ± 1 °C) over a period of 16 days, showing the inhibition of total bacteria growth and psychrophilic counts [97].

5.2. Meat Products

Meat is quite easily contaminated by some pathogens, in particular by *L. monocytogenes*. In fact, one of the main challenges in the meat industry is to avoid the re-contamination by this microorganism in ready-to-eat products [109]. It was reported that the possibilities of contamination in meat products in the U.S. and Canada ranged from 0.4%–71% and 0%–21%, respectively [110]. The meat industry is prone to propose new strategies to eliminate this problem and the consequent economical losses through the development of innovative antimicrobial edible coatings supported in biopolymer matrices. For example, mustard extracts at 0.5% (w/v) were added to k-carrageenan- (0.2%, w/v) and chitosan-based (2%, w/v) coatings prepared using 1.5% malic or acetic acid [99]. These formulations resulted in the reduction of the viability of five different strains of *L. monocytogenes* on inoculated vacuum-packed, cooked and cured roast chicken slices at 4 °C due to their ability to hydrolyze sinigrin (a glucosinolate present in mustard extracts) with the formation of allyl isothiocyanate. These authors expanded the applicability of this coating to *Campylobacter* bacteria. The importance of this microorganism is due to its prevalence since it has been reported as the second responsible of food-borne illnesses with 145,350 and 845,024 cases per year in Canada and the U.S., respectively [111]. Then, the mustard extract in concentrations up to 300 mg/g was added to k-carrageenan- (0.2%, w/v) and chitosan-based (2%, w/v) coatings prepared using acetic acid (1%, v/v), and the effect against *Campylobacter jejuni* (*C. jejuni*) on vacuum-packaged fresh chicken breasts stored at 4 °C was studied [100]. They concluded that coatings containing 200–300 mg/g of mustard extract reduced the population of *C. jejuni* on chicken breasts compared to those coated with only k-carrageenan/chitosan.

Other antimicrobial agents have been proposed to obtain new bio-based edible coatings active against *L. monocytogenes*. In this context, thyme (*Thymus vulgaris* L.) and propionic acid (both at 0.5%, w/w) have been reported as effective antimicrobial agents when they were added into edible coatings based on sodium alginate and sorbitol, both at 1% w/w. In fact, these active agents improved the shelf life and safety of fresh chicken against *E. coli* and *L. innocua* [101].

A mixture of lactic acid, nisin and lauric arginate was added at different concentrations to an oleic acid nanoemulsion as part of a waxy starch-based edible coating (4%, w/v) to reduce the growth of *Brochothrix thermosphacta*, *L. monocytogenes* and *Micrococcus luteus* in pork meat [22]. The optimum combination of antimicrobial agents was 17.5 mg/mL of lactic acid, 3.75 mg/mL of nisin and 0.0625 mg/mL of lauric arginate showing the highest antimicrobial effect.

Raw meat shows high water activity and plenty of nutrients to allow bacterial growth, resulting in major losses on quality at cold temperatures, so the use of active edible films is a promising alternative to improve the meat quality. However, deeper research is necessary since most of these films did not improve the meat shelf-life significantly. Another alternative is based on the use of bioactive edible coatings based on chitosan (2% w/v) with clove oil (0.05% v/v) and/or ethylene-diamine-tetra-acetate (10 mM) to reduce *E. coli* and *S. aureus* growth on refrigerated lean pork slices at 4 °C [50].

Dry-cured ham is another major product of the meat sector, and contamination results in a serious problem, particularly in the main producers of this foodstuff (Spain and the U.S.) [112]. *Tyrophagus putrescentiae* infestation in cured ham has been reduced by coating with xanthan gum with 10%–20% propylene glycol and carrageenan/propylene glycol alginate [102]. The contamination by *Salmonella* has been reduced on ham slices by the application of an antimicrobial coating based on defatted soybean meal (26.6%) and lactoperoxidase (81 U/mg) extracted from bovine milk (22.3%) blended with xanthan (13.2%), glycerol (6.0%) and water (31.9%) [103]. Low levels of *L. monocytogenes* on Frankfurt sausages from pork meat and ham slices were achieved when samples were coated with Na-alginate edible films (1.5%, w/v) immersed in three different Greek alcoholic beverages, namely "tsipouro" (41% v/v ethanol), "raki" or "tsikoudia" (39.6% v/v ethanol) used as antimicrobial agents and stored under chilling conditions [104]. However, the same study revealed that the inadequate reheating after storage may not be enough to cancel the risks from *L. monocytogenes* populations on roast beef samples. Nevertheless, these risks were reduced under chilling temperatures for 30 days by approximately 0.9–0.3 log CFU/cm^2 when coated with 5% (w/w) chitosan in an acidic solution containing 2% (v/v) of acetic, lactic and levulinic acids and 20% (v/v) of lauric arginate ester [105].

In conclusion, these studies demonstrated the effectiveness of the new antimicrobial edible coatings against pathogens inoculated on different meat products, such as fresh chicken and pork, as well as on ready-to-eat meat products, such as dry ham, cooked ham and roast beef.

5.3. Fruits and Vegetables

Fruits and vegetables are characterized by their high nutritional value, particularly being rich in vitamins, minerals and fibers. However, their high water activity promotes fast degradation since they are constituted by living tissues susceptible to enzymatic browning, off-flavors development, texture breakdown and microbial contamination [113]. These undesirable properties, as well as the continuous increase in the consumer's requirements have attracted the food industry's interest in improving protection of fresh fruits and vegetables. Some alternatives have been proposed to reduce spoilage and microbiological contamination, increasing their shelf-life. Research efforts in this field have resulted in the development of new antimicrobial edible coatings. Table 3 summarizes the most relevant studies reported in the literature on the applicability of new antimicrobial coatings for fruits and vegetables.

Table 3. Antimicrobial coatings developed for fruits and vegetables.

Food Applicability	Product	Matrix	Antimicrobial Agent	Ref.
Fruits	Watermelon (*C. lanatus*)	Sodium-alginate, pectin, and calcium lactate	Trans-cinnamaldehyde	[114]
	Cantaloupe melon	Chitosan and pectin	Trans-cinnamaldehyde	[115]
	Persimmon	Pectin, citric acid and calcium chloride	Nisin	[116]
	Strawberries	Pectin, pullulan, and chitosan	Sodium benzoate, Potassium sorbate	[117]
		Pectin and calcium chloride	Eugenol, Citral, Ascorbic acid	[118]
	Raspberries	Pectin and calcium chloride	Eugenol, Citral, Ascorbic acid	[119]
	Arbutus unedo L. fruit	Sodium alginate	Citral, Eugenol	[10]
	Pineapple	Sodium alginate	Lemongrass essential oil	[120]
	Fuji apples	Sodium alginate	Lemongrass essential oil	[35]
	Blueberry	Chitosan	Carvacrol, Cinnamaldehyde, *Trans*-cinnamaldehyde	[121]
	Avocado	Gum arabic, aloe vera and chitosan	Thyme oil	[122]

Table 3. *Cont.*

Food Applicability	Product	Matrix	Antimicrobial Agent	Ref.
Vegetables	Pepper	Chitosan	Lemongrass essential oil	[123]
		Pullulan	Leather bergenia leaves ethanolic extracts	[124]
		Pullulan	*Satureja hortensis* aqueous or ethanolic extracts	[125]
	Pumpkin	Xanthan gum, guar and chitosan	–	[126]
		Starch	Carvacrol	[127]
		Zein	Benzoic acid	[128]
	Cherry tomatoes	Zein	Cinnamon, Mustard essential oil, commercial wax	[129]
	Fresh cut broccoli	Chitosan	Bioactive compounds and essential oils	[130]
	Cauliflower florets	Maltodextrins and methylcellulose	Lactic acid, Citrus extract, Lemongrass essential oil	[131]
	Green beans	Modified chitosan	Mandarin essential oil	[132]
	Rucola	Modified chitosan	Lemon, mandarin, oregano or clove essential oils	[133]

Fresh-cut watermelon (*Citrullus lanatus*) was coated with three different solutions: sodium alginate, pectin and calcium lactate, and stored at 4 °C for 15 days. Glycerol was used as the plasticizer in all cases. Sodium alginate was added to the solution in three different concentrations (0.5, 1 and 2 g/100 g). Finally, *trans*-cinnamaldehyde encapsulated powder (2 g/100 g) was added to the solution as the active antimicrobial agent. Watermelon samples coated with sodium alginate showed the lowest counts on psychrotrophic and coliform microorganisms while also decreasing the growth of yeasts and molds [114]. Fresh-cut cantaloupe (*Cucumis melo L.*) was coated with chitosan (0.5, 1, 2 g/100 g, pectin (0.5, 1, 2 g/100 g), encapsulated in trans-cinnamaldehyde (1, 2, 3 g /100 g) and stored under chilling conditions at 4 °C [115]. The authors concluded that the highest concentration of the antimicrobial agent (3 g/100 g) was more effective (4.44 log cycles reduction) against the aerobic population than the uncoated sample by Day 15. However, coating with chitosan with no other active agents did not show significant inhibition for these microorganisms. Pectin biopolymers were also used to coat fresh-cut persimmon (*Diospyros kaki*) [116]. The edible coating was elaborated from a base solution of apple pectin at 10 g/kg emulsified with oleic acid and Tween80, whereas glycerol was added as the plasticizer. As anti-browning agents, 10 g/kg citric acid and 10 g/kg calcium chloride ($CaCl_2$) were added into the coating solution. Finally, nisin was added as the antimicrobial agent at 500 international units (IU) per mL. Persimmons were peeled, cut and dipped into the coating solution for 3 min. After dipping, fruit pieces were removed and dried before being placed on polypropylene trays and sealed with polypropylene-polyethylene terephthalate film. After eight days of incubation, the inhibition of the mesophilic aerobic bacteria growth was observed while coating also reduced the populations of *E. coli*, *Salmonella enteritidis* and *L. monocytogenes*.

Coating strategies are very common in strawberry (*Fragaria ananassa*) processing as a consequence of their short post-harvest life, high metabolism and microbial decay [134]. Edible active coatings based on pectin, pullulan and chitosan with sodium benzoate and potassium sorbate have been reported as potential antimicrobial coatings for strawberries [117]. Microbiological analyses showed that the application of coatings reduced the total aerobic counts, molds and yeasts growth, with chitosan offering the best results in microbial growth tests. Edible coatings based on sodium alginate (1%, w/v) and pectin (2%, w/v) enriched with essential oils (citral at 0.15% and 0.3%, w/v, and eugenol at 0.1% and 0.2%, w/v) also showed their antimicrobial effect against aerobic mesophilic and psychrophilic bacteria, molds and yeasts [118]. Similar results were reported for raspberry coated with the same formulations [119]. Edible coatings enriched with citral and eugenol were widely effective in reducing microbial spoilage. For example, essential oils or their constituents in alginate matrices are able to reduce microbial spoilage in fresh-cut pineapple [120] and fuji apple [35].

Other biopolymers used for coating fruit products have been also used as essential oil carriers in antimicrobial fruit active packaging. In this sense, chitosan has been reported as a coating for blueberries with the addition of three compounds with antimicrobial properties, carvacrol, cinnamaldehyde and *trans*-cinnamaldehyde (0.5%, w/v) [121]. Chitosan with this essential oils mixture was the most effective coating against mesophilic aerobic bacteria and also helped to reduce populations of bacteria and yeasts/molds. Antifungal effects against *Colletotrichum gloeosporioides* of gum arabic (10%, w/v), aloe vera (2%, w/v) and chitosan (1%, w/v) by themselves or in combination with thyme oil (1%, w/v) were studied on avocado fruit [122]. This study recommends the formulation of chitosan with thyme oil (3:1, v/v) as a potential antifungal coating for avocado storage.

Pepper has been widely coated onto different matrices since it is highly susceptible to chilling injury at temperatures below 7 °C [135]. However, at chilling temperatures, there is some enhancement in the rupture of the pepper surface and the consequent increase in susceptibility to contamination by different microorganisms, in particular by *Colletotrichum capsici*, the major causal agent of anthracnose. The addition of lemongrass essential oil at 0.5% and 1.0% (w/w) into 0.5 and 1.0% (w/w) of chitosan solution has been successful to control anthracnose in bell pepper [123]. In fact, the fungal growth was effectively controlled by 0.5% and 1.0% (w/w) lemongrass essential oil, whereas the application of 1.0% (w/w) of chitosan with 0.5% (w/w) of the lemongrass essential oil was effective as antimicrobial coating for bell peppers stored at room temperature for 21 days.

Bergenia crassifolia is another natural source to be considered in the protection of pepper. The antimicrobial effect of ethanolic extracts from bergenia leaves has been proven in pullulan coatings at different concentrations (0.4%, 1%, 2%, 5%, 10%, 20%, w/v) [124]. Samples coated with the antimicrobial solutions showed reductions in microbial growth by 1 log CFU/g when compared to control materials. A different coating for pepper to prevent the growth of Gram-positive and Gram-negative bacteria and *Penicillium expansum* has been recently described in the literature [125]. In this case, pullulan solution was prepared by dissolving pure pullulan (10%, w/v) and glycerol (5%, w/v) in distilled water. Then, aqueous or ethanol extracts of *Satureja hortensis* (20%, w/v) were added into the coating solution. Results showed that the formulation with the aqueous *Satureja hortensis* extract was more effective than ethanol against the growth of Gram-positive, Gram-negative bacteria and *Penicillium expansum*.

The development of different antimicrobial edible coatings for fresh-cut pumpkins (*Cucurbita moschata*) has increased in recent years [136]. Coatings based on different concentrations of xanthan gum, guar and chitosan reduced the growth of *Salmonella* ssp. [126]. Edible coatings with starch and carvacrol in minimally-processed pumpkin reduced the contamination by *E. coli*, *S. enterica serotype Typhimurium*, *Aeromonas hydrophila* and *S. aureus* [127]. Zein-based coatings with benzoic acid as the antimicrobial agent have been successfully tested on the quality of sliced pumpkin samples [128]. No mold growth was observed, and a final decrease in the total counts of mesophilic aerobic bacteria around the 1.0 log level was observed for coated samples. Zein has been also used as an edible matrix for coating cherry tomatoes [129]. The objective of this study was to investigate the effectiveness of zein-based coatings with propylene glycol (10%, w/v for both) in reducing populations of *S. enterica* serovar *Typhimurium* and preserving quality of cherry tomatoes. Then, a range of 5%–20% of cinnamon or mustard essential oils was added to the zein solution. On the other hand, a commercial wax formulation was also used in a different batch as an antibacterial agent. All of them were used to control the *S. enterica* dissemination in cherry tomatoes stored at 10 °C up to three weeks. As a result, the population of *S. Typhimurium* was reduced by 4.6 and 2.8 log CFU/g by the zein coatings with 20% cinnamon and 20% mustard oil, respectively. The same coating reduced populations of *S. Typhimurium* to levels below the detection limit. However, no antimicrobial activity was observed on the fruit coated with the commercial wax.

The antimicrobial properties of chitosan coatings (1%, v/v) enriched with four bioactive compounds (bee pollen, ethanol extract of propolis, pomegranate dried extract and resveratrol) and seven essential oils (tea tree, rosemary, clove, lemon, oregano, calendula and aloe vera) at different concentrations against mesophilic and psychrotrophic bacteria, *E. coli* and *L. monocytogenes* were studied on minimally-processed broccoli [130]. In vitro assays performed in tea tree, rosemary, pollen and propolis showed a remarkable inhibitory effects on *E. coli* and *L. monocytogenes*. Regarding *in vivo* analyses, in general terms, rosemary showed no significant effect on the reduction of the bacterial population. Chitosan coating with tea tree exerted a bacteriostatic effect on mesophilic and psychrotrophic bacteria counts with reductions around two order log lower than control sample up to seven days of storage. Chitosan coating with resveratrol or pomegranate produced a relevant reduction in mesophilic and psychrotrophic bacteria counts. Broccoli samples coated with chitosan and propolis showed a significant reduction in pathogen counts (1.0–2.0 log $CFU \cdot g^{-1}$), up to five days. When pollen was added to chitosan, a significant inhibitory effect in mesophilic and psychrotrophic bacteria counts (2.0–2.5 log $CFU \cdot g^{-1}$), compared to control samples, was observed. In a different work, cauliflower florets coated with maltodextrins (7.5 g/L) and methylcellulose (2.5 g/L) using lactic acid, citrus extract and lemongrass essential oil as antimicrobial agents at concentrations ranging from 0–34 mg/L were studied [131]. Complete inhibition of *L. innocua* after seven days of storage at 4 °C was reported. The same microorganism was reduced on green beans (*Phaseolus vulgaris* L.) coated with modified chitosan (3% N-palmitoyl chitosan, degree of palmitoylation 47%) containing 0.05% w/w nanoemulsion of mandarin essential oil [132]. Finally, modified chitosan (0.05% w/w) in 1% (v/v) lactic acid solution, enriched with 0.1% w/w of nanoemulsified lemon essential oil, was successfully used as antimicrobial coating of rucola during storage at 4 °C for three days and at 8 °C during 21 days [133]. A reduction of the initial microbial load around two log was obtained. After three days of storage under chilling conditions, the microbial load in coated samples remained constant with respect to Day 0, whereas the control showed a significant increase of about one log.

6. Market Analysis

The value of the global packaging market is expected to reach 910 billion euros by 2018. In particular, the worldwide consumption of flexible packaging is expected to reach 225 billion euros by 2020. In fact, the food industry covers more than two thirds of the current consumption of flexible packaging worldwide [137]. Particularly, novel edible coatings are emerging based on bio-based polymers with additives with specific functional properties to improve the shelf life of food products. In addition, coated food brings distinctive possibilities for the development of new products, processing improvement and general quality of the packaged food [11]. It has been observed that North America captured the highest market share in the global food coating ingredients market in 2015; this is majorly due to the increase in the confectionary market of the zone [138]. Europe accounted for the largest market of food coating ingredients for bakery and confectionery products in 2014 [139]. The market trend makes a rapid growth at a compound annual growth rate (CAGR) of 5.9% from 2014–2019 expectable; due to the changes in consumer lifestyles, increasing consumer disposable income and new developments to improve the organoleptic properties, to extend the shelf-life of food products and to provide the consumers safety [140].

7. Conclusions

This review underlines the most recent trends in the use of new edible coatings enriched with antimicrobial agents to reduce the growth of different microorganisms, such as Gram-negative and Gram-positive bacteria, molds and yeasts. The use of antimicrobials obtained from natural sources is one of the consequences of the rising consumer interest for healthy foods free of chemical additives. Among them, it is worth noting that organic acids and their salts (lauric, acetic, sorbic, citric, benzoic or propionic acids), spices and herb-derived compounds (essential oils and their main components), chitosan and natural antimicrobials obtained from bacteria, such as nisin, pediocin, natamycin or reuterin, have been recently proposed as antimicrobial agents for edible coating formulations.

The main coating techniques in food packaging are spraying, dipping or spreading. In this context, spraying attracts the industrial interest in packaging in contrast to dipping or spreading mainly due to two different factors: firstly, the potential cost reduction by applying this technique; secondly, the high quality of the final product that could be achieved when compared to those products obtained by using conventional techniques. However, coatings have still limited applications in food packaging due to their poor barrier to water vapor and low mechanical properties. Blending with different biopolymers, the addition of hydrophobic materials such as oils or waxes or chemical modification of the biopolymers structure have been proposed to overcome these drawbacks.

Meat and fish products, fruits and vegetables are the most susceptible food products to be coated with antimicrobial edible films. Chitosan, gelatin, methylcellulose, soy, whey, egg, wheat gluten, corn and collagen have been recently reported as coatings of fish products. Regarding meat, k-carrageenan, chitosan, sodium alginate, xanthan gum and soybean meal are the main matrices. On the other hand, fruits have been coated with sodium alginate, pectin, chitosan, pullulan and gum arabic, whereas chitosan, pullulan, xanthan gum, agar, zein, maltodextrins and methylcellulose were used as coatings of vegetables.

In conclusion, the current situation reviewed in this study underlines the necessity of focusing future research on the selection of the appropriate antimicrobial agents and the most adequate polymer matrices, to ensure good interactions among them and effectiveness against the target microorganisms. Their applicability to the food packaging industry needs further and deeper studies, since some of them showed high impact on the organoleptic characteristics of food products. It will be also necessary to study in detail the possible interactions between the coating films and the packaged food. As a general conclusion of this review, antimicrobial edible coatings are ready to suppose an effective alternative in active packaging materials to improve the safety of processed food products for commercial purposes.

Acknowledgments: The authors acknowledge the funding support of the Spanish Ministry of Economy and Competitiveness (MINECO, Ref. MAT2014-59242-C2-2-R).

Conflicts of Interest: The authors declare no conflict of interest.

References

1. Lucera, A.; Costa, C.; Conte, A.; Del Nobile, M.A. Food applications of natural antimicrobial compounds. *Front. Microbiol.* **2012**, *3*, 287. [CrossRef] [PubMed]
2. Mellinas, C.; Valdés, A.; Ramos, M.; Burgos, N.; Garrigós, M.C.; Jiménez, A. Active edible films: Current state and future trends. *J. Appl. Polym. Sci.* **2016**, *133*. [CrossRef]
3. Realini, C.E.; Marcos, B. Active and intelligent packaging systems for a modern society. *Meat Sci.* **2014**, *98*, 404–419. [CrossRef] [PubMed]
4. Gyawali, R.; Ibrahim, S.A. Natural products as antimicrobial agents. *Food Control* **2014**, *46*, 412–429. [CrossRef]
5. Tavassoli-Kafrani, E.; Shekarchizadeh, H.; Masoudpour-Behabadi, M. Development of edible films and coatings from alginates and carrageenans. *Carbohyd. Polym.* **2016**, *137*, 360–374. [CrossRef] [PubMed]
6. Shakila, R.J.; Jeevithan, E.; Arumugam, V.; Jeyasekaran, G. Suitability of antimicrobial grouper bone gelatin films as edible coatings for vacuum-packaged fish steaks. *J. Aquat. Food Prod. Technol.* **2016**, *25*, 724–734. [CrossRef]
7. Dhall, R.K. Advances in edible coatings for fresh fruits and vegetables: A review. *CRC Crit. Rev. Food Sci.* **2013**, *53*, 435–450. [CrossRef] [PubMed]
8. Karaca, H.; Pérez-Gago, M.B.; Taberner, V.; Palou, L. Evaluating food additives as antifungal agents against monilinia fructicola in vitro and in hydroxypropyl methylcellulose-lipid composite edible coatings for plums. *Int. J. Food Microbiol.* **2014**, *179*, 72–79. [CrossRef] [PubMed]
9. Campos, C.A.; Gerschenson, L.N.; Flores, S.K. Development of edible films and coatings with antimicrobial activity. *Food Bioprocess Technol.* **2011**, *4*, 849–875. [CrossRef]

10. Treviño-Garza, M.Z.; García, S.; Flores-González, M.S.; Arévalo-Niño, K. Edible active coatings based on pectin, pullulan, and chitosan increase quality and shelf life of strawberries (*Fragaria ananassa*). *J. Food Sci.* **2015**, *80*, M1823–M1830. [CrossRef] [PubMed]

11. Salgado, P.R.; Ortiz, C.M.; Musso, Y.S.; Di Giorgio, L.; Mauri, A.N. Edible films and coatings containing bioactives. *Curr. Opin. Food Sci.* **2015**, *5*, 86–92. [CrossRef]

12. Galus, S.; Kadzińska, J. Food applications of emulsion-based edible films and coatings. *Trends Food Sci. Technol.* **2015**, *45*, 273–283. [CrossRef]

13. Sánchez-Ortega, I.; García-Almendárez, B.E.; Santos-López, E.M.; Amaro-Reyes, A.; Barboza-Corona, J.E.; Regalado, C. Antimicrobial edible films and coatings for meat and meat products preservation. *Sci. World J.* **2014**, *2014*. [CrossRef] [PubMed]

14. Atarés, L.; Chiralt, A. Essential oils as additives in biodegradable films and coatings for active food packaging. *Trends Food Sci. Technol.* **2016**, *48*, 51–62. [CrossRef]

15. Donsì, F.; Marchese, E.; Maresca, P.; Pataro, G.; Vu, K.D.; Salmieri, S.; Lacroix, M.; Ferrari, G. Green beans preservation by combination of a modified chitosan based-coating containing nanoemulsion of mandarin essential oil with high pressure or pulsed light processing. *Postharvest Biol. Technol.* **2015**, *106*, 21–32. [CrossRef]

16. Hauser, C.; Thielmann, J.; Muranyi, P. Organic acids: Usage and potential in antimicrobial packaging. In *Antimicrobial Food Packaging*; Barros-Velazquez, J., Ed.; Elsevier: Amsterdam, The Netherlands, 2016; pp. 563–580.

17. Gharsallaoui, A.; Oulahal, N.; Joly, C.; Degraeve, P. Nisin as a food preservative: Part 1: Physicochemical properties, antimicrobial activity, and main uses. *CRC Crit. Rev. Food Sci.* **2016**, *56*, 1262–1274. [CrossRef] [PubMed]

18. Etayash, H.; Azmi, S.; Dangeti, R.; Kaur, K. Peptide bacteriocins–structure activity relationships. *Curr. Top. Med. Chem.* **2016**, *16*, 220–241. [CrossRef]

19. Elsabee, M.Z.; Abdou, E.S. Chitosan based edible films and coatings: A review. *Mater. Sci. Eng. C* **2013**, *33*, 1819–1841. [CrossRef] [PubMed]

20. Krašniewska, K.; Gniewosz, M.; Kosakowska, O.; Cis, A. Preservation of brussels sprouts by pullulan coating containing oregano essential oil. *J. Food Protect.* **2016**, *79*, 493–500. [CrossRef] [PubMed]

21. Yuceer, M.; Caner, C. Antimicrobial lysozyme-chitosan coatings affect functional properties and shelf life of chicken eggs during storage. *J. Sci. Food Agric.* **2014**, *94*, 153–162. [CrossRef] [PubMed]

22. Matiacevich, S.; Acevedo, N.; López, D. Characterization of edible active coating based on alginate-thyme oil-propionic acid for the preservation of fresh chicken breast fillets. *J. Food Process. Preserv.* **2015**, *39*, 2792–2801. [CrossRef]

23. Jin, T.Z.; Huang, M.; Niemira, B.A.; Cheng, L. Shelf life extension of fresh ginseng roots using sanitiser washing, edible antimicrobial coating and modified atmosphere packaging. *Int. J. Food Sci. Technol.* **2016**, *51*, 2132–2139. [CrossRef]

24. Raybaudi-Massilia, R.; Mosqueda-Melgar, J.; Soliva-Fortuny, R.; Martín-Belloso, O. Combinational edible antimicrobial films and coatings. In *Antimicrobial Food Packaging*; Barros-Velazquez, J., Ed.; Elsevier: Amsterdam, The Netherlands, 2016; pp. 633–646.

25. Valdes, A.; Mellinas, A.C.; Ramos, M.; Burgos, N.; Jimenez, A.; Garrigos, M.C. Use of herbs, spices and their bioactive compounds in active food packaging. *RSC Adv.* **2015**, *5*, 40324–40335. [CrossRef]

26. Tohidi, B.; Rahimmalek, M.; Arzani, A. Essential oil composition, total phenolic, flavonoid contents, and antioxidant activity of thymus species collected from different regions of iran. *Food Chem.* **2017**, *220*, 153–161. [CrossRef] [PubMed]

27. Calo, J.R.; Crandall, P.G.; O'Bryan, C.A.; Ricke, S.C. Essential oils as antimicrobials in food systems—A review. *Food Control* **2015**, *54*, 111–119. [CrossRef]

28. Ramos, M.; Jiménez, A.; Garrigós, M.C. Active nanocomposite in food contact materials. In *Nanoscience in Food and Agriculture 4. Sustainable Agriculture Reviews*; Ranjan, S., Dasgupta, N., Lichtfouse, E., Eds.; Springer International Publishing: Vienna, Austria, 2017; Volume 24, pp. 1–45.

29. Cao, L.; Si, J.Y.; Liu, Y.; Sun, H.; Jin, W.; Li, Z.; Zhao, X.H.; Pan, R.L. Essential oil composition, antimicrobial and antioxidant properties of mosla chinensis maxim. *Food Chem.* **2009**, *115*, 801–805. [CrossRef]

30. Ćavar Zeljković, S.; Maksimović, M. Chemical composition and bioactivity of essential oil from thymus species in balkan peninsula. *Phytochem. Rev.* **2015**, *14*, 335–352. [CrossRef]

31. Bastarrachea, L.; Dhawan, S.; Sablani, S. Engineering properties of polymeric-based antimicrobial films for food packaging: A review. *Food Eng. Rev.* **2011**, *3*, 79–93. [CrossRef]

32. Guerreiro, A.C.; Gago, C.M.L.; Miguel, M.G.C.; Faleiro, M.L.; Antunes, M.D.C. The influence of edible coatings enriched with citral and eugenol on the raspberry storage ability, nutritional and sensory quality. *Food Pack. Shelf Life* **2016**, *9*, 20–28. [CrossRef]

33. Hashemi, S.M.B.; Mousavi Khaneghah, A.; Ghaderi Ghahfarrokhi, M.; Eş, I. Basil-seed gum containing origanum vulgare subsp. Viride essential oil as edible coating for fresh cut apricots. *Postharvest Biol. Technol.* **2017**, *125*, 26–34. [CrossRef]

34. Jouki, M.; Yazdi, F.T.; Mortazavi, S.A.; Koocheki, A. Quince seed mucilage films incorporated with oregano essential oil: Physical, thermal, barrier, antioxidant and antibacterial properties. *Food Hydrocoll.* **2014**, *36*, 9–19. [CrossRef]

35. Salvia-Trujillo, L.; Rojas-Graü, M.A.; Soliva-Fortuny, R.; Martín-Belloso, O. Use of antimicrobial nanoemulsions as edible coatings: Impact on safety and quality attributes of fresh-cut fuji apples. *Postharvest Biol. Technol.* **2015**, *105*, 8–16. [CrossRef]

36. Gómez-Estaca, J.; López de Lacey, A.; López-Caballero, M.E.; Gómez-Guillén, M.C.; Montero, P. Biodegradable gelatin-chitosan films incorporated with essential oils as antimicrobial agents for fish preservation. *Food Microbiol.* **2010**, *27*, 889–896. [CrossRef] [PubMed]

37. Alparslan, Y.; Yapici, H.H.; Metin, C.; Baygar, T.; Günlü, A. Quality assessment of shrimps preserved with orange leaf essential oil incorporated gelatin. *LWT—Food Sci. Technol.* **2016**, *72*, 457–466. [CrossRef]

38. Emiroğlu, Z.K.; Yemiş, G.P.; Coşkun, B.K.; Candoğan, K. Antimicrobial activity of soy edible films incorporated with thyme and oregano essential oils on fresh ground beef patties. *Meat Sci.* **2010**, *86*, 283–288. [CrossRef] [PubMed]

39. Ravishankar, S.; Jaroni, D.; Zhu, L.; Olsen, C.; McHugh, T.; Friedman, M. Inactivation of listeria monocytogenes on ham and bologna using pectin-based apple, carrot, and hibiscus edible films containing carvacrol and cinnamaldehyde. *J. Food Sci.* **2012**, *77*, M377–M382. [CrossRef] [PubMed]

40. Zhang, Y.; Ma, Q.; Critzer, F.; Davidson, P.M.; Zhong, Q. Effect of alginate coatings with cinnamon bark oil and soybean oil on quality and microbiological safety of cantaloupe. *Int. J. Food Microbiol.* **2015**, *215*, 25–30. [CrossRef] [PubMed]

41. Molaee Aghaee, E.; Kamkar, A.; Akhondzadeh Basti, A. Antimicrobial effect of garlic essential oil (*Allium sativum* L.) in combination with chitosan biodegradable coating films. *J. Med. Plants* **2016**, *15*, 141–150.

42. Ngamakeue, N.; Chitprasert, P. Encapsulation of holy basil essential oil in gelatin: Effects of palmitic acid in carboxymethyl cellulose emulsion coating on antioxidant and antimicrobial activities. *Food Bioprocess Technol.* **2016**, *9*, 1735–1745. [CrossRef]

43. Jovanović, G.D.; Klaus, A.S.; Nikšić, M.P. Antimicrobial activity of chitosan coatings and films against listeria monocytogenes on black radish. *Rev. Argent. Microbiol.* **2016**, *48*, 128–136. [CrossRef] [PubMed]

44. Aider, M. Chitosan application for active bio-based films production and potential in the food industry: Review. *LWT—Food Sci. Technol.* **2010**, *43*, 837–842. [CrossRef]

45. Fortunati, E. Multifunctional films, blends, and nanocomposites based on chitosan: Use in antimicrobial packaging. In *Antimicrobial Food Packaging*; Barros-Velazquez, J., Ed.; Elsevier: Amsterdam, The Netherlands, 2016; pp. 467–477.

46. Carrión-Granda, X.; Fernández-Pan, I.; Jaime, I.; Rovira, J.; Maté, J.I. Improvement of the microbiological quality of ready-to-eat peeled shrimps (penaeus vannamei) by the use of chitosan coatings. *Int. J. Food Microbiol.* **2016**, *232*, 144–149. [CrossRef] [PubMed]

47. Mei, J.; Guo, Q.; Wu, Y.; Li, Y. Evaluation of chitosan-starch-based edible coating to improve the shelf life of bod ljong cheese. *J. Food Protect.* **2015**, *78*, 1327–1334. [CrossRef] [PubMed]

48. Duran, M.; Aday, M.S.; Zorba, N.N.D.; Temizkan, R.; Büyükcan, M.B.; Caner, C. Potential of antimicrobial active packaging 'containing natamycin, nisin, pomegranate and grape seed extract in chitosan coating' to extend shelf life of fresh strawberry. *Food Bioprod. Process.* **2016**, *98*, 354–363. [CrossRef]

49. Ndoti-Nembe, A.; Vu, K.D.; Han, J.; Doucet, N.; Lacroix, M. Antimicrobial effects of nisin, essential oil, and γ-irradiation treatments against high load of salmonella typhimurium on mini-carrots. *J. Food Sci.* **2015**, *80*, M1544–M1548. [CrossRef] [PubMed]

50. Sánchez-Ortega, I.; García-Almendárez, B.E.; Santos-López, E.M.; Reyes-González, L.R.; Regalado, C. Characterization and antimicrobial effect of starch-based edible coating suspensions. *Food Hydrocoll.* **2016**, *52*, 906–913. [CrossRef]

51. Guo, M.; Jin, T.Z.; Wang, L.; Scullen, O.J.; Sommers, C.H. Antimicrobial films and coatings for inactivation of listeria innocua on ready-to-eat deli turkey meat. *Food Control* **2014**, *40*, 64–70. [CrossRef]

52. Lin, L.S.; Wang, B.J.; Weng, Y.M. Quality preservation of commercial fish balls with antimicrobial zein coatings. *J. Food Qual.* **2011**, *34*, 81–87. [CrossRef]

53. Wu, C.; Hu, Y.; Chen, S.; Chen, J.; Liu, D.; Ye, X. Formation mechanism of nano-scale antibiotic and its preservation performance for silvery pomfret. *Food Control* **2016**, *69*, 331–338. [CrossRef]

54. Kang, H.-J.; Kim, S.-J.; You, Y.-S.; Lacroix, M.; Han, J. Inhibitory effect of soy protein coating formulations on walnut (juglans regia l.) kernels against lipid oxidation. *LWT—Food Sci. Technol.* **2013**, *51*, 393–396. [CrossRef]

55. Espitia, P.J.P.; Du, W.-X.; Avena-Bustillos, R.d.J.; Soares, N.d.F.F.; McHugh, T.H. Edible films from pectin: Physical-mechanical and antimicrobial properties—A review. *Food Hydrocoll.* **2014**, *35*, 287–296. [CrossRef]

56. Falguera, V.; Quintero, J.P.; Jiménez, A.; Muñoz, J.A.; Ibarz, A. Edible films and coatings: Structures, active functions and trends in their use. *Trends Food Sci. Technol.* **2011**, *22*, 292–303. [CrossRef]

57. Andrade, R.; Skurtys, O.; Osorio, F. Atomizing spray systems for application of edible coatings. *Comp. Rev. Food Sci. Food Safety* **2012**, *11*, 323–337. [CrossRef]

58. Martín-Belloso, O.; Rojas-Graü, M.A.; Soliva-Fortuny, R. Delivery of flavor and active ingredients using edible films and coatings. In *Edible Films and Coatings for Food Applications*; Embuscado, M.E., Huber, K.C., Eds.; Springer: New York, NY, USA, 2009; pp. 295–314.

59. Ustunol, Z. Edible films and coatings for meat and poultry. In *Edible Films and Coatings for Food Applications*; Embuscado, M.E., Huber, K.C., Eds.; Springer: New York, NY, USA, 2009; pp. 245–268.

60. Ramos, M.; Jiménez, A.; Peltzer, M.; Garrigós, M.C. Characterization and antimicrobial activity studies of polypropylene films with carvacrol and thymol for active packaging. *J. Food Eng.* **2012**, *109*, 513–519. [CrossRef]

61. Bosquez-Molina, E.; Guerrero-Legarreta, I.; Vernon-Carter, E.J. Moisture barrier properties and morphology of mesquite gum–candelilla wax based edible emulsion coatings. *Food Res. Int.* **2003**, *36*, 885–893. [CrossRef]

62. Dhanapal, A.; Sasikala, P.; Rajamani, L.; Kavitha, V.; Yazhini, G.; Banu, M.S. Edible films from polysaccharides. *Food Sci. Qual. Manag.* **2012**, *3*, 9–18.

63. Skurtys, O.; Acevedo, C.; Pedreschi, F.; Enronoe, J.; Osorio, F.; Aguiler, J.M. Food hydrocolloid edible films and coatings. In *Food Hydrocolloids: Characteristics, Properties and Structures*; Nova Science Publishers, Inc.: New York, NY, USA, 2010; pp. 6–9.

64. Nasr, G.; Yule, A.; Bendig, L. *Industrial Sprays and Atomization: Design, Analysis and Applications*; Lightning Source UK Ltd.: Milton Keynes, UK, 2002.

65. *Airless Spray Systems. The Efficient Choice for Many Liquid Painting Applications*; Nordson Corporation: Armhest, OH, USA, 2004.

66. Mannouch, S. Spray Gun Technique. Itw Devilbiss Industrial Training Centre. Available online: http://www.devilbiss.com/ (accessed on 6 January 2017).

67. Peretto, G.; Du, W.X.; Avena-Bustillos, R.J.; De J. Berrios, J.; Sambo, P.; McHugh, T.H. Electrostatic and conventional spraying of alginate-based edible coating with natural antimicrobials for preserving fresh strawberry quality. *Food Bioprocess Technol.* **2017**, *10*, 165–174. [CrossRef]

68. Chiu, P.E.; Lai, L.S. Antimicrobial activities of tapioca starch/decolorized hsian-tsao leaf gum coatings containing green tea extracts in fruit-based salads, romaine hearts and pork slices. *Int. J. Food Microbiol.* **2010**, *139*, 23–30. [CrossRef] [PubMed]

69. Lu, F.; Ding, Y.; Ye, X.; Liu, D. Cinnamon and nisin in alginate–calcium coating maintain quality of fresh northern snakehead fish fillets. *LWT—Food Sci. Technol.* **2010**, *43*, 1331–1335. [CrossRef]

70. Schneller, T.; Waser, R.; Kosec, M.; Payne, D. *Chemical Solution Deposition of Functional Oxide Thin Films*; Springer: Vienna, Austria, 2013.

71. Costa, C.; Conte, A.; Del Nobile, M.A. Effective preservation techniques to prolong the shelf life of ready-to-eat oysters. *J. Sci. Food Agric.* **2014**, *94*, 2661–2667. [CrossRef] [PubMed]

72. Hamzah, H.M.; Osman, A.; Tan, C.P.; Mohamad Ghazali, F. Carrageenan as an alternative coating for papaya (carica papaya l. Cv. Eksotika). *Postharvest Biol. Technol.* **2013**, *75*, 142–146. [CrossRef]

73. Mastromatteo, M.; Conte, A.; Del Nobile, M.A. Packaging strategies to prolong the shelf life of fresh carrots (daucus carota l.). *Innov. Food Sci. Emerg. Technol.* **2012**, *13*, 215–220. [CrossRef]

74. Méndez-Vilas, A. *Microbial Pathogens and Strategies for Combating Them: Science, Technology and Education*; Microbiology Book Series 1; Formatex Research Center: Badajoz, Spain, 2013.

75. Khan, M.I.; Nasef, M.M. Spreading behaviour of silicone oil and glycerol drops on coated papers. *Leonardo J. Sci.* **2009**, *14*, 18–30.

76. Kumar, G.; Prabhu, K.N. Review of non-reactive and reactive wetting of liquids on surfaces. *Adv. Colloid Interface Sci.* **2007**, *133*, 61–89. [CrossRef] [PubMed]

77. Šikalo, Š.; Marengo, M.; Tropea, C.; Ganić, E.N. Analysis of impact of droplets on horizontal surfaces. *Exp. Therm. Fluid Sci.* **2002**, *25*, 503–510. [CrossRef]

78. Silvestru, B.M.; Pâslaru, E.; Fras Zemljic, L.; Sdrobis, A.; Pricope, G.; Vasile, C. Chitosan coatings applied to polyethylene surface to obtain food-packaging materials. *Cellul. Chem. Technol.* **2014**, *48*, 565–575.

79. Nithya, V.; Murthy, P.S.; Halami, P.M. Development and application of active films for food packaging using antibacterial peptide of bacillus licheniformis me1. *J. Appl. Microbiol.* **2013**, *115*, 475–483. [CrossRef] [PubMed]

80. Min, S.; Krochta, J.M. Inhibition of penicillium commune by edible whey protein films incorporating lactoferrin, lacto-ferrin hydrolysate, and lactoperoxidase systems. *J. Food Sci.* **2005**, *70*, M87–M94. [CrossRef]

81. Lim, G.-O.; Jang, S.-A.; Song, K.B. Physical and antimicrobial properties of gelidium corneum/nano-clay composite film containing grapefruit seed extract or thymol. *J. Food Eng.* **2010**, *98*, 415–420. [CrossRef]

82. Ayranci, E.; Tunc, S. A method for the measurement of the oxygen permeability and the development of edible films to reduce the rate of oxidative reactions in fresh foods. *Food Chem.* **2003**, *80*, 423–431. [CrossRef]

83. Marques, P.T.; Lima, A.M.F.; Bianco, G.; Laurindo, J.B.; Borsali, R.; Le Meins, J.F.; Soldi, V. Thermal properties and stability of cassava starch films cross-linked with tetraethylene glycol diacrylate. *Polym. Degrad. Stabil.* **2006**, *91*, 726–732. [CrossRef]

84. Gutiérrez, T.J.; Tapia, M.S.; Pérez, E.; Famá, L. Structural and mechanical properties of edible films made from native and modified cush-cush yam and cassava starch. *Food Hydrocoll.* **2015**, *45*, 211–217. [CrossRef]

85. Azarakhsh, N.; Osman, A.; Ghazali, H.M.; Tan, C.P.; Mohd Adzahan, N. Effects of gellan-based edible coating on the quality of fresh-cut pineapple during cold storage. *Food Bioprocess Technol.* **2014**, *7*, 2144–2151. [CrossRef]

86. Gutiérrez, T.J.; Morales, N.J.; Pérez, E.; Tapia, M.S.; Famá, L. Physico-chemical properties of edible films derived from native and phosphated cush-cush yam and cassava starches. *Food Pack. Shelf Life* **2015**, *3*, 1–8. [CrossRef]

87. Schmid, M.; Pröls, S.; Kainz, D.M.; Hammann, F.; Grupa, U. Effect of thermally induced denaturation on molecular interaction-response relationships of whey protein isolate based films and coatings. *Prog. Org. Coat.* **2017**, *104*, 161–172. [CrossRef]

88. Shuang, C. Development and Characterization of Antimicrobial Food Coatings Based on Chitosan and Essential Oils. Ph.D. Thesis, University of Tennessee, Knoxville, TN, USA, 2004.

89. Yang, F.; Hu, S.; Lu, Y.; Yang, H.; Zhao, Y.; Li, L. Effects of coatings of polyethyleneimine and thyme essential oil combined with chitosan on sliced fresh channa argus during refrigerated storage. *J. Food Process Eng.* **2015**, *38*, 225–233. [CrossRef]

90. Alemán, A.; González, F.; Arancibia, M.Y.; López-Caballero, M.E.; Montero, P.; Gómez-Guillén, M.C. Comparative study between film and coating packaging based on shrimp concentrate obtained from marine industrial waste for fish sausage preservation. *Food Control* **2016**, *70*, 325–332. [CrossRef]

91. Jasour, M.S.; Ehsani, A.; Mehryar, L.; Naghibi, S.S. Chitosan coating incorporated with the lactoperoxidase system: An active edible coating for fish preservation. *J. Sci. Food Agric.* **2015**, *95*, 1373–1378. [CrossRef] [PubMed]

92. Shokri, S.; Ehsani, A.; Jasour, M.S. Efficacy of lactoperoxidase system-whey protein coating on shelf-life extension of rainbow trout fillets during cold storage (4 °C). *Food Bioprocess Technol.* **2014**, *8*, 54–62. [CrossRef]

93. Thaker, M.; Hanjabam, M.D.; Gudipati, V.; Kannuchamy, N. Protective effect of fish gelatin-based natural antimicrobial coatings on quality of indian salmon fillets during refrigerated storage. *J. Food Process Eng.* **2015**. [CrossRef]

94. Wu, C.; Fu, S.; Xiang, Y.; Yuan, C.; Hu, Y.; Chen, S.; Liu, D.; Ye, X. Effect of chitosan gallate coating on the quality maintenance of refrigerated (4 °C) silver pomfret (*pampus argentus*). *Food Bioprocess Technol.* **2016**, *9*, 1835–1843. [CrossRef]

95. Dursun, S.; Erkan, N. The effect of edible coating on the quality of smoked fish. *Ital. J. Food Sci.* **2014**, *26*, 370–382.

96. Choulitoudi, E.; Bravou, K.; Bimpilas, A.; Tsironi, T.; Tsimogiannis, D.; Taoukis, P.; Oreopoulou, V. Antimicrobial and antioxidant activity of *satureja thymbra* in gilthead seabream fillets edible coating. *Food Bioprod. Process.* **2016**, *100*, 570–577. [CrossRef]

97. Ariaii, P.; Tavakolipour, H.; Rezaei, M.; Elhami Rad, A.H.; Bahram, S. Effect of methylcellulose coating enriched with pimpinella affinis oil on the quality of silver carp fillet during refrigerator storage condition. *J. Food Process. Preserv.* **2015**, *39*, 1647–1655. [CrossRef]

98. Chen, B.J.; Zhou, Y.J.; Wei, X.Y.; Xie, H.J.; Hider, R.C.; Zhou, T. Edible antimicrobial coating incorporating a polymeric iron chelator and its application in the preservation of surimi product. *Food Bioprocess Technol.* **2016**, *9*, 1031–1039. [CrossRef]

99. Olaimat, A.N.; Holley, R.A. Inhibition of listeria monocytogenes on cooked cured chicken breasts by acidified coating containing allyl isothiocyanate or deodorized oriental mustard extract. *Food Microbiol.* **2016**, *57*, 90–95. [CrossRef] [PubMed]

100. Olaimat, A.N.; Fang, Y.; Holley, R.A. Inhibition of campylobacter jejuni on fresh chicken breasts by κ-carrageenan/chitosan-based coatings containing allyl isothiocyanate or deodorized oriental mustard extract. *Int. J. Food Microbiol.* **2014**, *187*, 77–82. [CrossRef] [PubMed]

101. He, S.; Yang, Q.; Ren, X.; Zi, J.; Lu, S.; Wang, S.; Zhang, Y.; Wang, Y. Antimicrobial efficiency of chitosan solutions and coatings incorporated with clove oil and/or ethylenediaminetetraacetate. *J. Food Safety* **2014**, *34*, 345–352. [CrossRef]

102. Zhao, Y.; Abbar, S.; Phillips, T.W.; Williams, J.B.; Smith, B.S.; Schilling, M.W. Developing food-grade coatings for dry-cured hams to protect against ham mite infestation. *Meat Sci.* **2016**, *113*, 73–79. [CrossRef] [PubMed]

103. Lee, H.; Kim, J.E.; Min, S.C. Quantitative risk assessments of the effect of an edible defatted soybean meal-based antimicrobial film on the survival of salmonella on ham. *J. Food Eng.* **2015**, *158*, 30–38. [CrossRef]

104. Kapetanakou, A.E.; Karyotis, D.; Skandamis, P.N. Control of listeria monocytogenes by applying ethanol-based antimicrobial edible films on ham slices and microwave-reheated frankfurters. *Food Microbiol.* **2016**, *54*, 80–90. [CrossRef]

105. Wang, L.; Zhao, L.; Yuan, J.; Jin, T.Z. Application of a novel antimicrobial coating on roast beef for inactivation and inhibition of listeria monocytogenes during storage. *Int. J. Food Microbiol.* **2015**, *211*, 66–72. [CrossRef] [PubMed]

106. Noorihashemabad, Z.; Mehdi Ojagh, S.; Alishahi, A. A comprehensive surviving on application and diversity of biofilms in seafood. *Int. J. Biosci.* **2015**, *6*, 15–30.

107. Neetoo, H.; Mahomoodally, F. Use of antimicrobial films and edible coatings incorporating chemical and biological preservatives to control growth of listeria monocytogenes on cold smoked salmon. *Biomed. Res. Int.* **2014**, *2014*, 534915. [CrossRef] [PubMed]

108. Neetoo, H.; Ye, M.; Chen, H. Bioactive alginate coatings to control listeria monocytogenes on cold-smoked salmon slices and fillets. *Int. J. Food Microbiol.* **2010**, *136*, 326–331. [CrossRef] [PubMed]

109. Yener, F.Y.G.; Korel, F.; Yemenicioğlu, A. Antimicrobial activity of lactoperoxidase system incorporated into cross-linked alginate films. *J. Food Sci.* **2009**, *74*, M73–M79. [CrossRef] [PubMed]

110. Todd, E.C.D.; Notermans, S. Surveillance of listeriosis and its causative pathogen, listeria monocytogenes. *Food Control* **2011**, *22*, 1484–1490. [CrossRef]

111. Thomas, M.K.; Murray, R.; Flockhart, L.; Pintar, K.; Pollari, F.; Fazil, A.; Nesbitt, A.; Marshall, B. Estimates of the burden of foodborne illness in canada for 30 specified pathogens and unspecified agents. *Foodborne Pathog. Dis.* **2012**, *10*, 639–648. [CrossRef] [PubMed]

112. Rentfrow, G.; Chaplin, R.; Suman, S.P. Technology of dry-cured ham production: Science enhancing art. *Anim. Front.* **2012**, *2*, 26–31. [CrossRef]

113. Manzocco, L.; Da Pieve, S.; Maifreni, M. Impact of uv-c light on safety and quality of fresh-cut melon. *Innov. Food Sci. Emerg. Technol.* **2011**, *12*, 13–17. [CrossRef]

114. Sipahi, R.E.; Castell-Perez, M.E.; Moreira, R.G.; Gomes, C.; Castillo, A. Improved multilayered antimicrobial alginate-based edible coating extends the shelf life of fresh-cut watermelon (citrullus lanatus). *LWT—Food Sci. Technol.* **2013**, *51*, 9–15. [CrossRef]

115. Martiñon, M.E.; Moreira, R.G.; Castell-Perez, M.E.; Gomes, C. Development of a multilayered antimicrobial edible coating for shelf-life extension of fresh-cut cantaloupe (*Cucumis melo* l.) stored at 4 °C. *LWT—Food Sci. Technol.* **2014**, *56*, 341–350. [CrossRef]

116. Sanchís, E.; Ghidelli, C.; Sheth, C.C.; Mateos, M.; Palou, L.; Pérez-Gago, M.B. Integration of antimicrobial pectin-based edible coating and active modified atmosphere packaging to preserve the quality and microbial safety of fresh-cut persimmon (*Diospyros kaki* thunb. Cv. Rojo brillante). *J. Sci. Food Agric.* **2016**, *97*, 252–260.

117. Guerreiro, A.C.; Gago, C.M.L.; Faleiro, M.L.; Miguel, M.G.C.; Antunes, M.D.C. The use of polysaccharide-based edible coatings enriched with essential oils to improve shelf-life of strawberries. *Postharvest Biol. Technol.* **2015**, *110*, 51–60. [CrossRef]

118. Guerreiro, A.C.; Gago, C.M.L.; Faleiro, M.L.; Miguel, M.G.C.; Antunes, M.D.C. Raspberry fresh fruit quality as affected by pectin- and alginate-based edible coatings enriched with essential oils. *Sci. Horticult.* **2015**, *194*, 138–146. [CrossRef]

119. Guerreiro, A.C.; Gago, C.M.L.; Faleiro, M.L.; Miguel, M.G.C.; Antunes, M.D.C. The effect of alginate-based edible coatings enriched with essential oils constituents on *Arbutus unedo* L. Fresh fruit storage. *Postharvest Biol. Technol.* **2015**, *100*, 226–233. [CrossRef]

120. Azarakhsh, N.; Osman, A.; Ghazali, H.M.; Tan, C.P.; Mohd Adzahan, N. Lemongrass essential oil incorporated into alginate-based edible coating for shelf-life extension and quality retention of fresh-cut pineapple. *Postharvest Biol. Technol.* **2014**, *88*, 1–7. [CrossRef]

121. Sun, X.; Narciso, J.; Wang, Z.; Ference, C.; Bai, J.; Zhou, K. Effects of chitosan-essential oil coatings on safety and quality of fresh blueberries. *J. Food Sci.* **2014**, *79*, M955–M960. [CrossRef] [PubMed]

122. Bill, M.; Sivakumar, D.; Korsten, L.; Thompson, A.K. The efficacy of combined application of edible coatings and thyme oil in inducing resistance components in avocado (persea americana mill.) against anthracnose during post-harvest storage. *Crop Prot.* **2014**, *64*, 159–167. [CrossRef]

123. Ali, A.; Noh, N.M.; Mustafa, M.A. Antimicrobial activity of chitosan enriched with lemongrass oil against anthracnose of bell pepper. *Food Packag. Shelf Life* **2015**, *3*, 56–61. [CrossRef]

124. Kraśniewska, K.; Gniewosz, M.; Synowiec, A.; Przybył, J.L.; Bączek, K.; Węglarz, Z. The application of pullulan coating enriched with extracts from bergenia crassifolia to control the growth of food microorganisms and improve the quality of peppers and apples. *Food Bioprod. Process.* **2015**, *94*, 422–433. [CrossRef]

125. Kraśniewska, K.; Gniewosz, M.; Synowiec, A.; Przybył, J.L.; Bączek, K.; Węglarz, Z. The use of pullulan coating enriched with plant extracts from *Satureja hortensis* L. To maintain pepper and apple quality and safety. *Postharvest Biol. Technol.* **2014**, *90*, 63–72. [CrossRef]

126. Cortez-Vega, W.R.; Brose Piotrowicz, I.B.; Prentice, C.; Borges, C.D. Influence of different edible coatings in minimally processed pumpkin (cucurbita moschata duch). *Int. Food Res. J.* **2014**, *21*, 2017–2023.

127. Santos, A.R.; da Silva, A.F.; Amaral, V.C.S.; Ribeiro, A.B.; de Abreu Filho, B.A.; Mikcha, J.M.G. Application of edible coating with starch and carvacrol in minimally processed pumpkin. *J. Food Sci. Technol.* **2016**, *53*, 1975–1983. [CrossRef] [PubMed]

128. Aksu, F.; Uran, H.; Dülger Altiner, D.; Sandikçi Atunarmaz, S. Effects of different packaging techniques on the microbiological and physicochemical properties of coated pumpkin slices. *Food Sci. Technol. (Camp.)* **2016**. [CrossRef]

129. Yun, J.; Fan, X.; Li, X.; Jin, T.Z.; Jia, X.; Mattheis, J.P. Natural surface coating to inactivate salmonella enterica serovar typhimurium and maintain quality of cherry tomatoes. *Int. J. Food Microbiol.* **2015**, *193*, 59–67. [CrossRef] [PubMed]

130. Alvarez, M.V.; Ponce, A.G.; Moreira, M.d.R. Antimicrobial efficiency of chitosan coating enriched with bioactive compounds to improve the safety of fresh cut broccoli. *LWT—Food Sci. Technol.* **2013**, *50*, 78–87. [CrossRef]

131. Boumail, A.; Salmieri, S.; St-Yves, F.; Lauzon, M.; Lacroix, M. Effect of antimicrobial coatings on microbiological, sensorial and physico-chemical properties of pre-cut cauliflowers. *Postharvest Biol. Technol.* **2016**, *116*, 1–7. [CrossRef]

132. Severino, R.; Vu, K.D.; Donsì, F.; Salmieri, S.; Ferrari, G.; Lacroix, M. Antibacterial and physical effects of modified chitosan based-coating containing nanoemulsion of mandarin essential oil and three non-thermal treatments against listeria innocua in green beans. *Int. J. Food Microbiol.* **2014**, *191*, 82–88. [CrossRef] [PubMed]

133. Sessa, M.; Ferrari, G.; Donsì, F. Novel edible coating containing essential oil nanoemulsions to prolong the shelf life of vegetable products. *Chem. Eng. Trans.* **2015**, *43*, 55–60.

134. Gol, N.B.; Patel, P.R.; Rao, T.V.R. Improvement of quality and shelf-life of strawberries with edible coatings enriched with chitosan. *Postharvest Biol. Technol.* **2013**, *85*, 185–195. [CrossRef]

135. Nunes, M.C.N. *Color Atlas of Postharvest Quality of Fruits And Vegetables*; John Wiley & Sons: New York, NY, USA, 2009.

136. Sasaki, F.F.; Del Aguila, J.S.; Gallo, C.R.; Ortega, E.M.M.; Jacomino, A.P.; Kluge, R.A. Physiological, qualitative and microbiological changes in minimally processed squash submitted to different cut types. *Hortic. Bras.* **2006**, *24*, 170–174.

137. Smithers Pira. The Future of Global Packaging to 2018. Available online: http://www.smitherspira. com/products/market-reports/packaging/global-world-packaging-industry-market-report/ (accessed on 12 April 2017).

138. Future Markets Insights. Food Coating Ingredients Market: Global Industry Analysis and Opportunity Assessment 2015–2025. Available online: http://www.futuremarketinsights.com/reports/food-coating-ingredients-market/ (accessed on 12 April 2017).

139. Research and Markets. Global Food Coating Ingredients Market Size, Share, Development, Growth and Demand Forecast to 2020—Industry Insights by Types, by Applications. Available online: http://www. researchandmarkets.com/research/hwnl29/global_food (accessed on 12 April 2017).

140. Markets and Markets. Food Coating Ingredients Market Worth $3.7 Billion by 2019. Available online: http://www.marketsandmarkets.com/PressReleases/food-coating-ingredients.asp (accessed on 12 April 2017).

3

Effects of Rare Earth Elements on Properties of Ni-Base Superalloy Powders and Coatings

Chunlian Hu [1] and Shanglin Hou [2,*]

[1]   Alloy Powder Co., Ltd., Lanzhou University of Technology, Lanzhou 730050, China; huchl2005@126.com
[2]   School of Science, Lanzhou University of Technology, Lanzhou 730050, China
*   Correspondence: houshanglin@163.com

Academic Editors: Niteen Jadhav and Andrew J. Vreugdenhil

Abstract: NiCrMoY alloy powders were prepared using inert gas atomization by incorporation of rare earth elements, such as Mo, Nb, and Y into Ni60A powders, the coatings were sprayed by oxy-acetylene flame spray and then remelted with high-frequency induction. The morphologies, hollow particle ratio, particle-size distribution, apparent density, flowability, and the oxygen content of the NiCrMoY alloy powders were investigated, and the microstructure and hardness of the coatings were evaluated by optical microscopy (OM). Due to incorporation of the rare earth elements of Mo, Nb, or Y, the majority of the NiCrMoY alloy particles are near-spherical, the minority of which have small satellites, the surface of the particles is smoother and hollow particles are fewer, the particles exhibit larger apparent density and lower flowability than those of particles without incorporation, i.e., Ni60A powders, and particle-size distribution exhibits a single peak and fits normal distribution. The microstructure of the NiCrMoY alloy coatings exhibits finer structure and Rockwell hardness HRC of 60–63 in which the bulk- and needle-like hard phases are formed.

Keywords: rare earth; microstructure; alloy powder; coating

1. Introduction

Due to excellent weldability properties, surface stability, corrosion resistance, and mechanical properties at high temperature, nickel-based superalloys are widely used for gasturbine components and other applications, such as the base materials for hot components, e.g., hot parts of aerospace turbine engines [1]. Nickel-based coatings can function either as overlay coatings or asbond coats in a thermal barrier coating system. They are usually applied using thermal spraying processes, such as low-pressure plasma spraying (LPPS), high velocity oxygen fuel spraying (HVOF), vacuum plasmas praying (VPS) and atmospheric plasma spraying (APS) [2]. All types of spraying processes use powder as feedstock, which typically results in a characteristic splat-structure [3].

Incorporation of rare earth (RE) elements into alloys may improve their high-temperature oxidation resistance or other mechanical properties. Chromium and aluminum are added to promote resistance to oxidation and hot corrosion. Incorporation of minor amounts of rare earth elements, such as Ce, Y, Zr, La, or their oxides enhance the bonding strength of the oxide layer [4] and improve the high-temperature oxidation resistance of alumina- and chromia-forming alloys [5]. Stringer [6] suggested the enhancement of oxide nucleation processes through the presence of rare earth elements. Antill and Peakall [7] reported that the beneficial effect of the rare earth elements was primarily to improve scale plasticity for accommodating stresses due to the difference in the thermal expansion coefficients between the alloy and the oxide scale. Tien and Pettit [8] concluded that the application of rare earth elements provide sites for vacancy condensation in an Fe–25Cr–4Al alloy, with consequent improvement of scale adhesion. It was reported that the rare earth elements, such as La, Y, Ce, and their oxides, can be used for reducing the oxidation rate and improving corrosion resistance of the superalloys [9]. It was also found that high-density, fine, and uniform structure is formed in coatings deposited by Ni-based alloy powders with the addition of rare earth elements so as to improve the wear and corrosion resistance of Ni-based alloys [10]. Recently He [11] reported the microstructure and hot corrosion resistance of Co–Si-modified aluminide coating on nickel-based superalloys, and interdiffusion between a polycrystalline nickel-based superalloy (René 80) and two MCrAlY bondcoats, each with a different chemical composition, is demonstrated in [12].

Ni-based self-fluxing alloy powder Ni60A is one of the most important protective coating materials owing to excellent high-temperature corrosion resistance, and the coating sprayed by Ni60A powders with the incorporation rare earth elements, such as W and Mo, has better wear resistance [13]. It is well known that the properties of the powders influence characteristics of sprayed coatings extensively [14–16]. For example, both oxygen content and flowability of powders may induce porosity, impurities, or cracks in the coatings [17].

In this work, NiCrMoY alloy powders were prepared by using inert gas atomization with incorporation of rare earth elements, such as Mo, Nb, and Y into Ni60A powders, the coatings were sprayed by oxy-acetylene flame spray and then remelted with high-frequency induction. The morphologies, hollow particle ratio, particle-size distribution, apparent density, flowability, and oxygen content of NiCrMoY alloy powders were investigated, and the microstructure and hardness of the NiCrMoY-sprayed coatings are presented. The research results show that the properties of NiCrMoY alloy powders are improved due to the incorporation of rare earth elements, and the microstructure of NiCrMoY alloy coatings exhibit fine structure and Rockwell hardness HRC of 60–63.

2. Experimental Procedure

2.1. Preparation and Property Test of Powders

The Ni60A and NiCrMoY alloy powders were prepared by double-stage coupling fast freezing and low-pressure gas atomizing with an atomizing gas pressure of 5–10 MPa, the melt overheating temperature is 100–150 °C, and the particles had a near spherical morphology with internal pores and a size of 40–110 μm. The chemical compositions of the Ni60A alloy and NiCrMoY alloy powders are shown in Table 1 which was measured at the Testing Center of the Shanghai Research Institute of Materials according to ASTM E1019-11 [18], ASTM E2594-09(2014) [19], ASTM E354-14 [20], and ISO4938:1988 [21].

Table 1. Chemical compositions of the NiCrMoY and Ni60A alloy powders (wt %).

Alloy	Chemical Compositions									
	C	B	Si	Cr	Fe	Mo	Cu	Nb	Y	Ni
Ni60A	0.98	2.91	3.96	16.4	3.2	–	–	–	–	Bal.
NiCrMoY	1.0	2.85	4.0	16.5	3.0	2.5	1.5	0.5	0.15	Bal.

The epoxy resin and the ethidenediamine were mixed proportionally with the Ni60A and NiCrMoY alloy powders, respectively, and then polished a cross-section to test the porosity of the powder after the epoxy resin solidified. The morphologies of powders were examined by a Reicher-Jung (Leica MeF3) optical microscope (OM) (Leica Microsystems, Wetzlar, Germany). A Mastersizer 2000 laser diffraction particle size analyzer (Malvern Instruments Ltd., Malvern, UK) was performed to analyze particle-size distribution. The apparent density and flowability of two kinds of powders were carried out with a FL4-1-type Hall flowmeter (Baishan Jiujiu Instruments Ltd., Baishan, Jilin, China). The oxygen content was measured with aTCH600 hydrogen-nitrogen-oxygen analyzer (LECO Corporation, St. Joseph, MI, USA).

2.2. Preparation and Property Test of Coatings

The Ni60A and NiCrMoY alloy powders were sprayed onto a degreased and grit-blasted mild 45# steel rod substrate of Φ 50 mm × 200 mm in size, to a thickness of 1.8 mm by oxy-acetylene flame spray and high-frequency induction remelting. The substrate was preheated and acetylene was used as the fuel gas. The spraying parameters are presented in Table 2. After being sprayed and cooled, the samples were cut with a size of 10 mm × 10 mm × 1.5 mm, and then polished using 400–1200# SiC waterproof abrasive paper, cleaned in alcohol, and corroded for 5–6 s in aqua regia.

Table 2. Oxy-acetylene flame spraying parameters.

Parameters	Gas	
	O_2	C_2H_2
Pressure (MPa)	1.2	0.1
Flow rate (m^3/h)	1.6–1.8	1.2–1.5
Spray rate (kg/h)	7.0	
Spray distance (mm)	150–200	
Thickness (mm)	250	
Remelting voltage (V)	550	
Remelting current (A)	280	

An optical microscope (OM) was used to characterize the microstructure of the Ni60A and NiCrMoY alloy sprayed coatings. The hardness of the sprayed coatings was evaluated with aHRMS-45 digital Rockwell hardness tester.

3. Results and Discussion

3.1. Properties of Powders

The morphologies of the NiCrMoY alloy powders and the Ni60A alloy powders are shown in Figure 1a,b, respectively. It can be seen that the NiCrMoY powders and Ni60A powders are near-spherical, but the sphericity of the NiCrMoY powders is better than that of the Ni60A powders, and a small number of the NiCrMoY powders have small satellites, while the Ni60A powders have more joint structures.

It is well know that morphology of powders depends on the surface tension of alloy melt, cooling speed, and shrinkage time. Better sphericity and the smoother surface of the powders are formed owing to the increasing surface tension, slow cooling speed, and long shrinkage time [22]. Due to incorporation of high melting point alloy elements, such as Mo and Nb, into the NiCrMoY powders, the melting point of NiCrMoY powders increases and particle surface tension becomes larger than the powders without addition, i.e., Ni60A alloy powders. Meanwhile the addition of Cu and rare earth Y makes the grains' surface smooth, improves the malleability [23], and there is enough time and energy to form a better spherical shape and smoother surface for the NiCrMoY alloy grains (as shown in Figure 1a) during the formation process of the powders by double-stage coupling fast freezing and low-pressure gas atomizing.

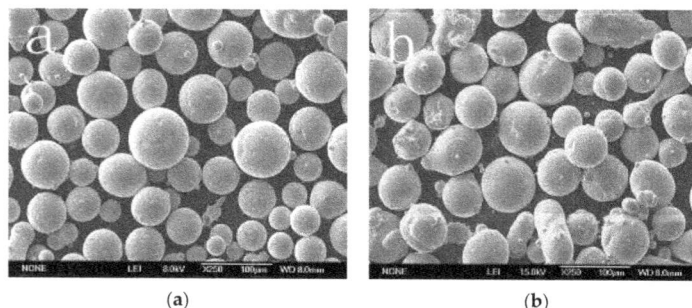

(a)   (b)

Figure 1. Morphologies of (**a**) the NiCrMoY Powders and (**b**) the Ni60A Powders.

When the particles come out of the nozzle and are atomized by low-pressure gas atomization, the large particles get cooled slowly and have a higher temperature than the small ones, thus, small particles adhere to the surface of the large ones to form joint structure and satellite. Moreover the Ni60A powders easily adhere to the small particles because of its low melting point, so that Ni60A alloy powders have a poor rate of sphere formation and have satellites, as shown in Figure 1b.

The porosity of the powders can be expressed by the hollow particle ratio of the particles to the ones without porosity in terms of unit area of the cross-section of the powder sample. Figure 2a,b show the cross-section of NiCrMoY alloy powders and Ni60A alloy powders, respectively, but the samples in Figure 2 were gradually ground and polished without corrosion. It can be seen that the hollow particle ratio of NiCrMoY alloy powders is 6.5%, which is lower than 12.5% for Ni60A alloy powders. This is because the rare earth elements interact with oxygen in the alloy melt to form tiny and dispersive rare earth compounds, and Y enhances the non-oxidizability and malleability of the alloy.

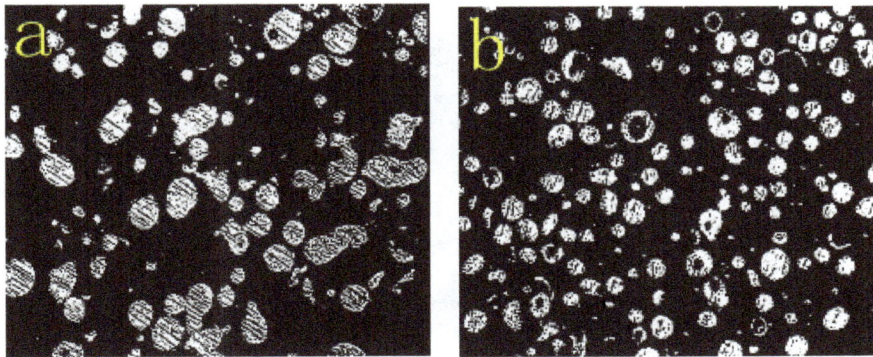

Figure 2. Cross-section of (**a**) NiCrMoY powders and (**b**) Ni60A powders.

Although deoxidization and degasification are carried out in the melting process, there is still small amounts of air existing in the alloy melt, and the temperature and pressure of the atomizing gas (nitrogen) in the atomization barrel increases rapidly during the atomization process, so the cooling velocity of alloy droplets slows so that more gas comes out from the alloy liquids. Meanwhile, due to the stirring induced by the high-pressure, some alloy powders contain nitrogen in the particle-forming process to form porosity in the particles [24]. However, the hollow particles are easy to burst and form pinholes in the coating layer when spraying. This is an important factor which influences the quality of the coatings, so the hollow particle ratio should be reduced as much as possible.

The particle-size distribution of the NiCrMoY and Ni60A alloy powders are carried out for an appropriate dispersing agent and the dispersion time is shown in Figure 3a,b, respectively. It can be seen from Figure 3a that the particle-size distribution of the NiCrMoY alloy powders shows a single peak and fits a normal distribution, and most of particle sizes are in the 38.59–118.15 μm range, and the median particle diameter d is 68.3 μm. Figure 3b shows the particle-size distribution of the Ni60A powders, which is bimodal, dispersive, and has a larger median particle diameter.

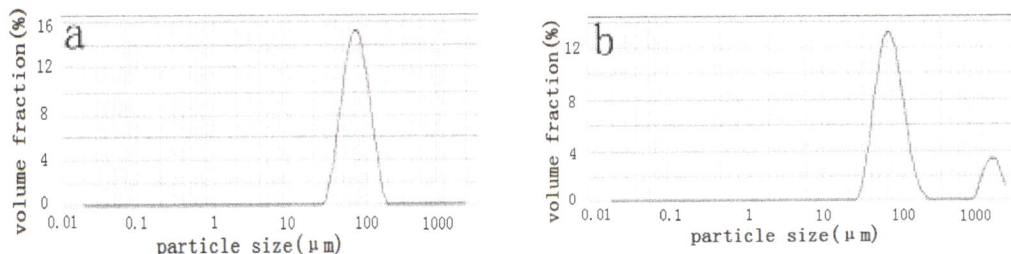

Figure 3. Particle size distribution of (**a**) NiCrMoY powders and (**b**) Ni60A powders.

The NiCrMoY alloy powder has a higher melting point and increasing surface tension due to the addition of the high melting-point alloy, such as Mo or Nb; thus, it requires more energy during gas atomization and the powder size increases compared to those without addition under the same atomizing parameters. However, because the Ni60A alloy powders have many joint structures and satellites, the particle size distribution of Ni60A powders is bimodal and dispersive, as shown in Figure 3b.

Figure 4 shows the apparent density and flowability of the NiCrMoY alloy powders and the Ni60A alloy powders. The apparent density and flowability of the NiCrMoY powders are 4.300 g/cm^3 and 14.07 s/50 g, respectively, while those of the Ni60A powders are 4.031 g/cm^3 and 15.05 s/50 g. Thus, it can be seen that the apparent density and flowability of the NiCrMoY alloy powders are better than those of Ni60A alloy powders.

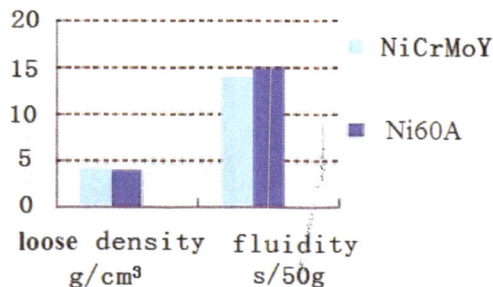

Figure 4. The apparent density and flowability of the powders.

The apparent density and flowability of powders depend on particle-size distribution, morphology, hollow particle ratio, and so on. The flowability of powders increases with better sphericity, and the apparent density of powders increases with the decreasing hollow particle ratio; high apparent density induces particle-size distribution dispersion in spite of large or small powders. This is in good agreement with the particle-size distribution, morphology, and the hollow particle ratio of the two kinds of powders mentioned above.

Figure 5 reveals the oxygen content of the NiCrMoY alloy powders and the Ni60A alloy powders, respectively. The oxygen content of the NiCrMoY alloy powders is 0.042%, lower than that of the 0.072% of the Ni60A powders. This is because rare earth Y improves the non-oxidizability of the alloy powders. The oxygen content of powders has a noticeable effect on the property of the sprayed coatings, and induces defects in coatings, so it should be reduced as much as possible.

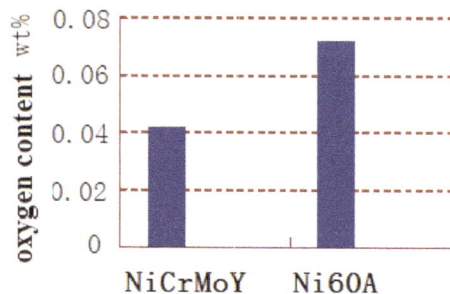

Figure 5. The oxygen content of the powders.

3.2. Properties of Coatings

The microstructure of coatings of the NiCrMoY alloy powders and the Ni60A alloy powders are shown in Figure 6a,b, respectively. As shown in Figure 6a, boride and carbide structures are distributed in the austenitic matrix of the Ni60A alloy coating. Not only boride and carbide structures be seen from Figure 6b, but also many needle-like hard phases that are uniformly distributed in the austenitic matrix of the NiCrMoY alloy coating. This is because Mo is a kind of refractory metal with large atomic radius, which can induce noticeable distortion in the crystal lattice of the nickel solid solution. It is reported that carbide is formed uniformly due to the incorporation of the rare earth elements so as to improve the mechanical property of the alloy, especially its shock property [25,26].

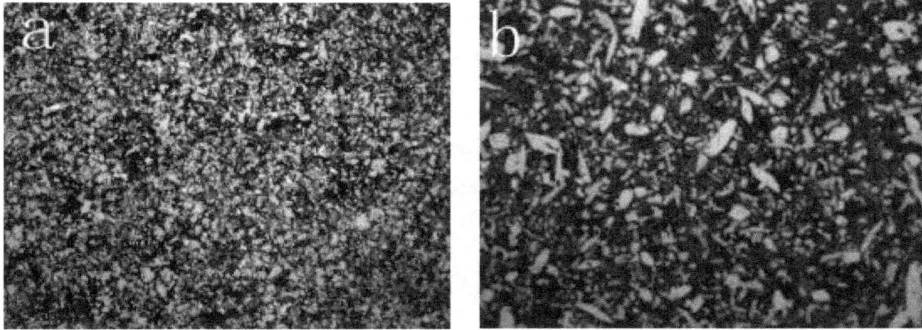

Figure 6. Microstructure of (**a**) the Ni60A coating and (**b**) the NiCrMoY coating.

The test results of hardness of the Ni60A coating and the NiCrMoY alloy coating are shown in Table 3. It can be seen that the hardness of the NiCrMoY alloy coating is higher. This is because the addition of the Mo element of the NiCrMoY coating causes grain refinement, increased toughness, decreased crack sensitivity, and enhanced high-temperature hardness and wear resistance. The addition of Nb strongly forms carbide and effectively refines grains. Thus, the appropriate incorporation of rare earth elements refines alloy structures, eliminates impurities, and forms the hard phases, such as carbide and boride, to prevent other new hard phases from forming. This causes the block- and needle-like hard phases to be uniformly distributed in alloy coatings, which increases the hardness of the alloy coatings

Table 3. Hardness of the Ni60A coating and NiCrMoY coating.

Coatings	HRC					
	1	2	3	4	5	Average
Ni60A	60.5	60.5	61.0	60.0	60.5	60.5
NiCrMoY	62.0	62.5	63.0	62.5	63.0	62.6

4. Conclusions

- Due to the incorporation of the rare earth elements Mo, Nb, or Y, the majority of the NiCrMoY alloy particles exhibit better sphericity, smoother surface, fewer joint structures, larger apparent density, and lower flowability than those of particles without incorporation.
- The particle-size distribution of NiCrMoY alloy powders shows a single peak and fits a normal distribution with a median particle diameter of 68.3 μm, an apparent density of 4.300 g/cm^3, and a flowability of 14.07 s.
- The microstructure of the NiCrMoY alloy coatings exhibits a finer structure and better hardness by the incorporation of appropriate amounts of Mo and Nb, and small amount of Y elements.

Acknowledgments: This work was financially supported by the National Natural Science Foundation of China (Grant Nos. 61665005 and 61367007) and the Natural Science Foundation of Gansu province of China (Grant No. 1112RJZA018).

Author Contributions: Chunlian Hu conceived, designed and performed the experiments; Shanglin Hou analyzed the data and wrote the paper.

Conflicts of Interest: The authors declare no conflict of interest.

References

1. Tsai, Y.L.; Wang, S.F.; Bor, H.Y.; Hsu, Y.F. Effects of Zr addition on the microstructureand mechanical behavior of a fine-grained nickel-based superalloy at elevated temperatures. *Mater. Sci. Eng. A* **2014**, *607*, 294–301. [CrossRef]
2. Scrivani, A.; Bardi, U.; Carrafiello, L.; Lavacchi, A.; Niccolai, F.; Rizzi, G. A comparative study of high velocity oxygen fuel, vacuum plasma spray, and axial plasma spray for the deposition of CoNiCrAlY bond coat alloy. *J. Therm. Spray Technol.* **2003**, *12*, 504–507. [CrossRef]
3. Safai, S.; Herman, H. Microstructural investigation of plasma-sprayed aluminum coatings. *Thin Solid Films* **1977**, *45*, 295–307. [CrossRef]
4. Christensen, R.J.; Tolpygo, V.K.; Clarke, D.R. The influence of the reactive element yttriumon the stress in alumina scales formed by oxidation. *Acta Mater.* **1997**, *45*, 1761–1766. [CrossRef]
5. Paul, A.; Elmrabet, S.; Odriozola, J.A. Low cost rare earth elements deposition method for enhancing the oxidation resistance at high temperature of Cr_2O_3 and Al_2O_3 forming alloys. *J. Alloys Compd.* **2001**, *323*, 70–73. [CrossRef]
6. Stringer, J.; Wallwork, G.R.; Wilcox, B.A.; Hed, A.Z. Effect of a thoria dispersion on high-temperature oxidation of chromium. *Corros. Sci.* **1972**, *12*, 625–636. [CrossRef]
7. Antill, J.; Peakall, K. Influence of an alloy addition of yttrium on the oxidation behavior of an austenitic and a ferritic SS in carbon dioxide. *J. Iron Steel Inst.* **1967**, *205*, 1136–1142.
8. Tien, J.K.; Pettit, F.S. Mechanism of oxide adherence on Fe–25Cr–4Al(Y or Sc) alloys. *Metall. Trans.* **1972**, *3*, 1587–1599. [CrossRef]
9. Thanneeru, R.; Patil, S.; Deshpande, S. Effect of trivalent rare earth dopants in nanocrystalline ceria coatings for high-temperature oxidation resistance. *Acta Mater.* **2007**, *55*, 3457–3466. [CrossRef]
10. Xiu, S.; Lei, W.; Yang, L. Effects of temperature and rare earth content on oxidation resistance of Ni-based superalloy. *Prog. Nat. Sci. Mater. Int.* **2011**, *21*, 227–235.
11. He, H.; Liu, Z.; Wang, W.; Zhou, C. Microstructure and hot corrosion behavior of Co–Si modified aluminide coating on nickel based superalloys. *Corros. Sci.* **2015**, *100*, 466–473. [CrossRef]
12. Elsaß, M.; Frommherz, M.; Scholz, A.; Oechsner, M. Interdiffusion in MCrAlY coated nickel-base superalloys. *Surf. Coat. Technol.* **2016**, *307*, 565–573. [CrossRef]
13. Tan, J.; Looney, L.; Hashmi, M. Component repair using HVOF thermal spraying. *J. Mater. Process. Technol.* **1999**, *92*, 203–208. [CrossRef]
14. Hu, C.; Hou, S. Failure analysis of plungers sprayed by Ni-based alloy on hydraulic feedback subsurface pump. *J. Chin. Soc. Corros. Prot.* **2012**, *32*, 80–84.
15. Dong, G.; Yan, B.; Deng, Q.; Yu, T. Effect of niobium on the microstructure and wear resistance of nickel-based alloy coating by laser cladding. *Rare Metal Mater. Eng.* **2011**, *40*, 973–977.
16. Tang, Y.; Yang, J. Influence of rare earth on wear ability of spray welding layer. *Hot Work. Technol.* **2001**, *1*, 32–33.
17. Lu, Z.; Tian, Y.; Zhu, C. Study on Ni60AA alloy powder coating by high frequency induction heating thermal. *Spray Technol.* **2012**, *4*, 44–46.
18. *ASTM E1019-11 Standard Test Methods for Determination of Carbon, Sulfur, Nitrogen, and Oxygen in Steel, Iron, Nickel, and Cobalt Alloys by Various Combustion and Fusion Techniques*; ASTM International: West Conshohocken, PA, USA, 2011.
19. *ASTM E2594-09 Standard Test Method for Analysis of Nickel Alloys by Inductively Coupled Plasma Atomic Emission Spectrometry (Performance-Based Method)*; ASTM International: West Conshohocken, PA, USA, 2014.
20. *ASTM E354-14 Standard Test Methods for Chemical Analysis of High-Temperature, Electrical, Magnetic, and Other Similar Iron, Nickel, and Cobalt Alloys*; ASTM International: West Conshohocken, PA, USA, 2014.
21. *ISO 4938 Steel and Iron-Determination of Nickel Content-Gravimetric or Titrimetric Method*; International Organization for Standardization: Geneva, Switzerland, 1988.
22. Wang, Y.; Shen, D.; Liao, B. Effect of rare earth on Ni-based spontaneous melting alloy by laser cladding. *Appl. Laser* **2003**, *3*, 139–141.
23. Zhang, C. Research on RE Micro-Alloy Effect in Self-Fusion Alloy. Master's Thesis, Shenyang University of Technology, Shenyang, Liaoning, China, 2005.

24. Liang, B.; Zhang, Z. Development and application in Ni-based powders containing rare earth elements. *J. Lanzhou Polytech. Coll.* **2007**, *1*, 28–30.
25. Ma, Y.; Huang, B.; Fan, J. Effect of rare earth Y on preparation of nanometer W-Ni-Fe composite powder. *Rare Metal Mater. Eng.* **2005**, *34*, 1135–1138.
26. Yuan, H.; Li, Z.; Xu, W. The study of Argon atomized superalloy powders. *Powder Metall. Ind.* **2010**, *4*, 1–5.

4

Defect-Free Large-Area (25 cm^2) Light Absorbing Perovskite Thin Films Made by Spray Coating

Mehran Habibi, Amin Rahimzadeh, Inas Bennouna and Morteza Eslamian *

University of Michigan-Shanghai Jiao Tong University Joint Institute, Shanghai 200240, China;
mhabibi82@sjtu.edu.cn (M.H.); amin.rahimzadeh@sjtu.edu.cn (A.R.); inas.bennouna@etu.univ-nantes.fr (I.B.)
* Correspondence: Morteza.Eslamian@sjtu.edu.cn or Morteza.Eslamian@gmail.com

Academic Editor: Alessandro Lavacchi

Abstract: In this work, we report on reproducible fabrication of defect-free large-area mixed halide perovskite (CH$_3$NH$_3$PbI$_{3-x}$Cl$_x$) thin films by scalable spray coating with the area of 25 cm^2. This is essential for the commercialization of the perovskite solar cell technology. Using an automated spray coater, the film thickness and roughness were optimized by controlling the solution concentration and substrate temperature. For the first time, the surface tension, contact angle, and viscosity of mixed halide perovskite dissolved in dimethylformamide (DMF) are reported as a function of the solution concentration. A low perovskite solution concentration of 10% was selected as an acceptable value to avoid crystallization dewetting. The determined optimum substrate temperature of 150 °C, followed by annealing at 100 °C render the highest perovskite precursor conversion, as well as the highest possible droplet spreading, desired to achieve a continuous thin film. The number of spray passes was also tuned to achieve a fully-covered film, for the condition of the spray nozzle used in this work. This work demonstrates that applying the optimum substrate temperature decreases the standard deviation of the film thickness and roughness, leading to an increase in the quality and reproducibility of the large-area spray-on films. The optimum perovskite solution concentration and the substrate temperature are universally applicable to other spray coating systems.

Keywords: mixed halide perovskite; large area perovskite; spray coating; perovskite solution physical properties; perovskite film optimization

---

1. Introduction

Within the past few years, a tremendous effort has been made to increase the power conversion efficiency (PCE) of perovskite solar cells (PSCs). In spite of achieving remarkable PCEs in the research labs, as high as 22.1% [1], two main obstacles still hinder the development of this technology: The device instability and the lack of knowledge and experience for large scale and large area device fabrication. Development of commercial methods for the fabrication of large area PSCs is one of the prerequisites for their commercialization, and it is as important as stabilizing the PSC performance. It is generally expected that, by increasing the film surface area, the defect and pinhole density would increase; therefore, research on the development of large area solar cells is essential. An ideal perovskite film must have a fully-covered monocrystalline structure with high uniformity and low roughness. Obtaining such ideal films is challenging if not impossible, due to the special behavior of the halide perovskite materials, which is the tendency to crystallize in a polycrystalline structure upon deposition, making the resulting thin films prone to dewetting due to crystallization (crystallization dewetting), and, therefore, the emergence of pinholes [2]. Perovskite crystal growth in all directions, including the direction normal to the film, tends to shrink and disintegrate the film, resulting in a decrease in the film coverage and an increase in the roughness. An ideal perovskite film must have a thickness within the range suitable for charge generation and transfer, as well. Thickness of the mixed halide perovskite films should be limited to 1 μm or so, dictated by the maximum diffusion length of the generated excitons in the perovskite structure [3]. Therefore, controlling the detrimental effect of the crystallization dewetting to achieve a fully-covered film, which also has desirable thickness and low roughness is quite challenging, especially when the film is deposited by a scalable method.

Solution-processed deposition of a thin film of perovskites may be performed using various casting methods, such as spin coating, dip-coating, doctor blading, spray coating, inkjet printing, screen printing, drop casting, slot-die coating, etc. Some of the aforementioned techniques, such as spin coating, in spite of providing precise controllability on the film morphology (thickness, coverage and roughness), are generally limited to batch processes and/or thin films with small effective surface areas, making them unsuitable for real-world applications. In the lab-scale and mostly using spin coating, various treatments are usually applied on the small-area perovskite films to reduce the roughness and increase the coverage and homogeneity and improve the crystalline structure. These methods include but are not limited to solvent engineering [4–6], manipulating the stoichiometry of the perovskite precursors (e.g., ratio of PbI_2/MAI solutions, where MAI stands for methylammonium iodide) [7,8], introducing additives to the perovskite solution (e.g., water, 1,8-diiodooctane(DIO)) [9,10], and controlling the annealing temperature and time [11]. However, preparation of large area perovskite films (>1 cm^2) with uniform characteristics across the film is harder to accomplish, compared with the films with small areas (≤ 0.1 cm^2). Enlarging the perovskite surface area causes a decline in the cell performance. Figure 1 compares the PCE of several PSCs made under identical conditions, but using various deposition methods and effective surface areas. The figure confirms a systematic drop in the PCE after enlarging the active area of similar cells, or as a result of module fabrication by connecting various small-area cells in series, in order to increase the effective area [7,12–14].

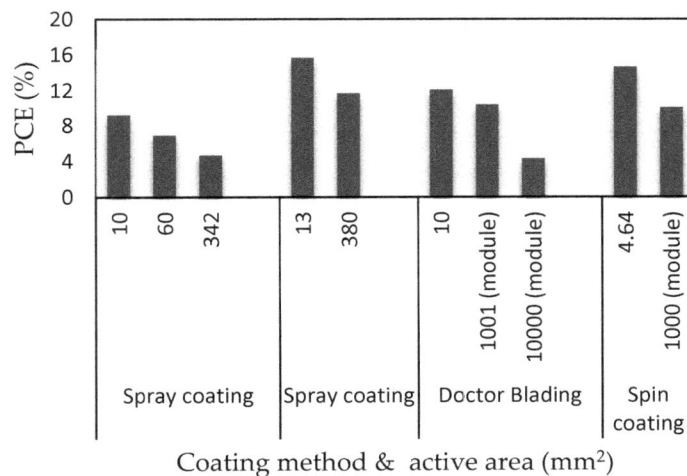

Figure 1. Degradation in the PCE of the PSCs made by various deposition techniques after enlarging the active area of individual cells, or by fabricating modules through connecting several small-area cells in series (data were taken from Refs. [7,12–14]). The precursor solutions associated with the mentioned coating processes are $MAPbI_3$, $MA(I_xBr_{1-x})_3$, PbI_2, and $MAPbI_3$, respectivley, and from left to right. In the doctor blading case, the perovskite layer was obtained by dip coating of the blade-coated PbI_2 film in the methylamonium iodide solution [12].

Following the aforementioned argument regarding the need for the fabrication of large-area solar cells using scalable methods, in this work, spray coating is used to produce large-area perovskite thin films with favorable morphological and light absorbing characteristics. This work focuses on the optimization of the perovskite light harvester layer only, and the fabrication of the entire device with large area is postponed to future works. Several advantages of spray coating compared to the other large scale casting techniques including the touch-free, low-cost and fast process, the possibility for deposition on flexible or rough substrates, and its capability for producing ultrathin films, assure its high potential for the roll-to-roll fabrication of solar cells on a large scale [7,15]. Spray coating may be also combined with shadow masks for pattern printing. Recently, it has been reported that spray-on films show better thermal stability compared to spun-on films [16]. Huang et al. [16] demonstrated that the prolonged annealing time required after spin coating may adversely affect the stability of spun-on films, whereas the spray-on perovskite films show high thermal stability, which originates from better crystallinity of spray-on perovskite films. In their study, they also observed better optoelectronic characteristics in spray-on perovskite films compared to spun-on counterparts, due to their higher carrier lifetime and better charge transfer capability. Despite the advantages of spray coating, fabrication of fully-covered and homogeneous thin (perovskite) films with desired low thickness and roughness is challenging. This is because the phenomenon of the liquid atomization and spraying is a random and stochastic process, which works based on transient impact of numerous droplets of different sizes across the wetted area. The droplets may first form a stable or unstable thin liquid film and then dry to form a thin solid film, or each individual droplet or patch of several merged droplets may dry to form a thin solid film. These uncertainties may cause unpredicted characteristics in the ensuing thin solid films [17]. The photovoltaic characteristics of a solar cell, such as its open-circuit voltage (V_{OC}) and fill factor (FF), are directly influenced by the quality of the film. Voids and pinholes in the perovskite films caused by a poor spray coating process may make short circuit pathways between the above and underneath layers of the perovskite, which may result in a decrease in the device shunt resistance and degradation of the device performance. Therefore, some pre-treatments, post-treatments, or additional processes have been suggested to achieve a desired spray-on thin film. For instance, Ramesh et al. [7] used a simple airbrush pen to spray $CH_3NH_3PbI_3$ perovskite precursor solution and tuned the ratio of the MAI to PbI_2 precursor solutions, spray flow rate, substrate temperature, and annealing temperature to achieve a desired film. In another work, Chandrasekhar et al. used electrostatic spray coating to spray the MAI solution onto a pre-cast PbI_2 film, where a more uniform and dense perovskite film with lager crystals was obtained, compared to that of the conventional spray coating [18]. Heo and coworkers [19] synthesized $CH_3NH_3PbI_{3-x}Cl_x$ powder and then dissolved it in a mixture of DMF (dimethylformamide) and GBL (g-butyrolactone). To adjust the perovskite crystal size in the spray-on film, they controlled the evaporation rate of DMF by changing the ratio GBL to DMF. Abdollahi Nejand et al. [20] sprayed a concentrated solution of $CH_3NH_3PbI_{3-x}Cl_x$, which caused the creation of columnar film of perovskite. Then, the film was exposed to low-pressure vapor of DMF to partially dissolve the crystals; the weakened film was then compacted by a cold-roll press. Through this method, the film coverage and the device performance were improved. Concurrent spraying of perovskite precursors using two spray nozzles is another suggested technique to control the film composition and achieve a pinhole-free film of perovskite [14]. In the literature, spraying of the MAI solution over a pre-cast PbI_2 film in a sequential deposition has been reported, as well [8,18,21–23]. Zabihi et al. sprayed perovskite precursor solutions sequentially, using two spray nozzles, on an ultrasonically vibrating substrate to form a mixed halide perovskite film [22]. The substrate vibration resulted in improved mixing and uniform deposition of the perovskite layer. In a recent work, we also used spray coating in a perovskite solar cell to fabricate a uniform PbI_2 layer and then converted it to single-halide $MAPbI_3$ perovskite film via pulsed-spray coating and drop casting of the MAI solution atop the PbI_2 layer [23].

Although spray deposition of a uniform and pinhole-free film of mixed halide perovskites (e.g., $CH_3NH_3PbI_{3-x}Cl_x$) in a PSC with planar structure is challenging, the mixed halide perovskites are more advantageous over single halide perovskites, e.g., $CH_3NH_3PbI_3$, due to larger charge carrier diffusion lengths (near 10 times), which results in a higher charge collection efficiency [19]. Therefore, in this work, single-step spray deposition of the mixed halide perovskite $CH_3NH_3PbI_{3-x}Cl_x$ is adopted. Then, the morphological and optoelectronic characteristics of the fabricated perovskite films with a square area of 25 cm^2 are optimized by systematically tuning the important process parameters, i.e., the solution concentration, substrate temperature, and the number of spray passes, while other parameters are pre-optimized and kept constant during the experiments. The best concentration of the solution of $CH_3NH_3PbI_{3-x}Cl_x$ dissolved in DMF is determined based on measuring the physical characteristics of the perovskite solution and also considering the detrimental effect of the crystallization dewetting that occurs at high solution concentrations [2]. The surface tension, contact angle and viscosity of the perovskite precursor solution are the main physical properties that govern the droplet impact and the coating process, and therefore are measured and reported in this work. The second important parameter, which is controlled to achieve a fully-covered film, is the substrate temperature. To begin with, a reasonable range of high temperatures is chosen due to the better spreading of droplets on high substrate temperatures (deduced by the measured contact angles versus temperature), and then the best substrate temperature is found based on the conversion of precursors to perovskite. Finally, the number of spray passes is tuned to achieve a fully-covered film with the lowest roughness and desired thickness. Standard deviation of the roughness and thickness data obtained from the spray-on films fabricated on a hotplate are lower than their counterparts sprayed on substrates kept at the ambient temperature. This fact reveals the strong feature of the high substrate temperature to increase the reproducibility of the spray-on films. It is important to distinguish between the substrate temperature in spray coating and the long-duration annealing temperature performed after the deposition process. In this work, all deposited samples were annealed on a hotplate at 100 °C for two hours.

2. Experimental

2.1. Materials and Methods

Lead chloride (PbCl$_2$, 98.5%), and N,Ndimethylformamide (DMF, 99.8%) were supplied by Sigma-Aldrich, St. Louis, MO, USA. Methylammonium iodide (MAI, 99.5%) was purchased from Xi'an Polymer Light Technology Corp. (Xi'an, China). MAI and PbCl$_2$ were mixed in 3:1 molar ratio and then dissolved in DMF in various concentrations (5% to 50% weight ratio: Ratio of the solid precursors mass to the total mass of the solvent and solid precursors in the solution). The MAPbI$_{3-x}$Cl$_x$ solution was heated to 60 °C and stirred on a magnetic stirrer overnight, and then cooled at room temperature. Small (1 × 1 cm^2) and large (6 × 6 cm^2) fluorine-doped tin oxide (FTO)-coated glass substrates with an average roughness of near 14.5 nm were cleaned by a mixture of deionized water and soap, acetone and isopropyl alcohol in an ultrasonic bath, sequentially. Then, the substrates were exposed to UV-Ozone irradiation for 15 min. Spray coating was performed by replacing the spray nozzle of an automatic spray coating system (Holmarc, Opto-Mechatronics Pvt. Ltd., Model HO-TH-04, Kochi, India) with a low flow rate commercial airbrush pen. This was done because the liquid container of the original ultrasonic nozzle of the Holmarc machine is excessively large, resulting in the wastage of the perovskite solution. The installed airbrush generates a fine mist, is easy to control, and is suitable for the fabrication of perovskite films. The speed and position of the spray nozzle in the x–y plane were controlled by software to move the nozzle in a continuous and raster movement to simulate an industrial spray coating process. The perovskite solution was atomized by the pressurized air at constant pressure and air flow rate. Parameters of the spray coating process are summarized in Table 1. Pictures of the spray coating machine and the spray nozzle are shown in Figure 2.

Table 1. Spray coating parameters for the fabrication of perovskite films.

Spray Parameters	Value
Nozzle to substrate distance (cm)	12
Nozzle speed (mm/s)	150
Flow rate (μL/s)	15
Air pressure (bar)	3.5
Number of passes	40, 70, 100
Substrate temperature (°C)	25–200

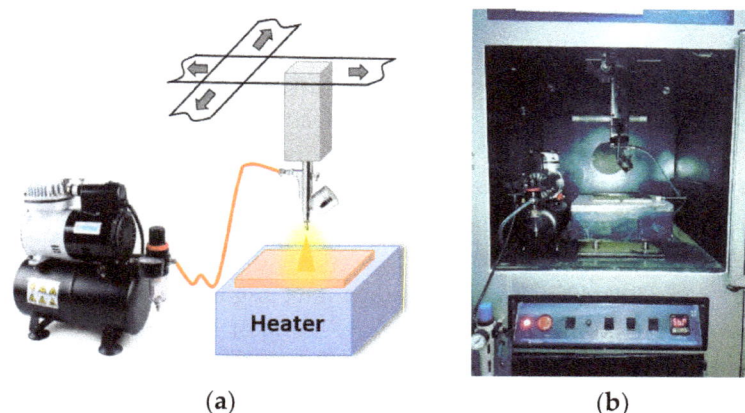

Figure 2. (**a**) schematic of the spray coating process and the aparatus. (**b**) picture of the Holmarc spray coating system, which accomodates the spray nozzle and the hotplate shown in (**a**).

The relative humidity of Shanghai during the spray deposition of the perovskite films was in the range of 35% to 50% in different days. Although the humidity has a detrimental effect on the perovskite quality, it mainly affects the perovskite film during and after annealing. Therefore, to minimize the effect of the humidity, the spray-on films, made in the ambient conditions in few minutes, were transferred to the glovebox after deposition and initial drying on the high temperature substrate, for complete drying and annealing on a hotplate at 100 °C, for two hours. In some of the tests (for determination of the crystallization dewetting), the samples were formed on small-size FTO-coated glass (1×1 cm^2) by spin coating at 2500 rpm for 20 s, and then the samples were annealed based on the same procedure mentioned above for the spray-on films. Perovskite films made on small area FTO-coated glass were encapsulated with poly (methyl methacrylate) (PMMA) to enhance the perovskite stability against humidity during the X-ray diffraction (XRD) and absorption tests.

2.2. Film and Device Characterization

Optical microscopy images of perovskite films were taken by a confocal laser scanning microscope (CLSM 700, Carl Zeiss AG, Oberkochen, Germany) and scanning electron microscopy (SEM) images were obtained using a Hitachi microscope, Model S-3400N, Tokyo, Japan. Average thickness and roughness of the films were measured by a stylus profilometer (KLA-Tencor P7, Milpitas, CA, USA). Roughness values were measured and averaged along six randomly chosen lines of 100 μm long, on two similar samples, in each case. Thickness values were also measured by the same instrument, in four randomly-chosen spots near the edge of the films with respect to the uncovered areas of the FTO-coated glass, and were averaged.

The conversion of the MAI and PbCl$_2$ precursors to mixed halide perovskite and the absorption of the perovskite films were evaluated by X-ray diffraction (XRD, model D5005, Bruker, Billerica, MA, USA) and UV-Visible absorption spectrophotometry (EV300, Thermo Fisher Scientific, Waltham, MA, USA), respectively. Physical properties of the perovskite solutions were measured, as well. The solution surface tension and contact angle were measured, using Theta Lite Optical Tensiometer (Biolin Scientific AB, Gothenburg, Sweden), and the viscosity was measured using a digital rotary viscometer (model NDJ-8S, Jiangsu Zhengji Instruments Co., Ltd., Jintan, China) in various concentrations, defined as the weight fraction of the solute (perovskite precursors) in the solution. A digital infrared camera (FLIR C2, Tallinn, Estonia) was used to confirm the substrate temperature, which is generated and maintained by an electric heater.

3. Results and Discussion

Knowledge of the concentration-dependent physical properties of the perovskite precursor solution, such as the contact angle, surface tension, and viscosity, would lead to better understanding of the droplet impact behavior during spray deposition, and also would help proper selection of the solution concentration. This is because the droplet impact dynamics is governed by the Ohnesorge number, a dimensionless group based on the liquid properties, as well as Reynolds and Weber numbers, which include the effect of the droplet momentum upon impact. Assuming constant impingement velocity and substrate texture and surface energy, concentration of the perovskite precursor solution and the substrate temperature control the spreading behavior of the perovskite droplets after impinging on the substrate, and therefore control the morphology of the resulting perovskite film. Physical properties of the precursor solution may also affect the thin film characteristics prepared by other casting methods. Various solution concentrations (9.5–33 wt.%) [20,24,25] have been used by others as the starting concentration for the fabrication of perovskite films, but, to the best of our knowledge, this has not been done systematically. Therefore, here we measured the contact angle, surface tension and viscosity of the mixed halide perovskite $MAPbI_{3-x}Cl_x$ dissolved in DMF solvent in a wide range of concentrations (5–50 wt.%). Table 2 lists the average measured properties along with the standard deviation of the measurements.

Table 2. Some of the physical properties of mixed halide perovskite solution ($MAPbI_{3-x}Cl_x$ in DMF) at various concentrations (wt.% of solute in the solution). MAI and $PbCl_2$ powders were mixed in 3:1 molar ratio and then dissolved in DMF to achieve various concentrations.

Concentration (wt.%)	Viscosity (mPa.s)	Surface Tension (mN/m)	Contact Angle (°)
0	0.887 ± 0.02	37.29 ± 0.02	12.39 ± 0.04
5	1.0 ± 0.0	36.81 ± 0.02	16.35 ± 0.24
10	1.0 ± 0.0	35.18 ± 0.02	20.79 ± 0.10
20	1.59 ± 0.016	34.39 ± 0.06	24.00 ±0.15
30	2.0 ± 0.0	31.44 ± 0.03	29.26 ± 0.11
40	2.0 ± 0.0	29.80 ± 0.03	30.40 ± 0.15
50	6.72 ± 0.135	28.87 ± 0.02	31.65 ± 0.21

Some studies, e.g., Mullins and Sekerka [26], suggest that some ionic solutions may act like a surfactant, in that, their surface tension may decrease with the solution concentration, and, in fact, our data of perovskite solutions at different concentrations comply with the reported observations for other similar solutions. Figure 3a shows a decrease in the surface tension of the perovskite solution with the solution concentration. On the other hand, examination of the perovskite solution droplets on glass substrates shows that the contact angle increases with concentration (Figure 3b). Young's equation, i.e., $\gamma\cos\theta = \gamma_{sg} - \gamma_{ls}$, relates the binary surface tensions with the equilibrium contact angle on a solid surface. In Young's equation, γ is the liquid–air surface tension or simply surface tension, γ_{sg} is the solid-gas surface tension, γ_{ls} is the liquid–solid surface tension, and θ is the equilibrium contact angle. Our results show that, as the solution concentration increases, γ decreases on one hand and contact angle increases on the other hand ($\cos\theta$ decreases), thus the solid–liquid surface tension, γ_{ls} must increase, since the solid–gas surface tension γ_{sg} remains constant for the same glass substrate next to the same surrounding gas (air). Some numerical works by Sear [27] and theoretical works by Djikaev and Ruckenstein [28] have shown the substrate effect on crystal nucleation. For instance, crystal nucleation is greater in the vicinity of the gas–liquid–solid triple line, since the concentration is the highest at the triple line, due to the coffee stain or coffee ring effect. Hence, to suppress the inhomogeneous crystal nucleation due to the coffee-ring effect, increasing the spreading of the solution droplets, as well as increasing the droplet drying rate, could be effective. In this work, this has been achieved by using treated FTO-coated glass substrates to increase wettability, as well as by choosing a low solution concentration (e.g., 10%), to achieve a low contact angle.

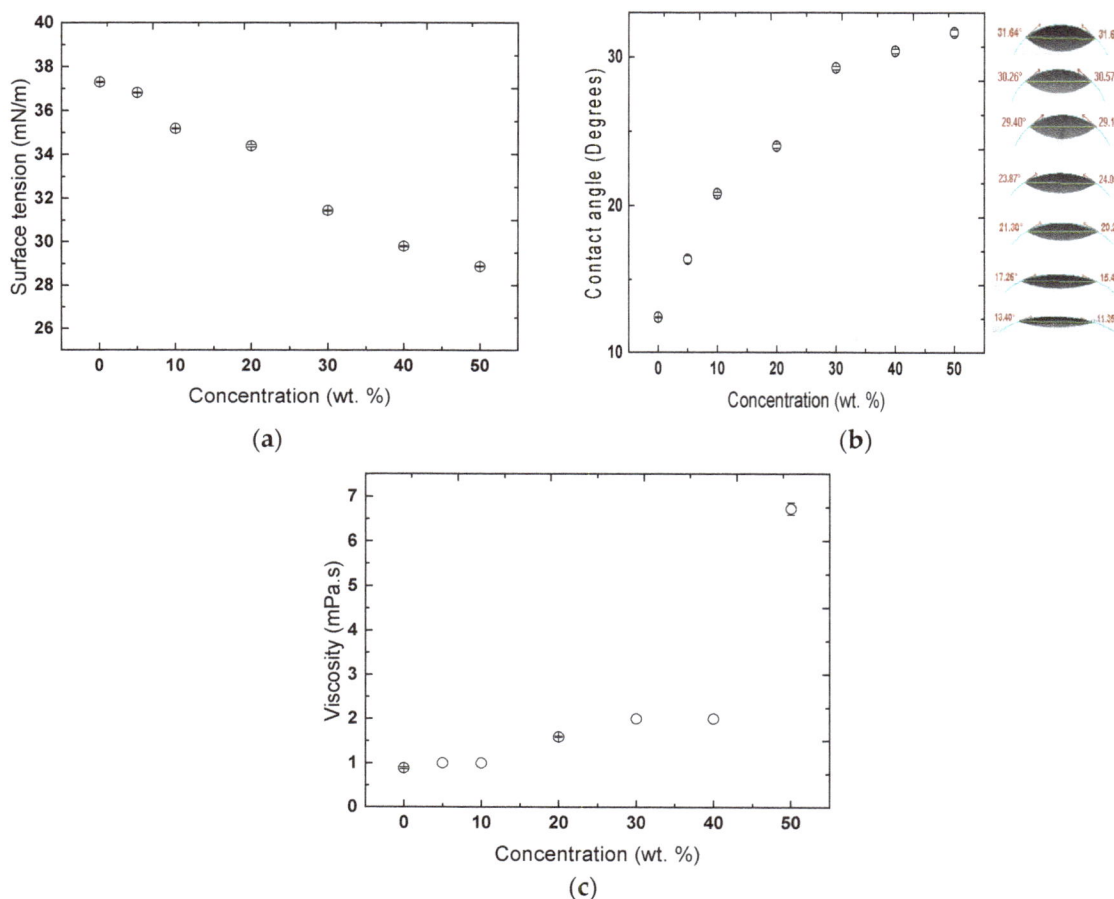

Figure 3. Variation of the physical properties of the mixed halide perovskite solution (MAPbI$_{3-x}$Cl$_x$ in DMF) versus solution concentration in wt.% of solute in solution. (**a**) liquid–air surface tension; (**b**) contact angle, and (**c**) viscosity. Note: The data and the associated errors are listed in Table 1.

Viscosity is another physical property of the liquid solution that affects the liquid film spreading and wetting, and is believed to increase nonlinearly as the solvent evaporates and the liquid film partially dries [29]. In droplet-based coating methods, viscosity, through the Ohnesorge number, affects droplet impact dynamics and spreading. As shown in Figure 3c, viscosity increases somewhat linearly at concentrations up to about 40 wt.%. Then, it sharply increases as the concentration increases to 50 wt.%, which corroborates the nonlinear behavior of the viscosity with the concentration. This is because, at high concentrations, the solution starts to show non-Newtonian behavior.

High concentration of the perovskite solution may cause the formation of larger crystals in the film and therefore surface dewetting, i.e., crystallization dewetting [2], which obviously leads to the formation of rough films with pinholes (Figure 4). All films in Figure 4 are spun under the same conditions, while the concentration changes from 10% to 30% and 50% from left to right, respectively. Since the data from Figure 3 clearly show that the solution droplets with lower concentration have smaller contact angles, and therefore higher surface wettability, and Figure 4 shows that smaller detrimental crystallization dewetting occurs when the perovskite thin films are made using a solution concentration of 10 wt.%, this concentration is selected as the favorable concentration for spray coating of the mixed halide perovskite MAPbI$_{3-x}$Cl$_x$ dissolved in DMF. At this low concentration, the crystallization dewetting is the minimum and spreading is the maximum, both favoring the formation of a continuous and defect-free film.

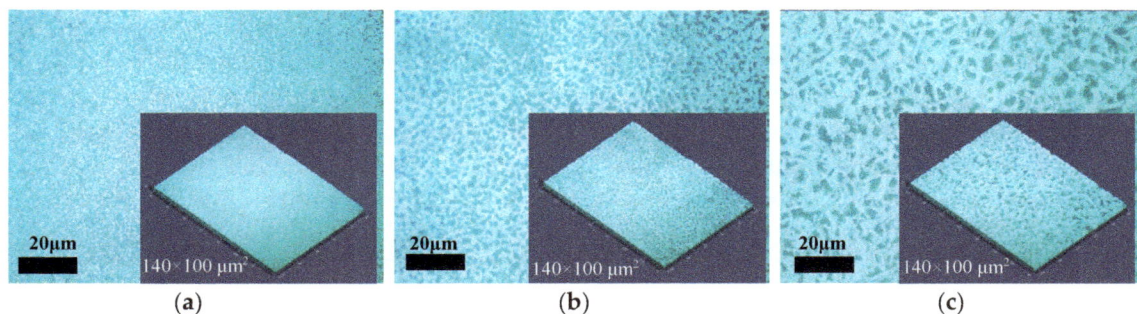

Figure 4. Optical images of spun-on perovskite films at different concentrations. The solution concentrations are (**a**) 10 wt.%; (**b**) 30 wt.%; and (**c**) 50 wt.%. This figure shows that high concentrations result in the formation of large crystals, and therefore, occurrence of crystallization dewetting.

Having determined the ideal solution concentration, now we will find the optimum substrate temperature. Spray deposition of perovskite solution on a low temperature substrate prolongs the drying time, which may adversely affect the film coverage [24]. This is due to the prolonged crystal growth at low temperatures and therefore the occurrence of crystallization dewetting. Figure 5 shows the range of the substrate temperatures (during the coating process) that have been used in recent published papers [3,7,8,13,16,18–22,24,25,30–37], in order to fabricate the perovskite layer of PSCs, mostly by a scalable technique. These techniques include spray coating, doctor blading, slot-die coating, roller coating, inkjet printing, and drop casting. In most of the methods, temperature of the substrate was raised to facilitate the coating process. However, in most of those works, a systematic procedure was not followed. Hence, in this work, the best substrate temperature for rapid drying and in situ heat treatment during spray deposition is selected based on two criteria; first, a range of temperatures are selected to achieve the best droplet spreading upon impingement on a hot substrate. To this end, the contact angles of 10 wt.% perovskite solution droplets were evaluated right after droplet release on glass substrates kept at various temperatures (25 °C to 200 °C with 25 °C intervals) (Figure 6a). The heated substrate increases the evaporation rate, which is beneficial for arresting the crystal growth and suppressing the crystallization dewetting. In addition, the data of Figure 6a show that a higher substrate temperature improves the film coverage by decreasing the contact angle, because of better droplet spreading and surface wettability. Therefore, based on the contact angle measurements, relatively high substrate temperatures seem to be more effective for a better coverage. It is noted that recent research on the thermal and thermodynamic stability of perovskites proves that high annealing temperatures for a long time results in the decomposition of the perovskite structure [38,39]; however, here in this work, the films are exposed to a high substrate temperature for a short period of time during the spray deposition, which lasts only for several seconds. It is also noted that excessive substrate temperatures above the Leidenfrost temperature must be avoided due to the formation of a vapor film over the substrate, which would interrupt the droplet spreading and the coating process.

In the second step, we further narrow down the choices of the substrate temperature by evaluating the conversion of the perovskite precursors to perovskite. We define the conversion ratio using the XRD patterns of the perovskite films, as the ratio of the PbI_2 peak intensity at 12.7° to the perovskite peak intensity at 14.2°. PbI_2 is an impurity in the perovskite film and has to be eliminated or minimized. Figure 6b illustrates the XRD patterns of four samples fabricated on substrates kept at 25, 100, 150 and 200 °C. It is noted that these are the substrate temperatures, and all samples were annealed at 100 °C after deposition. We observe that the intensity of the PbI_2 peak changes with the substrate temperature. Figure 6c shows the conversion ratio obtained at various substrate temperatures. It is observed that the maximum conversion occurs near 150 °C, which is near the boiling point of DMF, i.e., 153 °C. Low conversion ratio at temperatures lower than 150 °C is ascribed to initial incomplete conversion of precursors, whereas, at higher temperatures than the boiling point of DMF, it may be attributed to rapid evaporation of the solvent and the lack of enough solvent for a complete conversion. This observation is in agreement with the report of Mallajosyula and coworkers that considered substrate temperature of 145 °C during doctor blading of MAPbI$_3$ [37]. Therefore, according to the contact angle measurements of perovskite droplets at various substrate temperatures and also the conversion ratio data, 150 °C is chosen as the optimum substrate temperature, since, at this temperature, spreading of perovskite solution droplets and also conversion ratio of perovskite attain their maximum values.

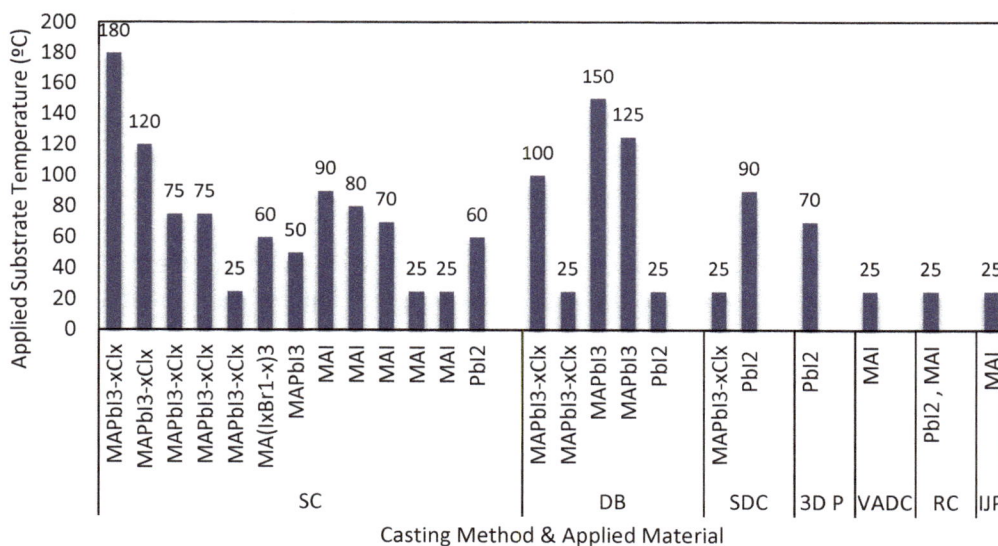

Figure 5. Employed substrte temeratures during the fabrication of perovsite films by various (large scale) techniques. (data taken from Refs. [3,7,8,13,16,18–22,24,25,30–37]). Abreviations: SC: Spray coating; DB: Doctor blading; SDC: Slot-die coating; 3D P: 3D printing based on SDC and N$_2$ gas-blowing; VADC: Vibration-assisted drop casting; RC: Roller coating; IJP: Inkjet printing.

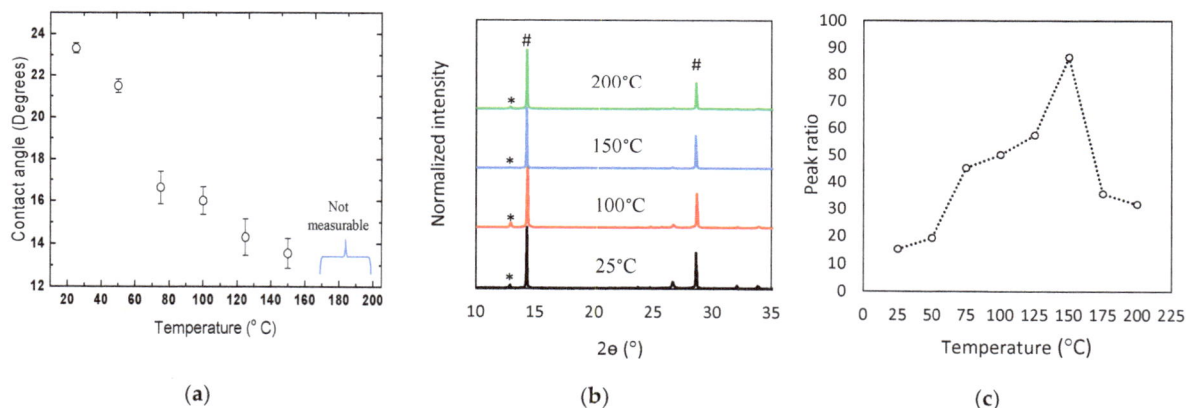

(a)   (b)   (c)

Figure 6. (a) contact angle of the perovskite solution droplets (MAPbI$_{3-x}$Cl$_x$ in DMF) versus substrate temperature; (b) XRD patterns of MAPbI$_{3-x}$Cl$_x$ perovskite films, spray deposited on FTO-coated glass substrates at various substrate temperatures (# and * represent the perovskite and PbI$_2$ peaks, respectively); (c) the quantified conversion ratios at various substrate temperatures. At temperatures higher than the boiling point of DMF (>153 °C), measurement of the contact angle was found to be not accurate, due to rapid solvent evaporation. It is noted that, after deposition, all perovskite films were annealed at 100 °C for 2 h.

The third and last parameter to optimize is the number of spray passes for the used spray nozzle and spray conditions (Table 1). The number of spray passes affects the film coverage, thickness, and roughness. In general, photons with long wavelengths are better absorbed in thicker films, which could translate to a higher current density in the active layer [25]; however, enhanced absorption occurs at the cost of increased charge recombination and the loss of the current density [25]. The limitation of charge diffusion length in perovskite films also confines the admissible span of thicknesses. In $MAPbI_{3-x}Cl_x$ mixed halide perovskite, the charge diffusion length has been measured to be around 1000 nm, while, in $MAPbI_3$ single halide perovskite, it is near 100 nm [3]. Therefore, there is a specific range of thicknesses in which the highest current from a perovskite film could be extracted. In this work, adjusting the thickness was performed by using multiple spray passes. A single spray pass is the travelling of the spray nozzle arm over the substrate with a specific nozzle velocity. Multiple-pass spraying is comprised of several consecutive single spray passes back and forth in a predesigned raster pattern to improve the coverage and achieve a desired thickness. Figure 7 shows the measured thickness and roughness values of the perovskite films, sprayed using 40, 70, and 100 passes, on the substrates kept at the ambient temperature. Coverage measurements revealed that, at the used spray flow rate and spray nozzle speed (Table 1), spraying using 40 and 70 passes was not enough to cover the film completely. Raising the passes to 100 resulted in full coverage of the substrate. Obviously, the film thickness increases by raising the number of spray passes; however, the 40-pass-spray-on film showed the highest roughness, which is due to the excessive number of pinholes. However, in the two other cases, increasing the number of passes from 70 to 100, resulted in an increase in the roughness again (Figure 7b). This is in agreement with the result of other reports, which declared that the roughness of spray-on perovskite films increases by increasing the thickness [25]. It is noted that in less-crystalline or amorphous films, such as polymeric films, the correlation between the number of spray passes and roughness may be somewhat different from what was observed here for perovskite films, simply because, in polymeric films, the phenomenon of crystallization dewetting and the formation of large grains is absent or it is insignificant [40].

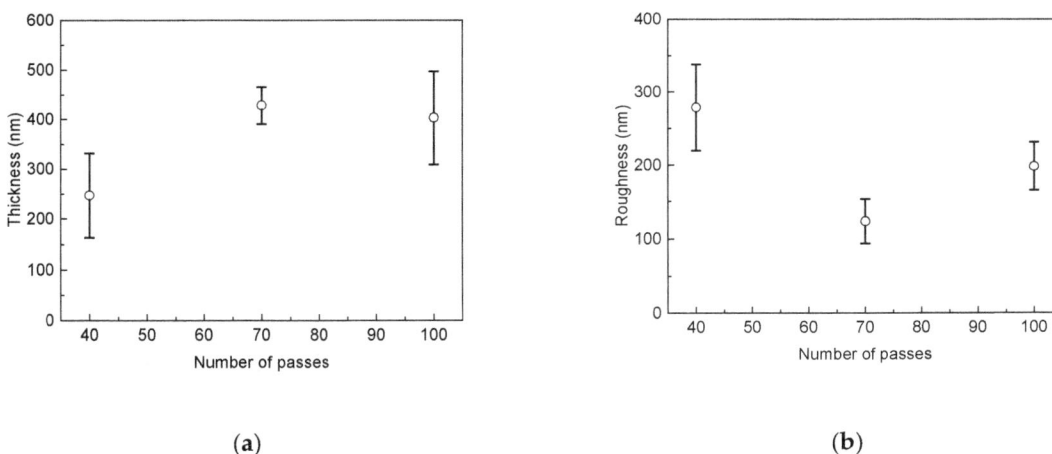

(a)   (b)

Figure 7. Measured (a) thickness and (b) roughness of spray-on perovskite films deposited on the substrates kept at the ambient temperature, at various number of spray passes.

To fabricate the best functional perovskite film, i.e., lowest roughness and proper thickness with negligible PbI_2 impurity, the spraying process was repeated using different passes (40, 70 and 100) on hotplates at the optimum temperature of 150 °C, which yields the best conversion ratio (Figure 8). Again, the full coverage is achieved at 100 passes, while the thickness and roughness decreased in all cases, compared with the samples fabricated on the substrates kept at the ambient temperature.

Figure 9 shows the SEM images of spray-on perovskite films deposited on the hotplates kept at 150 °C, while the number of passes varies. The average coverage, denoted as C on the images, obtained from the SEM images, reveals that a fully-covered film was obtained after 100 passes. The coverage was obtained by image processing using the ImageJ software (version 1.51j, National Institutes of Health (NIH), Bethesda, MD, USA).

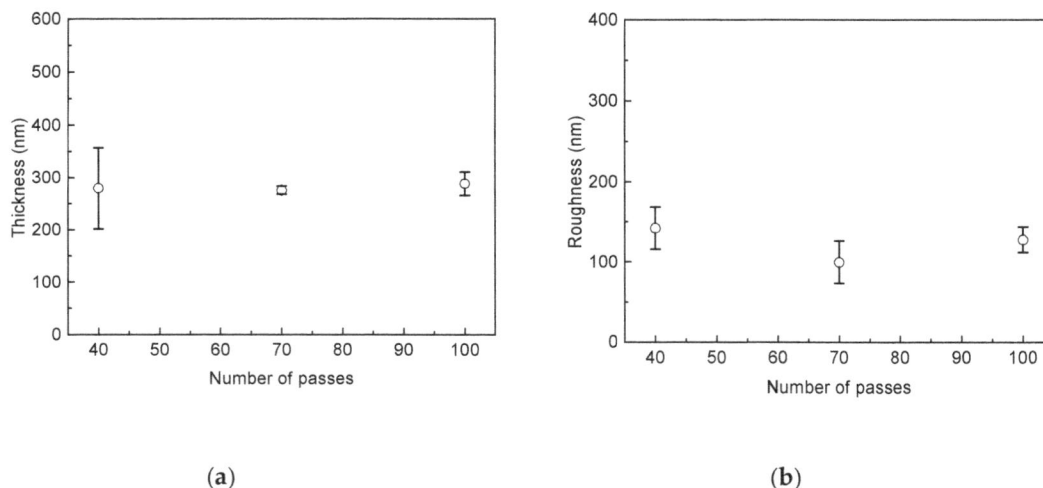

(a)   **(b)**

Figure 8. Measured (**a**) thickness and (**b**) roughness of spray-on perovskite films deposited on the substrates kept at 150 °C, at various number of spray passes.

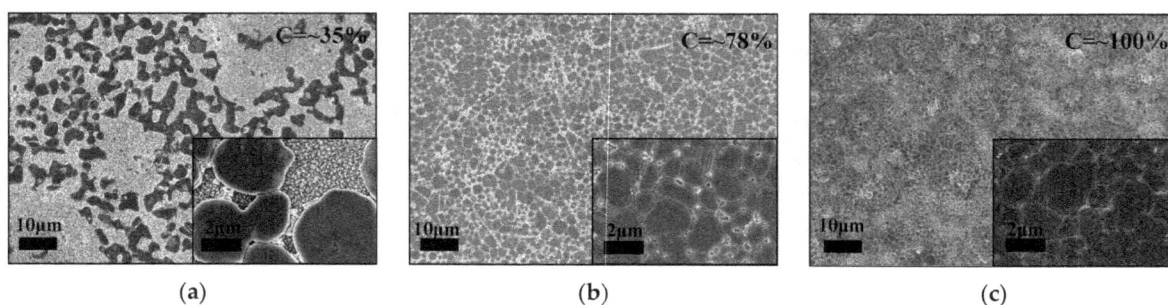

(a)   **(b)**   **(c)**

Figure 9. SEM images of spray-on perovskite films deposited on FTO-coated glass substrates at substrate temperatures of 150 °C. The number of spray passes are (**a**) 40; (**b**) 70; and (**c**) 100. Letter C refers to the percentage of coverage of perovskite layer on the substrates.

Interestingly, as depicted in Figure 10, the application of the high temperature substrate results in a decrease in the standard deviation of roughness and thickness data of spray-on films with respect to their counterparts fabricated at the ambient temperature. This is an important factor to increase the device reproducibility of large-area perovskite films. According to Figure 8a, the average thickness values of all three spray-on samples is less than 300 nm, which is acceptable for mixed halide perovskite films to perform well in a PSC [24]. In Figure 10, the spray-on films made using 70 passes show the minimum standard deviation in their thicknesses. The same films have the lowest roughness as well (Figure 8b). However, these films suffer from inadequate coverage (78%). Thus, the film made using 100 passes on the hotplate kept at 150 °C is considered as the best film.

The perovskite films, sprayed using 100 passes on the substrates kept at 25 and 150 °C, are compared using optical and SEM images shown in Figure 11a,b. The absorbance spectra of the same films are also shown in Figure 11c. According to the thickness values given in Figures 7a and 8a, the film sprayed using 100 passes on the substrate kept at the ambient temperature is thicker than the film sprayed on the hotplate at 150 °C, while its absorbance shown in Figure 11c is lower in a wide range of wavelengths. This may have originated from two sources. Firstly, high roughness in spray-on films may decrease the light absorbance [41]. Based on the roughness data of Figures 7b and 8b, the former film is rougher, i.e., its local variation of thickness is higher than the latter film. Inset optical images in Figure 11a,b also verify that the film fabricated on the hotplate is smoother than the film sprayed on the substrate kept at the ambient temperature, thus it can absorb the light more efficiently. Secondly, according to the conversion ratio data shown in Figure 6c, there is more

PbI$_2$ impurity in the perovskite film fabricated at 25 °C, with respect to the film deposited on the hotplate kept at 150 °C. Therefore, the film deposited on the hotplate is more functional and thus capable of better photon absorption, considering that the bandgap of the perovskite films is suitable for the absorption of photons in the visible range. Figure 11d is a picture of the fabricated film (5 × 5 cm^2), made using the optimum conditions of 100 spray passes deposited on the hotplate kept at 150 °C, using a perovskite solution concentration of 10 wt.%. The raster movement of the spray nozzle is also illustrated schematically on this picture. This spray-on film is semitransparent and thus has the capability to be used as a second absorber layer to make tandem perovskite–copper indium gallium diselenide (CIGS) or perovskite–silicon solar cells.

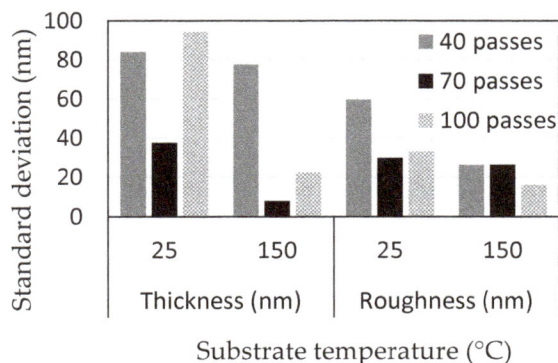

Figure 10. Standard deviation of the roughness and thickness data of perovskite films fabricated by various spray passes on the substrates kept at 25 and 150 °C. High substrate temperature improves reproducibility of the data.

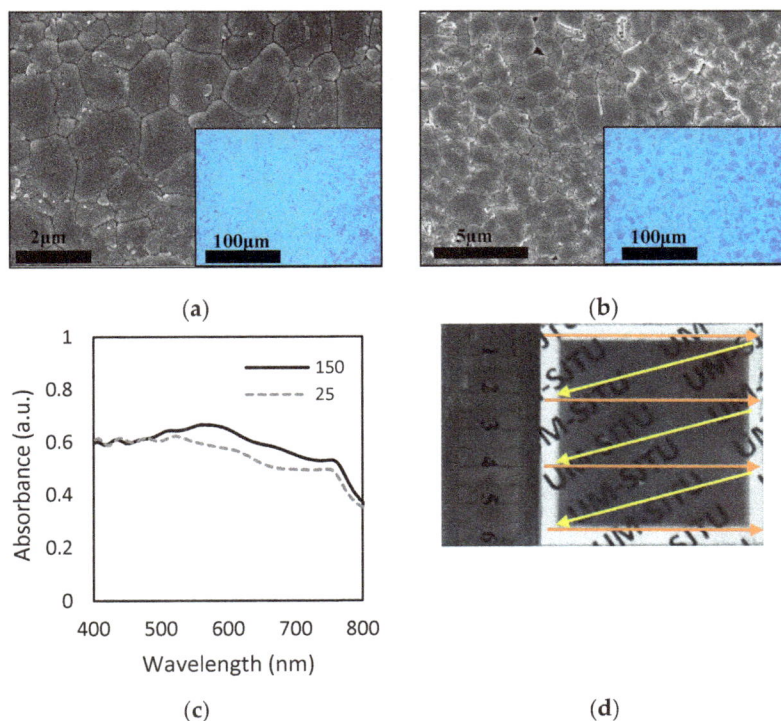

Figure 11. SEM and optical (inset) images of spray-on perovskite films made at substrate temperature of (**a**) 25 °C and (**b**) 150 °C. (**c**) absorbance spectra of perovskite films sprayed on FTO-coated glass kept at 150 °C; (**d**) the spray-on film deposited on the hotplate kept at 150 °C and the deposition pattern based on a zigzag movement of the spray nozzle over the substrate. The films were made at optimum condition of 100 spray passes using a 10 wt.% solution concentration.

4. Conclusions

A continuous, uniform, and pinhole-free perovskite thin film was fabricated by spray coating with an effective squared area of 25 cm^2. Optimum concentration of the mixed halide perovskite precursor (10 wt.% of MAPbI$_{3-x}$Cl$_x$ in DMF) and substrate temperature (150 °C) were determined as two essential parameters that strongly affect the morphology of the spray-on film. The optimum concentration and substrate temperature may be generalized to other spray systems. Then, for the spray nozzle used in this study, for the given spray flow rate and nozzle speed, the number of spray passes was optimized (100 spray passes). The combination of the best solution concentration, substrate temperature, and the number of spray passes produced a fully functional perovskite film with high surface area suitable for the commercialization of the technology. We also measured the contact angle, surface tension, and viscosity of the mixed halide perovskite solution in different concentrations, in order to find the optimum range of concentration. In order to choose the best substrate temperature, at first, we measured the contact angle of perovskite droplets on the hotplates at various temperatures. It was observed that a higher substrate temperature leads to a higher surface wettability (lower contact angle). At 150 °C, which is around the boiling point of DMF, we also observed the highest conversion ratio (efficient conversion of perovskite precursors to perovskite), which is a measure of the percentage of PbI$_2$ impurity in the perovskite film. Spraying of 10 wt.% mixed halide perovskite using 100 passes on a hotplate kept at 150 °C resulted in a fully-covered film with the lowest thickness and roughness, compared to the films made on the substrates kept at room temperature. Standard deviation of the thickness and roughness of the films made on hot substrates also decreased, which is an indication of increased reproducibility of the process. Although the fabricated large-area perovskite films demonstrated superior morphological and optoelectronic functionality, it is noted that their photovoltaic performance would be determined when used in a complete device. The fabrication of such large area devices is challenging and entails the optimization of all layers, which was beyond the scope of this work.

Acknowledgments: Research funding from the Shanghai Municipal Education Commission in the framework of the Oriental scholar and distinguished professor designation and funding from the National Natural Science Foundation of China (NSFC) is acknowledged.

Author Contributions: Mehran Habibi and Morteza Eslamian conceived the work; Mehran Habibi designed and performed the experiments regarding the optimization of perovskite layers, conducted characterization tests, and discussed the results. Amin Rahimzadeh performed the tests concerning the physical properties of perovskite solutions and interpreted the results. Inas Bennouna assisted with the experiments and characterizations; and Mehran Habibi, Amin Rahimzadeh and Morteza Eslamian wrote the paper. All authors read and approved the paper.

Conflicts of Interest: The authors declare no conflict of interest.

References

1.   Efficiency chart. Available online: http://www.nrel.gov/pv/assets/images/efficiency_chart.jpg (accessed on 10 March 2017).
2.   Habibi, M.; Rahimzadeh, A.; Eslamian, M. On dewetting of thin films due to crystallization (crystallization dewetting). *Eur. Phys. J. E Soft Matter.* **2016**, *39*, 30. [CrossRef] [PubMed]
3.   Habibi, M.; Zabihi, F.; Ahmadian-Yazdi, M.R.; Eslamian, M. Progress in emerging solution-processed thin film solar cells–Part II: Perovskite solar cells. *Renew. Sustain. Energy Rev.* **2016**, *62*, 1012–1031. [CrossRef]
4.   Cai, B.; Zhang, W.-H.; Qiu, J. Solvent engineering of spin-coating solutions for planar-structured high-efficiency perovskite solar cells. *Chin. J. Catal.* **2015**, *36*, 1183–1190. [CrossRef]
5.   Jeon, N.J.; Noh, J.H.; Kim, Y.C.; Yang, W.S.; Ryu, S.; Seok, S.I. Solvent engineering for high-performance inorganic-organic hybrid perovskite solar cells. *Nat. Mater.* **2014**, *13*, 897–903. [CrossRef] [PubMed]
6.   Ahmadian-Yazdi, M.; Zabihi, F.; Habibi, M.; Eslamian, M. Effects of Process Parameters on the Characteristics of Mixed-Halide Perovskite Solar Cells Fabricated by One-Step and Two-Step Sequential Coating. *Nanoscale Res. Lett.* **2016**, *11*, 408. [CrossRef] [PubMed]

7.   Ramesh, M.; Boopathi, K.M.; Huang, T.Y.; Huang, Y.C.; Tsao, C.S.; Chu, C.W. Using an airbrush pen for layer-by-layer growth of continuous perovskite thin films for hybrid solar cells. *ACS Appl. Mater. Interfaces* **2015**, *7*, 2359–2366. [CrossRef] [PubMed]

8.   Mohammadian, N.; Alizadeh, A.H.; Moshaii, A.; Gharibzadeh, S.; Alizadeh, A.; Mohammadpour, R.; Fathi, D. A two-step spin-spray deposition processing route for production of halide perovskite solar cell. *Thin Solid Films* **2016**, *616*, 754–759. [CrossRef]

9.   Gong, X.; Li, M.; Shi, X.-B.; Ma, H.; Wang, Z.-K.; Liao, L.-S. Controllable Perovskite Crystallization by Water Additive for High-Performance Solar Cells. *Adv. Funct. Mater.* **2015**, *25*, 6671–6678. [CrossRef]

10. Liang, P.W.; Liao, C.Y.; Chueh, C.C.; Zuo, F.; Williams, S.T.; Xin, X.K.; Lin, J.; Jen, A.K. Additive enhanced crystallization of solution-processed perovskite for highly efficient planar-heterojunction solar cells. *Adv. Mater.* **2014**, *26*, 3748–3754. [CrossRef] [PubMed]

11. Lau, C.F.J.; Deng, X.; Ma, Q.; Zheng, J.; Yun, J.S.; Green, M.A.; Huang, S.; Ho-Baillie, A.W.Y. CsPbIBr$_2$Perovskite Solar Cell by Spray-Assisted Deposition. *ACS Energy Lett.* **2016**, *1*, 573–577. [CrossRef]

12. Razza, S.; Di Giacomo, F.; Matteocci, F.; Cinà, L.; Palma, A.L.; Casaluci, S.; Cameron, P.; D'Epifanio, A.; Licoccia, S.; Reale, A.; et al. Perovskite solar cells and large area modules (100 cm^2) based on an air flow-assisted PbI$_2$ blade coating deposition process. *J. Power Sources* **2015**, *277*, 286–291. [CrossRef]

13. Yeo, J.-S.; Lee, C.-H.; Jang, D.; Lee, S.; Jo, S.M.; Joh, H.-I.; Kim, D.-Y. Reduced graphene oxide-assisted crystallization of perovskite via solution-process for efficient and stable planar solar cells with module-scales. *Nano Energy* **2016**, *30*, 667–676. [CrossRef]

14. Tait, J.G.; Manghooli, S.; Qiu, W.; Rakocevic, L.; Kootstra, L.; Jaysankar, M.; Masse de la Huerta, C.A.; Paetzold, U.W.; Gehlhaar, R.; Cheyns, D.; et al. Rapid composition screening for perovskite photovoltaics via concurrently pumped ultrasonic spray coating. *J. Mater. Chem. A* **2016**, *4*, 3792–3797. [CrossRef]

15. Markus, H.; Henrik, F.D.; Krebs, F.C. Development of Lab-to-Fab Production Equipment Across Several Length Scales for Printed Energy Technologies, Including Solar Cells. *Energy Technol.* **2015**, *3*, 293–304.

16. Huang, H.; Shi, J.; Zhu, L.; Li, D.; Luo, Y.; Meng, Q. Two-step ultrasonic spray deposition of CH$_3$NH$_3$PbI$_3$ for efficient and large-area perovskite solar cell. *Nano Energy* **2016**, *27*, 352–358. [CrossRef]

17. Eslamian, M. Spray-on thin film PV solar cells: Advances, potentials and challenges. *Coatings* **2014**, *4*, 60–84. [CrossRef]

18. Chandrasekhar, P.S.; Kumar, N.; Swami, S.K.; Dutta, V.; Komarala, V.K. Fabrication of perovskite films using an electrostatic assisted spray technique: the effect of the electric field on morphology, crystallinity and solar cell performance. *Nanoscale* **2016**, *8*, 6792–6800. [CrossRef] [PubMed]

19. Heo, J.H.; Lee, M.H.; Jang, M.H.; Im, S.H. Highly efficient CH3NH3PbI$_{3-x}$Cl$_x$ mixed halide perovskite solar cells prepared by re-dissolution and crystal grain growth via spray coating. *J. Mater. Chem. A* **2016**, *4*, 17636–17642. [CrossRef]

20. Abdollahi, N.B.; Gharibzadeh, S.; Ahmadi, V.; Shahverdi, H.R. New Scalable Cold-Roll Pressing for Post-treatment of Perovskite Microstructure in Perovskite Solar Cells. *J. Phys. Chem. C* **2016**, *120*, 2520–2528. [CrossRef]

21. Jung, Y.S.; Hwang, K.; Scholes, F.H.; Watkins, S.E.; Kim, D.Y.; Vak, D. Differentially pumped spray deposition as a rapid screening tool for organic and perovskite solar cells. *Sci. Rep.* **2016**, *6*, 20357. [CrossRef] [PubMed]

22. Zabihi, F.; Ahmadian-Yazdi, M.R.; Eslamian, M. Fundamental Study on the Fabrication of Inverted Planar Perovskite Solar Cells Using Two-Step Sequential Substrate Vibration-Assisted Spray Coating (2S-SVASC). *Nanoscale Res. Lett.* **2016**, *11*, 71. [CrossRef] [PubMed]

23. Habibi, M.; Ahmadian-Yazdi, M.R.; Eslamian, M. Optimization of spray coating for the fabrication of planar perovskite solar cells. *At. Sprays* **2017**. under review.

24. Barrows, A.T.; Pearson, A.J.; Kwak, C.K.; Dunbar, A.D.F.; Buckley, A.R.; Lidzey, D.G. Efficient planar heterojunction mixed-halide perovskite solar cells deposited via spray-deposition. *Energy Environ. Sci.* **2014**, *7*, 2944. [CrossRef]

25. Das, S.; Yang, B.; Gu, G.; Joshi, P.C.; Ivanov, I.N.; Rouleau, C.M.; Aytug, T.; Geohegan, D.B.; Xiao, K. High-Performance Flexible Perovskite Solar Cells by Using a Combination of Ultrasonic Spray-Coating and Low Thermal Budget Photonic Curing. *ACS Photonics* **2015**, *2*, 680–686. [CrossRef]

26. Mullins, W.W.; Sekerka, R.F. Morphological Stability of a Particle Growing by Diffusion or Heat Flow. *J. Appl. Phys.* **1963**, *34*, 323–329. [CrossRef]

27. Sear, R.P. Nucleation at contact lines where fluid–fluid interfaces meet solid surfaces. *J. Phys. Condens. Matter.* **2007**, *19*, 466106. [CrossRef]

28. Djikaev, Y.S.; Ruckenstein, E. Thermodynamics of Heterogeneous Crystal Nucleation in Contact and Immersion Modes. *J. Phys. Chem. A* **2008**, *112*, 11677–11687. [CrossRef] [PubMed]

29. Rahimzadeh, A.; Eslamian, M. Stability of thin liquid films subjected to ultrasonic vibration and characteristics of the resulting thin solid films. *Chem. Eng. Sci.* **2017**, *158*, 587–598. [CrossRef]

30. Deng, Y.; Peng, E.; Shao, Y.; Xiao, Z.; Dong, Q.; Huang, J. Scalable fabrication of efficient organolead trihalide perovskite solar cells with doctor-bladed active layers. *Energy Environ. Sci.* **2015**, *8*, 1544–1550. [CrossRef]

31. Wei, Z.; Chen, H.; Yan, K.; Yang, S. Inkjet printing and instant chemical transformation of a CH3NH3PbI3/nanocarbon electrode and interface for planar perovskite solar cells. *Angew. Chem.* **2014**, *53*, 13239–13243. [CrossRef] [PubMed]

32. Hwang, K.; Jung, Y.S.; Heo, Y.J.; Scholes, F.H.; Watkins, S.E.; Subbiah, J.; Jones, D.J.; Kim, D.Y.; Vak, D. Toward large scale roll-to-roll production of fully printed perovskite solar cells. *Adv. Mater.* **2015**, *27*, 1241–1247. [CrossRef] [PubMed]

33. Schmidt, T.M.; Larsen-Olsen, T.T.; Carlé, J.E.; Angmo, D.; Krebs, F.C. Upscaling of Perovskite Solar Cells: Fully Ambient Roll Processing of Flexible Perovskite Solar Cells with Printed Back Electrodes. *Adv. Energy Mater.* **2015**, *5*, 69. [CrossRef]

34. Kim, J.H.; Williams, S.T.; Cho, N.; Chueh, C.-C.; Jen, A.K.Y. Enhanced Environmental Stability of Planar Heterojunction Perovskite Solar Cells Based on Blade-Coating. *Adv. Energy Mater.* **2015**, *5*, 1401229. [CrossRef]

35. Back, H.; Kim, J.; Kim, G.; Kyun, K.T.; Kang, H.; Kong, J.; Lee, H.S.; Lee, K. Interfacial modification of hole transport layers for efficient large-area perovskite solar cells achieved via blade-coating. *Sol. Energy Mater. Sol. Cells* **2016**, *144*, 309–315. [CrossRef]

36. Park, S.-M.; Noh, Y.-J.; Jin, S.-H.; Na, S.-I. Efficient planar heterojunction perovskite solar cells fabricated via roller-coating. *Solar Energy Mater. Solar Cells* **2016**, *155*, 14–19. [CrossRef]

37. Mallajosyula, A.T.; Fernando, K.; Bhatt, S.; Singh, A.; Alphenaar, B.W.; Blancon, J.-C.; Nie, W.; Gupta, G.; Mohite, A.D. Large-area hysteresis-free perovskite solar cells via temperature controlled doctor blading under ambient environment. *Appl. Mater. Today* **2016**, *3*, 96–102. [CrossRef]

38. Dualeh, A.; Gao, P.; Seok, S.I.; Nazeeruddin, M.K.; Grätzel, M. Thermal behavior of methylammonium lead-trihalide perovskite photovoltaic light harvesters. *Chem. Mater.* **2014**, *26*, 6160–6164. [CrossRef]

39. Brunetti, B.; Cavallo, C.; Ciccioli, A.; Gigli, G.; Latini, A. On the thermal and thermodynamic (in)stability of methylammonium lead halide perovskites. *Sci. Rep.* **2016**, *6*, 31896. [CrossRef] [PubMed]

40. Xie, Y.; Gao, S.; Eslamian, M. Fundamental study on the effect of spray parameters on characteristics of P3HT:PCBM active layers made by spray coating. *Coatings* **2015**, *5*, 488–510. [CrossRef]

41. Wengeler, L.; Schmitt, M.; Peters, K.; Scharfer, P.; Schabel, W. Comparison of large scale coating techniques for organic and hybrid films in polymer based solar cells. *Chem. Eng. Process. Process Intensif.* **2013**, *68*, 38–44. [CrossRef]

5

Studies on the Effect of Arc Current Mode and Substrate Rotation Configuration on the Structure and Corrosion Behavior of PVD TiN Coatings

Liam Ward [1,*,†]**, Antony Pilkington** [2,3,†] **and Steve Dowey** [3,4,†]

[1]   School of Engineering, RMIT University, Melbourne 3001, Australia
[2]   School of Science, RMIT University, Melbourne 3001, Australia; antony.pilkington@gmail.com
[3]   Defence Materials Technology Centre, Melbourne 3122, Australia
[4]   Sutton Tools Pty Ltd., Melbourne 3074, Australia; sdowey@sutton.com.au
*   Correspondance: liam.ward@rmit.edu.au
†   These authors contributed equally to this work.

Academic Editor: Alessandro Lavacchi

Abstract: Thin, hard cathodic arc evaporated (CAE) metal nitride coatings are known to contain defects such as macro-particles, pinholes, voids and increased porosity, leading to reduced corrosion resistance. The focus of this research investigation was to compare the structure and corrosion behaviour of cathodic arc evaporated (CAE) TiN coatings deposited on AISI 1020 low carbon steel substrates using a pulsed current arc and a more conventional constant current arc source (DC). The effects of a double (2R) and triple (3R) substrate rotation configuration were also studied. Coating morphology and chemical composition were characterised using optical, SEM imaging and XRD analysis. Focus variation microscopy (FVM), an optical 3D measurement technique, was used to measure surface roughness. Corrosion studies were carried out using potentiodynamic scanning in 3.5% NaCl. Tafel extrapolation was carried out to determine E_{corr} and I_{corr} values for the coated samples. In general, increased surface roughness, and to a certain extent, corrosion resistance, were associated with thicker coatings deposited using 2R, compared to 3R rotation configuration. The arc source mode (continuous or pulsed) was shown to have little effect on the corrosion behavior. Corrosion behavior was controlled by the presence of defects, pinholes and macro-particles at lower anodic potentials, while the formation of large pitted regions and aggressive corrosion of the underlying substrate was observed at higher anodic potentials.

Keywords: cathodic arc evaporation; pulsed arc current; metal nitride; characterisation; potentiodynamic scanning

1. Introduction

Thin, hard physical vapour deposited (PVD) nitride and carbo-nitride coatings are primarily used in many engineering tribological applications (metal cutting and forming, for example) as a result of their reduced friction and improved wear properties. PVD coatings can offer additional benefits involving improved corrosion resistance and oxidation resistance when deposited on a substrate with low to medium corrosion resistant properties. The high chemical stability and relative inertness of these coatings, combined with the potential to form protective oxide layers in selected environments, are contributing factors to such benefits.

PVD coatings offer major environmental processing benefits compared to conventional galvanic corrosion barrier layer production techniques such as electroplating. The factors that have limited the applications of PVD coatings as complete barriers to corrosion [1] are the presence of coating defects such as pinholes, pores, voids, cracks and macro-particles. The presence of these defects can result in accelerated corrosion of the underlying substrate, due to localised corrosion such as galvanic, pitting and crevice corrosion [2]. The concept of using an inert coating to protect a more active substrate relies heavily on the integrity of the coating to completely isolate the substrate from the corrosive medium. Adhesion of the PVD coating to the substrate with low interfacial stress is important in anti-corrosion applications together with low porosity in the coating morphology [3]. PVD coatings are used in applications where both wear resistance and corrosion protection is important. Examples are in coated tooling for injection moulding of polymers [4,5], automotive trims, architectural and decorative facades [6], plumbing fittings [7] and protection of orthopaedic implants [8].

PVD TiN was the focus of attention for many years as a corrosion-resistant coating, with the development of techniques such as the optimisation of the process deposition parameters, the incorporation of alloying elements within the coating, substrate choice and the deposition of intermediate layers investigated as methods for improving the corrosion resistance [9–13].

The cathodic arc evaporation PVD process has achieved widespread industrial application for the deposition of tribological and corrosion-resistant ceramic coatings due to its relative simplicity of process control, adaptability to large deposition systems (scaleability) and facility to deposit multi-component and multilayer coatings from alloy cathodes such as TiAl or AlCr. A disadvantage inherent in the cathodic arc evaporation process is the ejection of molten droplets from the molten cathode material at cathode spots. These droplets become incorporated into the deposited film and lead to localised porosity which then acts as corrosion initiation sites for active substrates in corrosive media [14]. A low level of macro-particle incorporation is promoted by the use of cathodes with a high melting point, the use of a low arc current, magnetic or electrostatic deflection of the cathode spot motion, ducted or filtered arc shielding and the formation of a cathode compound layer by reaction with gaseous reagents with high melting points (cathode poisoning) [15]. Many of these techniques adversely affect the deposition rate.

Previous studies [16] have shown that the use of a pulsed current arc source compared with a more conventional continuous direct current (DC) source has resulted in decreased size and number of macro-particles. This, in turn, may lead to an improvement in the corrosion behaviour of these coatings.

Control of the cathodic arc process becomes more difficult as the arc current is reduced to the arc splitting threshold; hence, low current offers limited scope for producing lower roughness coatings in practical applications. However it is known that current modulation can be used under conditions where a lower threshold arc current can be used with the addition of short, higher current pulse to maintain a stable operation and deposition rate [16].

The focus of this paper is a comparative study of the aqueous corrosion behaviour of TiN deposited by the cathodic arc evaporation PVD technique onto AISI 1020 (mild steel) substrates. Process variables studied include the deposition of the coatings using the cathodic arc current in continuous (C) and pulsed (P) mode and as a function of coating thickness through substrate rotation (double–2R and triple–3R rotation). Anodic and cathodic potentiodynamic scans were conducted in 3.5% NaCl at ambient (room) temperature. The PVD coating morphology was characterised using SEM imaging prior to and after corrosion testing.

2. Materials and Methods

2.1. Sample Preparation

Coupons for the deposition and corrosion studies were machined from AISI 1020 low carbon steel bright drawn bar (18 mm diameter × 10 mm thick). Sequential grinding with abrasive papers was followed by polishing with diamond slurries down to 1 µm grit. The polished coupons were ultrasonically cleaned immediately prior to coating deposition in a multi-stage aqueous cleaning system incorporating ultrasonic degreasing in pure ethanol and hot N_2 gas drying.

Reference (un-coated) substrates were provided for corrosion studies by polishing additional coupons back to the substrate metal surface after a PVD coating thermal cycle. This heat treatment provided stress relief for the AISI 1020 coupon material to prevent residual drawing stresses affecting the corrosion behaviour.

2.2. Coating Design and Deposition

PVD coatings of TiN were prepared using cathodic arc evaporation in an INNOVA 1.1 PVD system (Balzers). Ti (grade 2, assay 99.5 at. %) was used as the cathode material. Cathodes of 150 mm diameter were arranged equidistant at two levels in the walls and door of the 1100 mm diameter vacuum chamber to provide an effective coating deposition height of 1000 mm. A total of 6 cathodes (2 levels with 3 cathodes in each level) were utilised for TiN deposition. The process gasses used were Ar (5N) and N_2 (5N), which were admitted to the deposition process through mass flow controllers (MFC). The PVD coatings were manufactured under computer control according to a deposition sequence involving vacuum preheating to 450 °C, Ar ion etching at 2.1×10^{-2} mbar and −200 V bias. All coatings were deposited at −100 V DC bias, 2.2 mbar working pressure at 450 °C. Coating deposition time was one hour. The coating deposition parameters are summarised in Table 1. Parameters were selected to deposit pulsed arc coatings with similar compositions to the DC arc coatings with thicknesses in the range 1–3 µm.

Table 1. Deposition details of conventional and pulsed TiN coatings.

Coating	Arc Current (A)	Pulse Duty Cycle, Frequency
TiN 2R C	180	n/a
TiN 3R C	180	n/a
TiN 2R P	230/128.8	10%, 1 kHz
TiN 3R P	230/128.8	10%, 1 kHz

The samples were mounted vertically in a multiple spindle carousel system on the cathode centreline for the coating deposition. Two thicknesses of PVD coating were prepared in each run by mounting sets of coupons in double (2R) or triple (3R) rotation fixtures. For 2R configuration, the closest approach of the sample to any cathode surface was 280 mm, while for 3R configuration, this value was 230 mm.

Two arc current regimes were investigated to provide coatings deposited with DC (C) and a novel pulsed cathodic arc (P) process. The pulse waveform, as shown schematically in Figure 1, was configured to provide a low level unconditionally stable holding current of 128.8 with a superposed higher current pulse of 230 A with a temporal duty cycle of 10% and period 1 ms corresponding to the maximum frequency of 1 kHz available from the arc power supplies (Fronius DPS 2500, Fronius, Pettenbach, Austria). Coating deposition was made under pressure control to provide monolithic coatings.

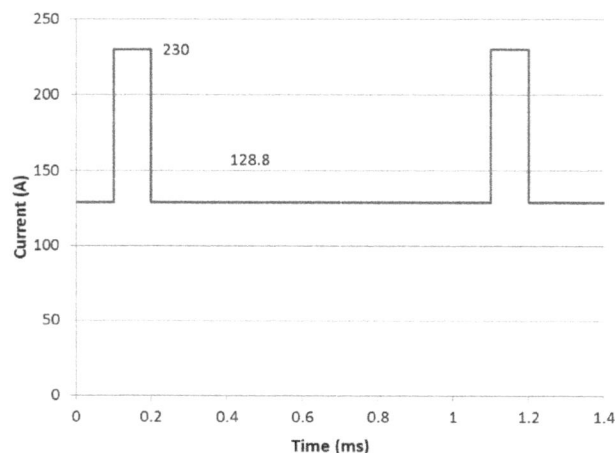

Figure 1. Schematic diagram showing 1 kHz pulsed arc waveform.

2.3. Coating Characterisation

A number of techniques were adopted with the intent of characterising the surface structure and morphology, determining the effects of the corrosion tests on the coating system integrity and the mode or extent of any coating failure.

The coating thickness was measured using the ball cratering technique with a 25 mm diameter AISI 52100 ball bearing and a 1 μm diamond slurry abrasive. Crater measurements were made from optical images of three separate craters per coating type.

The surface roughness of the deposited coatings was obtained using focus variation microscopy (FVM) a non-contact optical 3D measurement technique (Alicona Infinite Focus Microscope, Alicona Imaging GmbH Raaba/Graz, Austria) with a ×50 optic. The FVM datasets were processed using the IF4.5 surface roughness measurement software module.

X-ray diffraction (Siemens D4 Endeavour Diffractometer, Siemens, Munich, Germany) was employed to determine phase compositions of the TiN coatings deposited on AISI 1020 carbon steel. Cu kα radiation with X-ray source parameters 40 mA incident electron beam current and 40 kV acceleration voltage were adopted. The Bragg-Brentano (θ-2θ) geometry was used in the 2θ range of 20°–120° with a step size of 0.02° and 2 s collection time. Bruker EVA software (version 4) tools were used to strip kα2 content and smooth the raw spectra background. The Powder Diffraction File (PDF) library spectra were used as a means of identifying crystalline phases observed in the spectra.

The surface morphology of these coatings was investigated using scanning electron microscope (SEM) imaging in an FEI Quanta 200 SEM prior (FEI, Hillsbor, OR, USA) to and after corrosion testing.

2.4. Corrosion Testing

A Voltalab 21 Potentiostat (Radiometer Analytical, Lyon, France) was used to carry out the anodic/cathodic potentiodynamic scans. A conventional three-electrode cell configuration, utilising a saturated calomel electrode (SCE) as the reference electrode, a platinum wire counter electrode and the working electrode (sample) was used. All electrochemical tests were carried out in 3.5% NaCl at room temperature (20 °C) without aeration. Specimens were initially stabilised at the free corroding potential for 30 minutes prior to conducting potentiodynamic scans. Anodic/cathodic scans were conducted in the range of -1000 mV to $+500$ mV at a scan rate of 20 mV/min, in order to determine the overall anodic/cathodic corrosion characteristics of the coated systems and reference substrate. The results are presented as potentiodynamic polarisation curves in the form E vs. log$I$ plots. Tafel Extrapolation was used to determine the corrosion rates (E_{corr} and I_{corr} values) of the PVD metal nitride systems, based upon three scans per test configuration.

3. Results

3.1. Coating Thickness Results

Results of the coating thicknesses from the ball cratering technique are shown in Table 2. Thickness correlations were observed in terms of the effects of arc current source mode and substrate rotation configuration. The results show clearly that for all coating combinations, coatings deposited using triple substrate rotation were 53%–67% thinner (thickness of 1.3 µm) than those deposited using double rotation (thickness range of 2.3–2.8 µm).

Table 2. Coating thickness, surface roughness and evaporation rate values for conventional and pulsed TiN coatings.

Coating	Coating Thickness (µm)	Deposition Rate (nm/s)	Evaporation Rate (mg/s)	Evaporation Efficiency (g/C)	Surface Roughness Sq (nm)	Surface Roughness S10z (µm)
AISI 1020	–	–	–	–	14.7 ± 3.9	0.234 ± 0.02
TiN 2R C	2.8 ± 0.1	0.78	3.47	2.17×10^{-5}	180.83 ± 15.47	3.52 ± 0.39
TiN 3R C	1.3 ± 0.1	0.36	–	–	126.36 ± 12.93	1.93 ± 0.46
TiN 2R P	2.3 ± 0.1	0.64	2.97	2.05×10^{-5}	187.37 ± 2.51	3.06 ± 0.41
TiN 3R P	1.3 ± 0.1	0.37	–	–	137.74 ± 4.33	1.91 ± 0.34

For 2R rotation configuration, coatings deposited using a pulsed arc current source in general were slightly thinner (17%) than those deposited using a continuous DC arc source, although coating thickness was the same for 3R rotation configuration. Evaporation rates were greater for coatings deposited using a continuous DC compared to using a pulsed arc current source. The process efficiency defined as cathode mass loss/charge passed (C) is lower for a pulsed arc current.

3.2. Surface Roughness Results

Surface roughness results from the focus variation microscopy (FVM), provided as both the root mean square height of the area selected (Sq) and the 10-point height of the selected area (S10z), are provided in Table 2. Both sets of data are provided as Sq is the more common method for a quantitative comparison of overall surface roughness, while S10z is a more sensitive indication of the contribution to surface roughness associated with the presence of macro-particles i.e. maximum macro-particle height.

The Sq results highlight a number of important features, namely (i) all coatings showed considerably greater roughness than for the mild steel substrate (Sq = 14.7 nm; S10z = 0.23 µm); (ii) all coatings deposited using triple substrate rotation exhibited lower Sq values (126.4–137.7 nm) than their counterparts deposited using double substrate rotation (180.8–187.4 nm); and (iii) for any given substrate rotation configuration, very little difference was observed in Sq values for both continuous arc current and pulsed arc current deposited coatings. These results suggest that the surface roughness of these TiN arc deposited coatings is influenced predominantly by substrate rotation configuration and is independent of the arc current mode. The Sq data suggests that surface roughness increases with layer thickness in the coating thickness range of 1.3–2.8 µm.

The S10z values correlate closely with the Sq values, confirming that the substrate rotation mode influences the surface roughness parameter S10z. In addition, the higher S10z values observed for the thicker coatings indicates that an increase in size of macro-particle defects observed as the coating thickness increases may be due to preferential coating growth on macro-particles.

3.3. Phase Composition Results

XRD spectra for the pulsed and continuous TiN coatings deposited on AISI 1020 carbon steel, using double substrate rotation (2R), are shown in Figure 2. Note that the counts are plotted as arbitrary units (Y axis) as a function of 2θ (X axis). The spectra show that little difference is observed between pulsed and DC arc deposited coatings. Reflections due to FCC TiN (111), (200), (220) and (222) orientations are associated with the presence of TiN (Osbornite, PDF 01-087-0628). Ti_2N also features in the spectra (PDF 00-017-0386) but the (112) principle peak at 36.86° is masked by the TiN (111) peak. The minor peak at 2θ value of 32.9°, present in the spectra for TiN, is not identified against reference spectra for Ti, TiN or Ti_2N. It may be weak evidence for the presence of some Ti_2O_3 (reflection (104) from PDF 00-043-1033), since O is well known to be incorporated into PVD coatings and O is the main contamination on a molar basis (0.0156 at.%) in the grade Ti2 titanium cathode material. Slight differences observed in the (222) and (220) reflections can be attributed to differences in the plasma characteristics and metal ion energies associated with a pulsed current source.

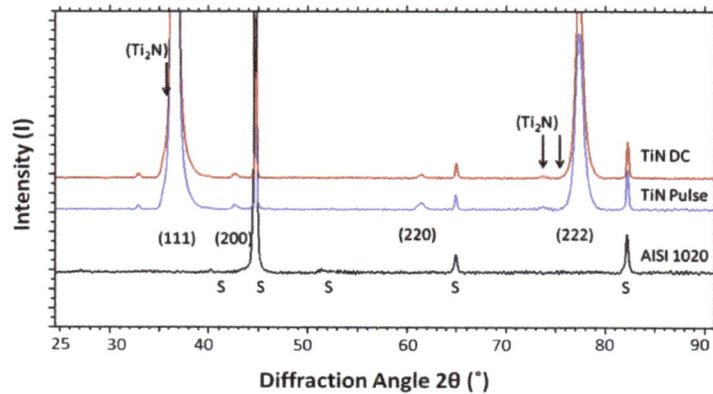

Figure 2. XRD spectra for cathodic arc evaporated TiN coatings.

3.4. Potentiodynamic Scanning Results

Potentiodynamic scans for TiN coatings deposited on 1020 mild steel using a continuous and pulsed arc current source are shown in Figure 3. All results are compared to the base material (1020 mild steel) serving as a control. C, P, 2R and 3R have been defined previously in the Introduction. Full potentiodynamic scans result in the generation of typical semi logarithmic curves with well-defined minima at E_{corr} and distinct anodic and cathodic regions.

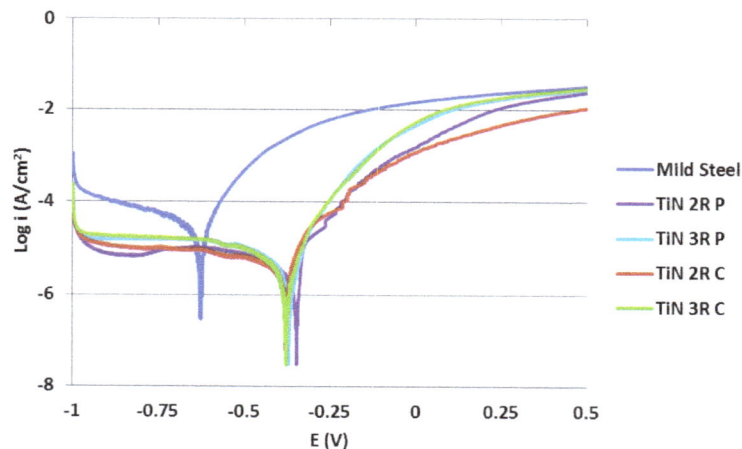

Figure 3. Potentiodynamic scans for continuous and pulsed arc current TiN coatings in 3.5% NaCl.

Analysis of the curves reveal a positive shift in the E_{corr} values (shift to the right) and lowering of the anodic/cathodic curves (indicative of lower I_{corr} values) can be observed for all TiN coatings compared with the base material, indicating an improvement in the corrosion resistance on the application of a PVD coating. Further analysis shows the cathodic regions of the curves for all coatings appear to be suppressed, showing lower cathodic current values compared with the uncoated carbon steel. However, there is very little difference in cathodic trends for all of the four coating systems analysed. This would suggest that the coated systems would show reduced corrosion behavior if the reactions are cathodically controlled. Further the forms of the anodic/cathodic curves are similar for all four coatings. At higher anodic potentials approaching +500 mV, the corrosion current values for the coatings deposited using triple rotation (3R) approach that of the uncoated substrate compared with the lower current values observed for the coatings deposited using double rotation (2R). This suggests that the corrosion behaviour of both the triple rotation (3R) coatings and uncoated samples are similar in this potential range, although the double rotation (2R) coatings show improved corrosion resistance.

3.5. Tafel Extrapolation Results

A summary of the E_{corr} and I_{corr} values, showing the mean and common standard deviation based on population, obtained from Tafel Extrapolation analysis of the polarisation curves for TiN coatings, are shown in Table 3.

Table 3. E_{corr} and I_{corr} values for conventional and pulsed TiN coatings.

Coating	E_{corr} (mV)	I_{corr} ($\mu A/cm^2$)
1020 Carbon Steel	-626 ± 3	15.57 ± 6.04
TiN 2R C	-350 ± 41	1.82 ± 1.08
TiN 3R C	-399 ± 32	2.59 ± 1.18
TiN 2R P	-357 ± 13	1.83 ± 0.36
TiN 3R P	-376 ± 6	2.80 ± 0.99

A positive shift in the E_{corr} value was observed compared with the 1020 carbon steel base material for all the coatings tested. Here, E_{corr} values ranged from -350 to -399 mV for the TiN coatings, compared with -626 mV for the 1020 carbon steel. Analysis of the different coating categories revealed more positive E_{corr} values were observed for the double rotation TiN coatings (-350 and -357 mV for 2R C and 2R P, respectively) compared with the triple rotation TiN coatings (-399 and -376 mV for 3R C and 3R P, respectively). This would suggest that the coatings deposited under double rotation conditions are slightly more thermodynamically stable in 3.5% NaCl than the coatings deposited under triple rotation conditions.

All coatings showed a reduction in the I_{corr} values ranging from 1.82 to 2.80 $\mu A/cm^2$, compared with 15.57 $\mu A/cm^2$ for the mild steel substrate. The results indicate a slight reduction in the corrosion current for the double rotation TiN (2R) coatings, compared with TiN deposited using triple rotation (3R) configuration, based upon the average values. However, these findings are inconclusive based upon the errors associated with these values suggesting little difference in corrosion current values for all coating conditions. Further work is required in this area to ascertain if improved corrosion resistance is associated with double rotation compared with triple rotation conditions.

Further analysis of the Tafel regions of the potentiodynamic scans show that the Tafel constants for the cathodic portions of the curves (βc) varied between -111 to -149 mV/decade for all TiN coatings, suggesting that oxygen reduction on the sample surface is likely to be the dominant cathodic process. The reduction of dissolved oxygen and water are expected to be the main cathodic reactions as the potentiodynamic scans were conducted under aerated conditions. For the uncoated sample, a slight increase in the βc value (-174 mV/decade) was observed. Tafel constants for the anodic portions of the curves (βa) varied between 26 and 102 mV/decade. The lower values observed for the anodic Tafel regions (βa) compared with the cathodic Tafel regions (βc) confirms that the corrosion processes are under cathodic (oxygen reduction) control.

3.6. Structural and Morphological Characterisation of Corroded and Uncorroded Coatings

3.6.1. SEM Image of TiN Fracture Cross Sections

SEM fracture cross sections of the TiN coatings deposited using double (2R) and triple (3R) rotation configurations in both continuous and pulsed arc current mode are shown in Figure 4a–d. Here, a dense, fibrous structure was evident in the coating cross section for all four coating conditions. Little differences were observed in the cross section morphologies for all coatings. The coatings deposited using 2R rotation (Figure 4a,b) were thicker than the coatings deposited using 3R rotation (Figure 4c,d)), confirming the thickness variations provided in Table 2. The coating surfaces (Figure 4b–d), showing a slightly dimpled morphology, contained both nodular flaws and circular craters associated with larger macro-particle pull-out, associated with the poor adhesion of these nodules to the coating. These nodular surface defects have overgrowth of coating material and are not simply spherical as-deposited macro-particles. These features are consistent with morphologies of arc deposited coatings. Overall, little differences in the morphology of the TiN coatings were observed, for both DC and pulsed arc deposited coatings, deposited using both double (2R) and triple (3R) rotation configurations.

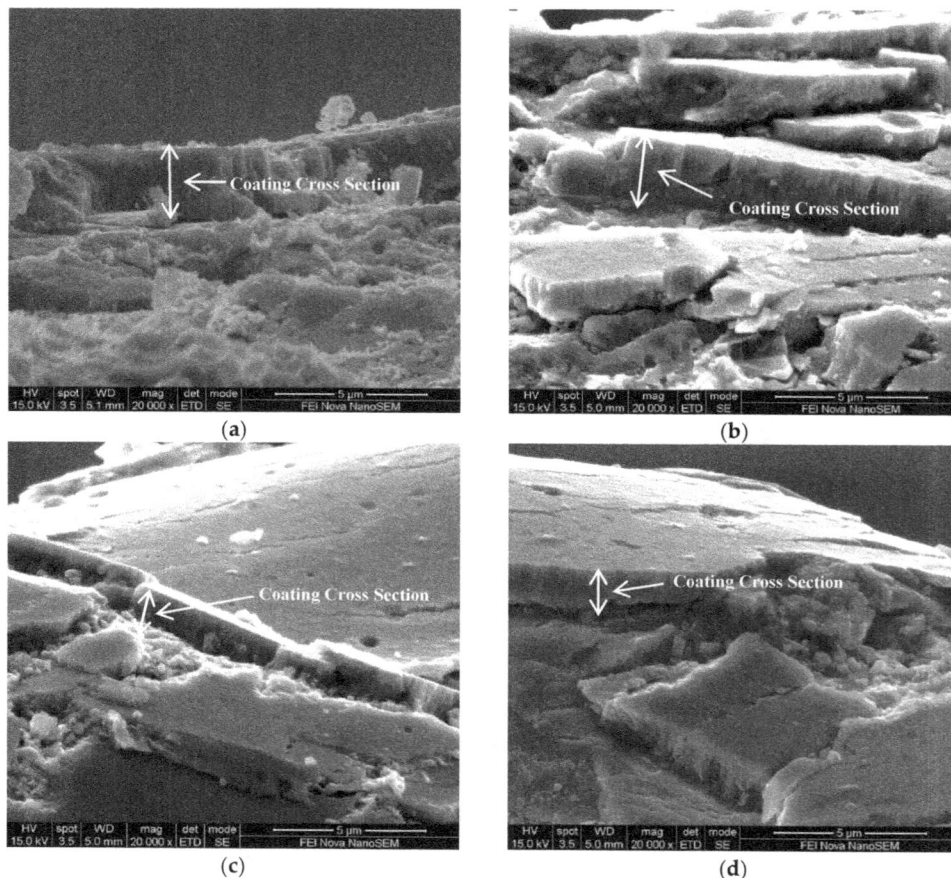

Figure 4. SEM fracture cross sections of (**a**) TiN 2R C; (**b**) TiN 2R P; (**c**) TiN 3R C; (**d**) TiN 3R P arc deposited coatings.

3.6.2. SEM and Optical Images of Corroded TiN Coatings

Scanning electron micrographs of the as deposited and corroded surfaces of the pulsed TiN coatings are shown in Figure 5a,b, respectively. The as deposited morphology is typical for cathodic arc PVD coatings, with surface defects due to macro-particles having characteristic diameters of 1–10 μm and nodular growth defects. Depressions or sockets are evident as circular regions with darker contrast in the SEM images, which are due to detached macro-particles. Little variation was observed in the surface characteristics of the as deposited coating surfaces deposited using conventional DC current (not shown) and pulsed arc current (Figure 5a), although there appeared to be a fewer number of defects present in the pulsed arc coating. After corrosion, however, a reduced number of macro-particles can be observed, while a greater number of sockets are further evident, as shown in Figure 5b.

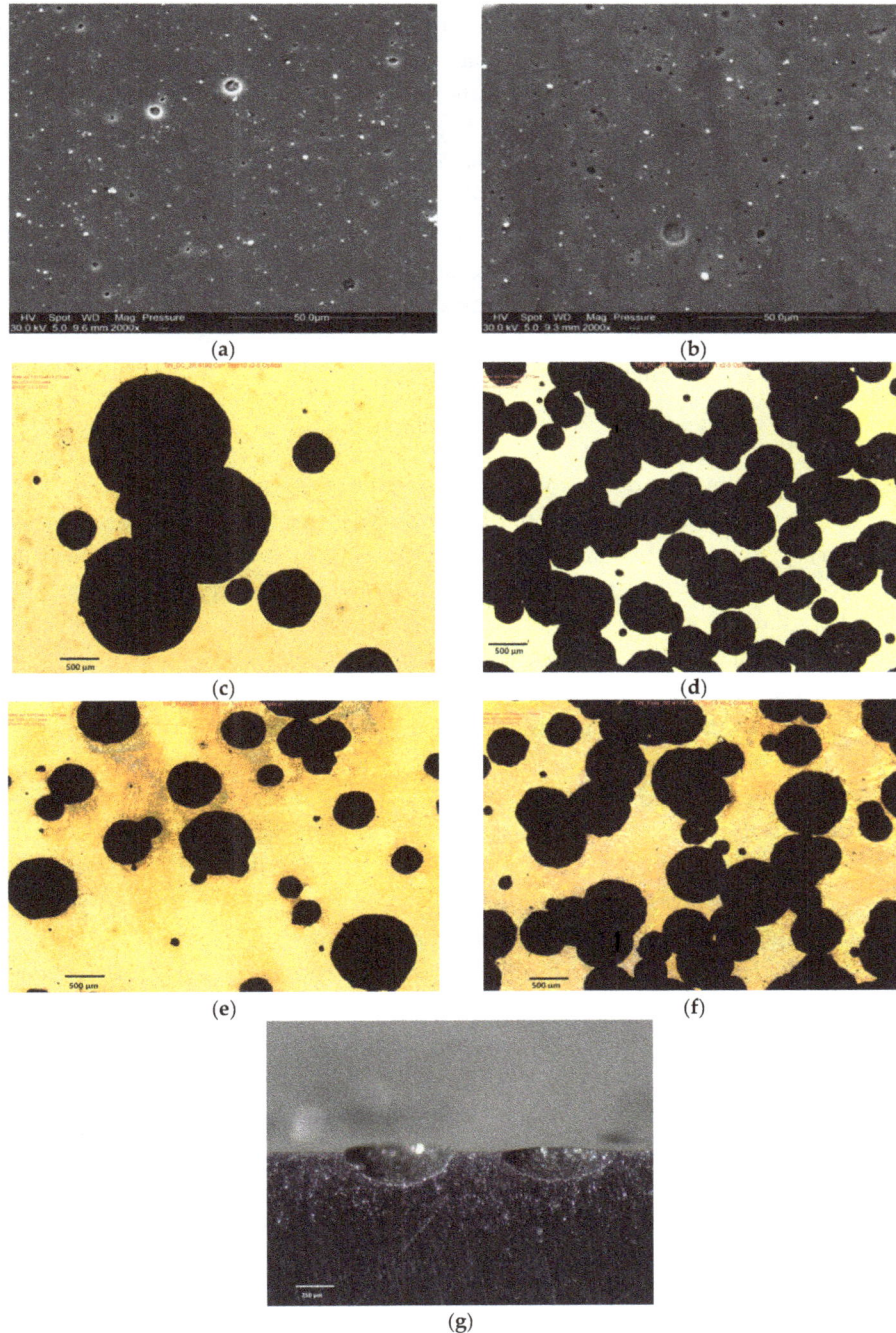

Figure 5. Micrographs showing (**a**) SEM image of the as deposited surface of TiN 3R P; (**b**) SEM image of the corroded surface of TiN 3R P; (**c**) optical image of the corroded surface of TiN 2R C; (**d**) optical image of the corroded surface of TiN 3R C; (**e**) optical image of the corroded surface of TiN 2R P; (**f**) optical image of the corroded surface of TiN 3R P; (**g**) optical image of corroded cross section of TiN 2R C showing pits.

Optical micrographs of the TiN coating surface topography after corrosion testing are shown in Figure 5c–f. All images were taken at the same magnification with micron bar scales shown on the photographs. Large areas of coating spallation with pitting damage were observed for both continuous and pulsed arc deposited coatings, indicating exposure of the underlying substrate. For continuous current coatings with double rotation, fewer number of pits were observed, however the diameter of these spalled regions varied from 400 μm to 2 mm, as shown in Figure 5c. In contrast, coatings deposited with triple rotation (Figure 5d) showed much higher pit density on the surface, although the diameters of these pits were lower, with typical diameters ranging from 125 to 700 μm. The pit density on these samples was such that many pits had merged together forming elongated regions of spallation. For pulsed arc coatings deposited with double rotation (Figure 5e), fewer pits with diameters ranging from 100 μm to 1 mm were observed compared with pulsed arc coatings deposited under triple rotation conditions, as shown in Figure 5f. Here, the pitting was intensified with a greater number of larger pits (1 mm diameter) merging together to form elongated regions of spallation.

Optical micrograph of a metallographic cross section of corroded TiN 2R C sample (×90 magnification) is shown in Figure 5g. Here the cross section of typical pits induced as a result of the corrosion process reveal that the pits tend to be hemispherical in nature, the pit depth at the centre being approximately one third of the pit diameter.

4. Discussion

It is widely accepted that one of the major limitations in the use of PVD metal nitride coatings as a corrosion-resistant barrier is the presence of defects such as voids and pinholes which can lead to a direct attack of the substrate by the corrosive media. This, in turn can induce a variety of problems such as galvanic coupling between the substrate and the coating, pitting, enhanced dissolution of the substrate and spallation of the coating, promoting the formation of large pits in the surface [17]. Improved performance can be achieved through the optimisation of the process deposition parameters, such as the substrate bias voltage, chamber gas pressure and sputtering target power to produce dense, pore-free, well adherent coatings with improved corrosion properties of TiN and TiAlN coatings [1].

Similarly, enhanced corrosion resistance was observed for cathodic arc deposited AlCrN and AlCrON coatings when deposited with greater film thickness and an optimised $N_2:O_2$ gas ratio of 75:25 [18]. Post deposition treatments, such as ion implantation, have been adopted as strategies for improving the corrosion resistance of nitride-based coatings [1]. In particular, coatings deposited using the cathodic arc evaporation process are known to contain large macro-particles, and hence a large driving force in the development of these coatings as corrosion-resistant barriers is to reduce the number of macro-particles present. One of the many factors that has been studied and is known to influence macro-particle formation is the arc current characteristic. Increased macro-particle density with increasing continuous DC current [19] has been reported. This was associated with increasing heat input. It was originally envisaged that increased corrosion resistance might be achieved through the use of pulsed arc technology, which may provide enhanced barrier layer performance for such coatings [16,20,21], thus providing the focus for this study.

4.1. Correlations Between the Process Deposition Conditions, Film Thickness and Surface Roughness of TiN Coatings

The study has shown that relationships exist between substrate configuration, film thickness, surface roughness and to some degree, corrosion behavior. The use of two substrate rotation configurations in this study, namely double rotation (2R) planetary and triple rotation (3R) planetary configurations, was observed to influence both the film thickness and surface roughness, which was shown to have some effect on the resultant corrosion properties. The surface roughness of the coating was found to be considerably rougher than the carbon steel substrate, which increased with increasing film thickness. This can be attributed to increasing size and number of macro-particles introduced into the coating as the thickness increases. 3R rotation configuration resulted in films of lower thickness (lower deposition rates) due to shielding from the arc evaporation source by adjacent test-pieces and jigs. Consequently, increased surface roughness was associated with the production of thicker films.

Buschel and Grimm [16] reported that a significant reduction in the number and size of droplets incorporated into cathodic arc deposited TiN and TiAlN coatings was observed when using a pulsed arc source. This was associated with increased pulse current and decreased pulse time. In contrast, the findings from the current study have shown no correlation between the surface roughness (S10z) values and the current mode for Ti cathodes at an average current of 180 A. The pulsed current source parameters used for deposition of these Ti-based coatings have not reduced the number of macro-particles formed, according to the surface roughness data.

For TiN, the crystallographic orientations, namely the (111), (200) and (220) observed from XRD analysis of the coatings, are typical of those observed elsewhere [22], although peaks associated with the (311) orientation were not observed in the current study. The similar nature of the observed orientations for both pulsed and continuous current mode would suggest that any differences in the ion energies and production of species in the plasma associated with higher peak currents during pulsing were not significant enough to induce changes in the crystallographic structure of the TiN coating.

4.2. Corrosion Behaviour of TiN Coatings

The potentiodynamic scans of TiN coatings shown in Figure 3 suggests that all the coatings have shown improved corrosion protection to the mild steel substrate, thus acting as a barrier layer between the substrate and the environment. This is evidenced by the more positive shifts in the E_{corr} values and lower I_{corr} values of the coatings compared to the carbon steel substrate. These results are in agreement with similar findings for TiAlN and AlCrN coatings [17].

The relative resistance of the coatings is influenced by the substrate rotation configuration and the arc current mode. The more positive E_{corr} values observed for coatings deposited using double rotation (2R) configuration compared with the triple rotation (3R) configuration, can be attributed to the increased film thickness associated with these coatings. Here, the slightly thicker coatings deposited using double rotation provided more noble characteristics associated with the coating, compared with the slightly thinner 3R coatings, where the more active E_{corr} values may be attributed to increased contributions from the underlying substrate.

Figure 6 shows a plot of the average I_{corr} versus coating thickness. While the trend indicates an increased coating thickness is accompanied by a reduction in the corrosion rate (increasing average I_{corr} value), as shown by the trend line, the errors associated with the I_{corr} values make this finding inconclusive. Further work is required in this area to systematically study the effects of film thickness on the corrosion behavior of TiN coatings, as functions of current mode (pulsed or continuous) and rotation configuration (2R and 3R). The latter is particularly important for investigating the energy available for nucleation and growth of the coating and the resultant morphological changes that may occur, which may influence the corrosion behavior.

However, it should be emphasised that the reported I_{corr} values only serve as an indication of relative behaviour within the free corroding potential region. At higher anodic potentials, approaching +500 mV as shown in Figure 3, the anodic curves of three coatings are identical to the anodic curve of the carbon steel. This suggests that at higher oxidising potentials, the coatings are no longer protective, resulting in coating breakdown and severe localised corrosion of the substrate at these breakdown points. Under these conditions, the anodic corrosion characteristic of the coating system is more closely aligned with the corrosion characteristics of the mild steel substrate. It should be noted that the anodic curves for the double rotation continuous coatings did not approach that of the carbon steel, double rotation pulsed and both triple rotation coatings, but exhibited lower current values at +500 mV. This suggests improved corrosion resistance of the thicker, continuous current deposited coatings at higher anodic potentials.

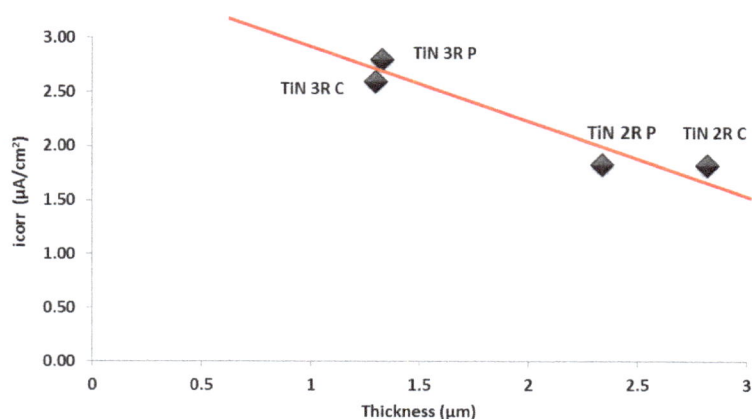

Figure 6. Plot of I_{corr} against physical vapour deposited (PVD) coating thickness

4.3. Morphological Effects on the Corrosion Behaviour of TiN Coatings

The optical and scanning electron micrographs provide an insight into the mechanisms that control the extent of corrosion occurring in the coatings. The reduced number of macro-particles and the increased number of voids observed on the coating surfaces after corrosion suggests that macro-particle detachment during the corrosion process is important in the corrosion mechanism (Figure 5a,b). The voids produced due to macro-particle detachment may act as initiation sites for pitting of the substrate since regions of the substrate are then exposed to the corrosive media. Therefore, the presence of defects, pinholes and macro-particles, before and after the corrosion process, become the limiting factors which control the corrosion behaviour of the coating system at lower anodic potentials.

After corrosion testing, the extremely large corroded regions exhibiting severe pitting with corrosion localised at the exposed substrate, as shown in the optical micrographs (Figure 5c–f), suggest that at the higher potentials approaching +500 mV the observed corrosion characteristics are possibly influenced by the presence of large areas of exposed substrate due to coating spallation being observed on the corroded coating surfaces, confirming that the corrosion behaviour of the coating system is predominantly controlled by the corrosion behaviour of the mild steel substrate as pitting commences. Further, it is postulated that the mechanism for production of these large corroded regions is existing pinholes, which can act as initiation sites for accelerated, localised (pitting) corrosion to occur and macro-particles can become detached during corrosion, providing large voids for further corrosion to initiate and propagate. The production of these corrosion sites within any one region can then result in the removal of the underlying substrate to the point that the steel can no longer support the coating, which in turn, cracks and spalls away. This then allows the coating system to corrode as freely as the mild steel alone [18].

Cross sections of the pits observed in Figure 5g are consistent with the formation of wide shallow pits, where the pit depth at the centre (maximum) is approximately one third of the pit diameter. This would suggest that metal dissolution is higher at the walls of the pit than at the bottom of the pit. The absence of any pit morphologies showing undercutting or subsurface pitting suggests that once the pit has initiated and has become stable, then the coating is being removed at the same rate as the pit is growing, thus ensuring that all of the pit area is exposed to the corrosive media.

It is suggested that coating removal within the pit region may occur either by chemical dissolution, mechanical removal or a combination of both. Pit formation may be initiated by chemical dissolution of the coating either at an existing defect or elsewhere on the surface, thus exposing more of the underlying steel substrate to attack. Conversely, the dissolution of the steel substrate at existing defects may result in regions of the coating which cannot support themselves and thus are prone to cracking and spallation. Removal of the coating would have to occur at the same rate that a pit grows in order to maintain the hemispherical morphology. It is unclear which of these mechanisms dominate, thus providing a basis for future studies in this area.

The increased number of pits observed on the surface of the 3R C and 3R P TiN coatings indicates the increased likelihood of breakdown and susceptibility to pitting under high anodic polarization conditions, which is governed predominantly by the substrate rotation configuration which promotes thinner films. While pit size/pit density was virtually independent of the arc current mode at triple rotation configuration, at double rotation configuration, fewer, larger pits were observed under continuous arc current mode (2R C) compared to pulsed arc current mode (2R P). The presence of fewer pits for 2R C coatings plus lower anodic currents at +500 mV serves as an indication of the more protective nature of these coatings under strong anodic polarization conditions.

The effect of coating integrity on the effectiveness of corrosion barrier layer systems has been demonstrated for other deposition techniques on mild steel, e.g., in the case of tungsten carbide (WC) based cermet coatings [23]. The poor corrosion protection was attributed to a poor coating structure featuring high porosity levels and the presence of micro-cracks within the coating. It was further observed that these micro-cracks provided a direct path from the environment to the substrate, inducing galvanic coupling effects between the substrate and the coating, resulting in enhanced dissolution and increased void/micro-crack formation. In the current study, if macro-particle detachment resulted in exposed regions of the substrate, enhanced dissolution of the substrate may have occurred due to galvanic effects between the substrate and the coating.

It is likely that for the thicker films deposited under continuous arc current conditions, there is a lower probability that the substrate will become exposed to the media and thus a lower probability of stable pits initiating at these sites, thus explaining why thicker films have fewer areas of large pits. However, the reduced corrosion resistance of the pulsed coatings cannot be explained by this mechanism.

Overall, while reduced corrosion can be controlled by having reduced number of pinholes and macro-particles during the deposition of the coatings at the microstructural level, enhanced corrosion resistance is favoured by structures which can restrict the initiation, formation and growth of the large regions of coating spallation and underlying substrate corrosion. The differences between the microstructures, composition and stress state of the experimental coatings deposited using a pulsed arc current source will be the subject for further study.

In addition, the main purpose of using potentiodynamic scanning and Tafel extrapolation was to provide an overall comparative study of the electrochemical corrosion behaviour of the coated steel samples, compared with the uncoated steel samples, when exposed to a saline environment. As a consequence of these scans, particularly at higher anodic potentials, breakdown of the coating at isolated regions and attack of the underlying substrate occurred, giving the appearance of localized corrosion. In order to assess the localized corrosion behavior of these coating systems, more appropriate electrochemical techniques should be employed and this forms the basis for future work in this area.

As shown in this study, increased surface roughness, and to a certain extent, corrosion resistance, was associated with thicker coatings deposited using 2R rotation configuration compared to 3R. In order to determine inherent effects associated with the arc current mode and substrate rotation configuration (outside of film thickness) on surface roughness and corrosion resistance, coatings for all test conditions should be deposited with equal film thickness and evaluated accordingly. As part of these studies, focus should be given to the formation, growth and distribution of defects within the coating in order to assess their role and influence on the corrosion behavior of the coatings.

5. Conclusions

- The deposition of TiN coatings resulted in less negative E_{corr} and lower I_{corr} values, determined from Tafel Extrapolation, when compared with the mild steel substrate after potentiodynamic corrosion testing in 3.5% NaCl.

- Coatings deposited under double rotation (2R) configuration showed more noble (more positive E_{corr} values) characteristics than those deposited under triple rotation (3R) configuration. This was attributed to the increased film thickness of the 2R coatings. Any correlations between substrate rotation configuration, arc current mode and observed I_{corr} values were inconclusive from this study and further work is required in this area.

- At high anodic potentials approaching +500 mV, the corrosion behaviour of most coating systems (except 2R C) was characteristic of uncoated mild steel. This suggests that the limiting factors controlling the corrosion behaviour of the coating system at lower anodic potentials were the presence of defects, pinholes and macro-particles. The initiation, formation and growth of large pitted regions with further coating spallation resulted in underlying substrate corrosion which became the dominant factors at higher anodic potentials.

- An increased number of pits were observed for thinner coatings deposited by triple rotation (both 3R C and 3R P). Fewer, but larger, pits were observed for 2R C coatings, indicating an increased resistance to breakdown at higher anodic potentials.

- Surface roughness from FVM measurements was found to increase with increasing film thickness, although no correlation was found with the type of arc current (continuous or pulsed) in the study.

- The effects of coating microstructure, composition, residual stress and film thickness on the reduced corrosion performance of coatings deposited using pulsed arc current techniques requires further investigation. This will include the deposition of coatings with similar film thicknesses to determine the effects of arc current mode and substrate rotation configuration on the resultant properties, with a focus on coating formation, growth and distribution of defects and their influence on the corrosion behavior.

Acknowledgments: This body of work was undertaken by the DMTC with researchers from RMIT University and Sutton Tools Pty. Ltd. The DMTC was established and is supported by the Australian Government's Defence Future Capability Technology Centre (DFCTC) initiative. The industrial DMTC partner Sutton Tools Pty. Ltd. provided sponsorship, research materials and supported the coating deposition facilities. The authors acknowledge the facilities, and the scientific and technical assistance, of the Australian Microscopy & Microanalysis Research Facility at the RMIT Microscopy & Microanalysis Facility, at RMIT University.

Author Contributions: L. Ward, A. Pilkington and S. Dowey conceived and designed the experiments; L. Ward, A. Pilkington and S. Dowey performed the experiments; L. Ward, A. Pilkington and S. Dowey analyzed the data; S. Dowey contributed reagents/materials/analysis tools; L. Ward and A. Pilkington wrote the paper.

Conflicts of Interest: The authors declare no conflict of interest. The funding sponsors had no role in the design of the study; in the collection, analyses, or interpretation of data; in the writing of the manuscript, and in the decision to publish the results.

References

1. Ward, L.P. Studies on the Corrosion behaviour of a selection of metal nitride coatings and the effect of surface modification. In Proceedings of the Corrosion and Prevention Conference 2009 (CAP09), Coffs Harbour, NSW, Australia, 15–18 November 2009.

2. Liu, C.; Bi, Q.; Leyland, A.; Matthews, A. An electrochemical impedance spectroscopy study of the corrosion behaviour of PVD coated steels in 0.5 N NaCl aqueous solution: Part I. Establishment of equivalent circuits for EIS data modelling. *Corr. Sci.* **2003**, *45*, 1243–1256. [CrossRef]

3. Navinsek, B.; Panjan, P.; Milosev, I. PVD coatings as an environmentally clean alternative to electroplating and electroless processes. *Surf. Coat. Technol.* **1999**, *116–119*, 476–487. [CrossRef]

4. Cunha, L.; Andritschky, M.; Rebouta, L.; Pischow, K. Corrosion of CrN and TiAlN coatings in chloride-containing atmospheres. *Surf. Coat. Technol.* **1999**, *116–119*, 1152–1160. [CrossRef]

5. Cunha, L.; Andritschky, M.; Pischow, K.; Wang, Z. Microstructure of CrN coatings produced by PVD techniques. *Thin Solid Films* **1999**, *355–356*, 465–470. [CrossRef]

6. Constantin, R.; Miremad, B. Performance of hard coatings, made by balanced and unbalanced magnetron sputtering, for decorative applications. *Surf. Coat. Technol.* **1999**, *120–121*, 728–733. [CrossRef]

7. Wang, Q.; Zhou, F.; Zhou, Z.; Li, L.K.-Y.; Yan, J. Electrochemical performance of TiCN coatings with low carbon concentration in simulated body fluid. *Surf. Coat. Technol.* **2014**, *253*, 199–204. [CrossRef]

8. Antunes, R.A.; de Assis, S.L.; Lorenzetti, S.G.; Higa, O.Z.; Costa, I. Comparison of in vitro corrosion behaviour and biocompatibility of Ti-13Zr-13Nb and passivated 316L stainless steel coated with TiCN. In Proceedings of the COBEM, 18th International Conference of Mechanical Engineering, Ouro Preto, MC, Brazil, 6–11 November 2005.

9. Massiani, Y.; Medjahed, A.; Crousier, JP.; Gravier, P.; Rebatel, I. Corrosion of sputtered titanium nitride films deposited on iron and stainless steel. *Surf. Coat. Technol.* **1991**, *45*, 115–120. [CrossRef]

10. Piippo, J.; Elsener, B.; Bohni, H. Electrochemical characterisation of TiN coatings. *Surf. Coat. Technol.* **1993**, *61*, 43–46.

11. Brown, R.; Alias, M.N.; Fontana, R. Effect of composition and thickness on corrosion behaviour of TiN and ZrN thin films. *Surf. Coat. Technol.* **1993**, *62*, 467–473. [CrossRef]

12. Kazuhisa, A.; Shigeyo, W.; Masahiro, S.; Isao, S.; Yokio, I.; Patrick, J.; William, S. Characterization of anodic oxide film formed on TiN coating in neutral borate buffer solution. *Corr. Sci.* **1998**, *40*, 1363–1377.

13. Massiani, Y.; Medjahed, A.; Gravier, P.; Argeme, L.; Fedrizzi, L. Electrochemical study of titanium nitride films obtained by reactive sputtering. *Thin Solid Films* **1990**, *191*, 305–316. [CrossRef]

14. Wang, H.W.; Stack, M.M.; Lyon, S.B.; Hovsepian, P.; Munz, W.-D. Preferential wear of droplet defects for combined cathodic arc-unbalanced magnetron sputtering CrN/NbN superlattice coatings during erosion in alkaline slurries. *J. Mater. Sci. Lett.* **2001**, *20*, 1995–1997. [CrossRef]

15. Pilkington, A. Development and Evaluation of Plasma-Assisted Alumina Based Coatings. Ph.D. Thesis, University of Sheffield, Sheffield, UK, 2014.

16. Büschel, M.; Grimm, W. Influence of the pulsing of the current of a vacuum arc on rate and droplets. *Surf. Coat. Technol.* **2001**, *142–144*, 665–668. [CrossRef]

17. Ward, L.P.; Pilkington, A.; Dowey, S.J.; Doyle, E.D. Corrosion Properties of Cathodic Arc Evaporated Coatings. In Proceedings of the Corrosion and Prevention Conference 2014, CAP14, Darwin, NT, Australia, 21–24 September 2014.

18. Ward, L.P.; Pilkington, A.; Dowey, S.J.; Toton, J.T.; Doyle, E.D. Studies on the corrosion behaviour of AlCrON coatings. In Proceedings of the Corrosion and Prevention Conference 2013, CAP13, Brisbane, QLD, Australia, 12–15 Nobember 2013.

19. Hu, Y.; Li, L.; Dai, H.; Li, X.; Cai, X.; Chu, P.K. Effects of pulse parameters on macro-particle production in pulsed cathodic vacuum arc deposition. *Surf. Coat. Technol.* **2007**, *201*, 6542–6544. [CrossRef]

20. Ramm, J.; Ante, M.; Bachmann, T.; Widrig, B.; Brändle, H.; Döbeli, M. Pulse enhanced electron emission (P3e™) arc evaporation and the synthesis of wear resistant Al–Cr–O coatings in corundum structure. *Surf. Coat. Technol.* **2007**, *202*, 876–883. [CrossRef]

21. Fuchs, H.; Engers, B.; Hettkamp, E.; Mecke, H.; Schultz, J. Deposition rate and thickness uniformity of thin films deposited by a pulsed cathodic arc process. *Surf. Coat. Technol.* **2001**, *142–144*, 655–660. [CrossRef]

22. Devia, D.M.; Restrepo-Parra, E.; Arango, P.J. Comparative study of titanium carbide and nitride coatings grown by cathodic vacuum arc technique. *App. Surf. Sci.* **2011**, *258*, 1164–1174. [CrossRef]

23. Ward, L.P.; Biddle, G.; Hinton, B.; Gerrard, D. Characterisation and evaluation of the corrosion behaviour of modified HVOF sprayed WC based coatings deposited on AISI 1020 mild steel substrates. In Proceedings of the Corrosion and Prevention Conference 2006 (CAP06), Hobart, Tasmania, Australia, 19–22 November 2006.

6

Interfacial Mechanics Analysis of a Brittle Coating–Ductile Substrate System Involved in Thermoelastic Contact

Chi Zhang, Le Gu *, Chongyang Nie, Chuanwei Zhang and Liqin Wang

Research Lab of Space & Aerospace Tribology, Harbin Insititute of Technology, Harbin 150001, China; zhc_hit@163.com (C.Z.); ncy422@163.com (C.N.); zhchwei@hit.edu.cn (C.Z.); lqwang@hit.edu.cn (L.W.)
* Correspondence: gule@hit.edu.cn

Academic Editor: Alessandro Lavacchi

Abstract: In this paper, interfacial stress analysis for a brittle coating/ductile substrate system, which is involved in a sliding contact with a rigid ball, is presented. By combining interface mechanics theory and the image point method, stress and displacement responses within a coated material for normal load, tangential load, and thermal load are obtained; further, the Green's functions are established. The effects of coating thickness, friction coefficient, and a coating's thermoelastic properties on the interfacial shear stress, τ_{xz}, and transverse stress, σ_{xx}, distributions are discussed in detail. A phenomenon, where interfacial shear stress tends to be relieved by frictional heating, is found in the case of a coating material's thermal expansion coefficient being less than a substrate material's thermal expansion coefficient. Additionally, numerical results show that distribution of interfacial stress can be altered and, therefore, interfacial damage can be modified by adjusting a coating's structural parameters and thermoelastic properties.

Keywords: contact mechanics; Green's function; interfacial mechanics; brittle coating

1. Introduction

Today, the key elements, such as bearings and gears, are under great pressure to meet the legislative demands of long life and high operational speeds [1–3]. These challenges can be achieved by employing high hardness, anti-corrosion, and wear-resistant brittle coatings, such as metal nitride coatings, metal oxide coatings, diamond-like carbon (DLC), etc., to protect working surfaces, as well as adopting high strength steel [4,5]. However, the huge mismatch between coatings and steel substrates in thermoelastic properties always leads to interfacial stress concentrations, which are induced thermally and mechanically, thus, increasing the risk of coating delamination. To avoid blindness in design and application, an interfacial stress analysis of coated solids in dry sliding contacts is essential.

Thus far, many researchers have established Hertzian contact models for coated, elastic semi-infinite space. Linearly, elastic theory uses various forms of transform integration techniques to produce solutions. Gu et al. [6] studied the stress response of coated materials under surface concentrate loading using the image point method. By comparison of calculation results and scratch tests, the notion that the load bearing capacities of DLC-coated Si_3N_4 was 1.5–3 times larger than that of DLC-coated M50. O'Sullivan and King studied the contact of coated materials using Papkovich–Neuber potentials [7]. Wang and Liu [8,9] applied fast Fourier transformation (FFT) to speed up calculations based on the work of O'Sullivan and King. A combination of FFT and Papkovich–Neuber potentials were further extended to study sliding contacts and partial slip contact problems for solids coated with a monolayer or a gradient layer [10,11]. The finite element method (FEM) was also used in studying interfacial adhesion strengths between coatings and substrates [11–15]. The Hertzian contact problem of coated materials has been extensively studied during past few decades; and a great number of useful conclusions have been drawn. However, the engineering applications of coatings still have a certain degree of blindness due to the small amount of research on the stress responses of coated materials involved in a thermoelastic contact. Ke et al. [16] discussed the effect of the friction coefficient and sliding velocity on the temperature distribution of semi-infinite materials coated with a gradient material. Choi et al. [17] established a two-dimensional thermoelastic contact model for a rigid flat punch sliding over the surface of a graded coating/substrate system. However, little attention has been paid to the interfacial stress distributions of coated solids involved in thermoelastic contacts. As a matter of fact, a significant amount of frictional heat is generated on the surface of coated materials involved in dry sliding contacts. Thermally induced deformation, subsequently, changes the Hertzian stress distribution of coated materials from the surface to several micrometers in depth, especially for hard coatings, which are distinguished from substrates in regard to thermoelastic properties. Thus, to fulfill a thorough optimization of a hard coating–substrate system, it is necessary to establish an interfacial stress analysis model for thermoelastic contacts.

In this paper, a stress analysis model for coated materials under thermoelastic contact conditions is built, based on interface mechanics. The Green's functions of the stress and displacement fields of coated materials for a normal force, tangential force, and thermal loading are established. With the aim of exploring ways to reduce the risks of delamination of hard coatings from steel substrates, the effects of the friction coefficient, coating thickness, and thermoelastic properties on interfacial shear stress τ_{xz} and transverse stress σ_{xx} distributions are investigated, as these two stresses are believed to be strongly related to the propagation of interface cracks in hard coating–substrate systems [18–20].

2. Theory

2.1. Description of the Thermoelastic Contact Model

As shown in Figure 1a, a rigid insulated ball is subjected to force F_z and slides on the surface of a coated, isotropic, elastic half space with a constant speed, V, in the positive x-direction.

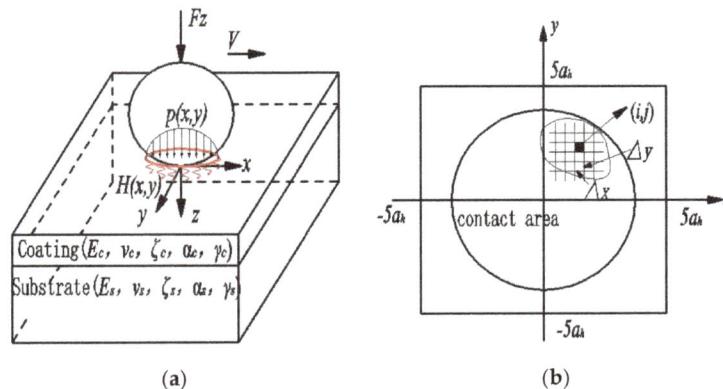

Figure 1. Schematic of the thermoelastic contact problem. (**a**) A rigid insulated ball sliding over the surface of a coated solid; (**b**) Contact area meshing.

The moving coordinate system (x,y,z) is attached to the ball with plane xOy coinciding with the surface of the coated material; that is, the contact area will remain stationary with respect to the movement of the ball. The normal pressure distribution is $p(x,y)$ N·m^{-2}; then, the friction force distribution, $q(x,y)$ N·m^{-2}, and the distribution of frictional heat flux, $H(x,y)$ W·m^{-2}, are generated within the contact region, with f being the friction coefficient. Their relationships are as follows:

$$q(x,y) = f_p(x,y), \quad H(x,y) = f_{vp}(x,y) \tag{1}$$

Consider this thermoelastic contact problem at room temperature in ambient humidity. Because of the thermal conductive coefficient of the air being always less than 10^{-3} times of that parameter for the coating material according to Reference [21], the free surface is assumed to be thermally insult. Due to punch sliding with a constant velocity, the temperature will very quickly approach a steady-state value, thus, the convection term in the heat equation is neglected. The coating and substrate are both isotropic materials, with E_c, v_c, ζ_c, α_c, and γ_c being the elastic modulus, Poisson ratio, thermal conductive coefficient, thermal expansion coefficient, and thermal diffusion coefficient for the coating material, respectively, and E_s, v_s, ζ_s, α_s, and γ_s being the same parameters for the substrate material.

According to Reference [22], as shown in Figure 1b, in order to obtain contact pressure $p(x,y)$, an area with $[-5a_h, 5a_h] \times [-5a_h, 5a_h]$ in the x and y directions is uniformly divided into $N_x \times N_y$ surface elements, centered on the grid nodes. Here, a_h is the Hertz contact radius without coating and thermal effects, and N_x and N_y are the number of elements in the x and y directions. As being proved in Reference [23], the contact pressure distribution can be approximated by a piecewise constant function that is uniform within each surface element with errors being less than 0.01% in the case of N_x and N_y being equal to 32. A more finely meshed surface network, with $N_x \times N_y$ being equal to 128×128, is adopted in this paper. With the displacements of the element being represented by its central grid; the contact problem can be expressed as a discretized form:

$$
\begin{aligned}
& p(i,j) > 0, g(i,j) = 0, (i,j) \in \Omega_c \\
& p(i,j) = 0, g(i,j) > 0, (i,j) \notin \Omega_c \\
& \Delta_x \Delta_y \sum_{j=0}^{N_y-1} \sum_{i=0}^{N_x-1} p_{(i,j)} = F_z \\
& g_{(i,j)} = g^0{}_{(i,j)} + u^{pz}{}_{(i,j)} + u^{tz}{}_{(i,j)} - \delta
\end{aligned}
\tag{2}
$$

For arbitrarily surface element (i,j), $p(i,j)$ is the normal pressure, $g(i,j)$ is the gap between the surfaces of two bodies in contact after external contact load and thermal load was applied, $g^0(i,j)$ is the initial gap before external mechanical load and heat flux applied on the ball, $u^{pz}(i,j)$ is the surface displacement of the coated material, which was induced mechanically, $u^{tz}(i,j)$ is the surface displacement of the coated material, which was induced thermally, Δ_x and Δ_y are the discretization lengths in the x and y directions, δ is the normal approach, and Ω_c is the contact area.

As illustrated in Figure 2, the ball radius is noted as d. The central point of elements (i,j) is noted as Q, with $x_{i,j}$ and $y_{i,j}$ being the positions of point Q in the xOy plane.

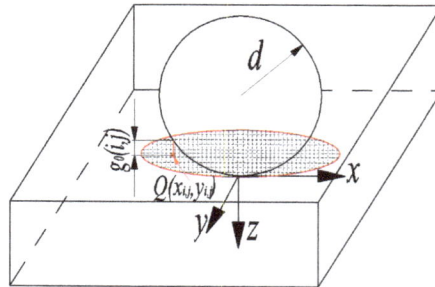

Figure 2. Schematic description of the initial gap between surface elements and ball.

The initial gap $g^0(i,j)$ can be described as the distance between the central point Q and the outer surface of the ball in the z direction:

$$
\begin{aligned}
& x_{i,j}^2/d^2 + y_{i,j}^2/d^2 + (z_{i,j} - d)^2/d^2 = 1 \\
& z_{i,j} = g^0(i,j)
\end{aligned}
\tag{3}
$$

Due to $z_{i,j} \ll d$ within the contact zone, the relationship between $g^0(i,j)$ and $x_{i,j}$, $y_{i,j}$ can be described as

$$g^0(i,j) = x_{i,j}^2/2d + y_{i,j}^2/2d \qquad (4)$$

Total normal displacement $u^z(i,j)$ can be obtained by the accumulation effects of the normal contact pressure and thermal loading of each surface element on the z-displacement of element (i,j).

$$u^z{}_{(i,j)} = u^{pz}{}_{(i,j)} + u^{tz}{}_{(i,j)} = \Delta_x\Delta_y \sum_{k=0}^{N_y-1} \sum_{l=0}^{N_x-1} G^p_{z-(i-l,j-k)} p_{(l,k)} + \Delta_x\Delta_y \sum_{k=0}^{N_y-1} \sum_{l=0}^{N_x-1} G^t_{z-(i-l,j-k)} H_{(l,k)} \qquad (5)$$

where G_z^p and G_z^t are Green's functions of displacement in the z-direction for a normal load and a thermal load. By applying the complementary energy principle to the thermal contact problem, we have

$$V^* = \frac{1}{2}\Delta_x\Delta_y \sum_{i=0}^{N_y-1} \sum_{j=0}^{N_x-1} p_{(l,k)} u^z{}_{(i,j)} + \Delta_x\Delta_y \sum_{i=0}^{N_y-1} \sum_{j=0}^{N_x-1} p_{(l,k)}(g^0{}_{(i,j)} - \delta) \qquad (6)$$

With the aid of the quadratic programming method, the true contact pressure, $p_{(i,j)}$, can be obtained, at which point contact complementary energy V^* is minimal and $p_{(i,j)} > 0$ is within the contact area.

After gaining contact pressure distribution and heat flux, the stress field can be obtained as follows:

$$\sigma_{AA-(x_0,y_0,z_0)} = \Delta_x\Delta_y \sum_{k=0}^{N_y-1} \sum_{l=0}^{N_x-1} G^p_{AA} p_{(l,k)} + \Delta_x\Delta_y \sum_{k=0}^{N_y-1} \sum_{l=0}^{N_x-1} G^v_{AA} q_{(l,k)} + \Delta_x\Delta_y \sum_{k=0}^{N_y-1} \sum_{l=0}^{N_x-1} G^t_{AA} H_{(l,k)} \qquad (7)$$

where σ_{AA} is an arbitrary stress component of point (x_0,y_0,z_0) within a semi-infinite space (AA may refer to xx, yy, xy, and so on). $p_{(l,k)}$, $q_{(l,k)}$, and $H_{(l,k)}$ are contact pressure distribution, tangential distribution, and heat flux distribution on surface elements (l,k), respectively. G_{AA}^p, G_{AA}^v, and G_{AA}^t are the values of Green's functions for components AA for a normal load, tangential load, and thermal load at point (x_0,y_0,z_0), respectively.

For an elastic solid with a coating bonded to its surface, the explicit expressions for G_{AA}^p, G_{AA}^v, and G_{AA}^t are not available. However, based on interface mechanics, they can be deduced and expressed in the accumulated form by employing the image point method [24,25]. The detailed deducing is presented in Section 2.2, Section 2.3, and Section 2.4. In this paper, three orders of image points are used to achieve a high accuracy as is proven in Section 3.1.

2.2. Green's Function for a Point Normal Load Acting on the Surface of a Coated Isotropic Thermoelastic Material

In the analytical model of interface mechanics (Figure 3), a straight interface is formed along the border of the coating and substrate. From Figure 3, we can see that coating I, with thickness h, is used to cover substrate II, and is connected only by the interface. Above the coating, there is a free surface, and a force, marked p_z, is applied at point O_1 on said surface and is along the z-axis. Considering the symmetries of this problem, a cylindrical coordinate (r,θ,z) is chosen with the r-axis being along the interface, and the z-axis being perpendicular to the r-axis and passing through O_1. The two axes intersect at O, namely the global point of origin. Coordinate plane $rO\theta$ coincides with the interface.

The interface stress, shown in Figure 3, can be calculated using the images method from the complex variable function. The boundary conditions and the constraints of the model are equivalent; furthermore, equations were composed using the relationship between stress and strain. As a method of images, the interface and the surface are imagined as mirrors that reflect points O or O_1, and will generate infinite points of mirror images. These images influence the interface stresses as a superposition forms, in order to fulfill the boundary conditions and constraints. Namely, the conditions of the interface continuum and free surface are satisfied and interface stresses are gained by superposing the stress solutions of these image points. The image points in the coating space are marked as O_k and those in the substrate space are marked as C_k. Local coordinate systems are established and originate from each image point.

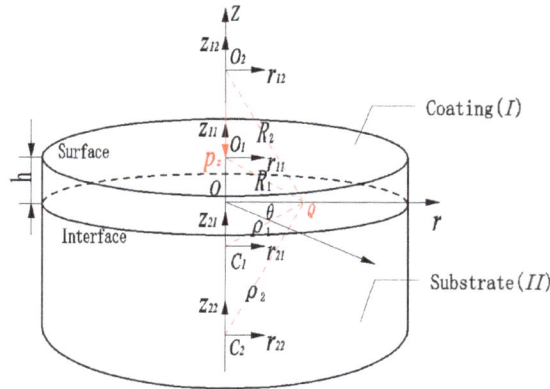

Figure 3. A point normal force p_z acting on the surface of coated material.

The relationship between the global coordinate and the local coordinates can be written as:

$$\begin{cases} R_k = \sqrt{r^2 + z_k^2}, z_k = z - (2k-1)h \\ \rho_k = \sqrt{r^2 + s_k^2}, s_k = z + (2k-1)h \end{cases} \quad k = 1,2,3....\infty. \tag{8}$$

Love's stress and displacement functions can be written as:

$$\begin{cases} 2\mu u_r = \frac{\partial \phi}{\partial r} + \frac{\partial \Psi}{\partial r}, 2\mu u_z = \frac{\partial \phi}{\partial z} - (3-4v)\Psi + z\frac{\partial \Psi}{\partial z} \\ \sigma_r = -\frac{1}{r}\frac{\partial \phi}{\partial r} - \frac{\partial^2 \phi}{\partial z^2} - \frac{z}{r}\frac{\partial \Psi}{\partial r} - z\frac{\partial^2 \Psi}{\partial r^2} - 2v\frac{\partial \Psi}{\partial z} \\ \sigma_\theta = \frac{1}{r}\frac{\partial \phi}{\partial r} + \frac{z}{r}\frac{\partial \Psi}{\partial r} - 2v\frac{\partial \Psi}{\partial z} \\ \sigma_z = \frac{\partial^2 \phi}{\partial z^2} - 2(1-v)\frac{\partial \Psi}{\partial z} + z\frac{\partial^2 \Psi}{\partial z^2} \\ \tau_{rz} = \frac{\partial^2 \phi}{\partial r \partial z} - (1-2v)\frac{\partial \Psi}{\partial r} + z\frac{\partial^2 \Psi}{\partial r \partial z} \end{cases} \tag{9}$$

where μ is the shear elasticity modulus, and its relationships with E and v is in the form of $\mu = 0.5E/(1+v)$. u is the displacement component and σ is the stress component in each directions. φ and Ψ are harmonic functions. For coating materials, the harmonic functions are noted as φ^1 and Ψ^1. For coating materials, the harmonic functions are noted as φ^2 and Ψ^2. Due to the singular point O_1 on the surface of the coating, the harmonic functions for coating material have a relationship with z_i and s_i at the same time. However, because no singular point exists in the substrate material, the harmonic functions for the substrate material only have a relationship with z_i. Thus, the harmonic functions can be written as

$$\begin{cases} \phi^1 = \sum\limits_{i=1}^{\infty} \phi_i^1(r,z_i) + \sum\limits_{i=1}^{\infty} \delta_i^1(r,s_i), \phi^2 = \sum\limits_{i=1}^{\infty} \phi_i^2(r,z_i) \\ \Psi^1 = \sum\limits_{i=1}^{\infty} \Omega_i^1(r,z_i) + \sum\limits_{i=1}^{\infty} \eta_i^1(r,s_i), \Psi^2 = \sum\limits_{i=1}^{\infty} \Omega_i^2(r,z_i) \end{cases} \tag{10}$$

where i is the sequence number of the image point.

One of the boundary conditions can be established by satisfying the conditions of the interface continuum.

$$u_r^1 = u_r^2, u_z^1 = u_z^2, \sigma_z^1 = \sigma_z^2, \tau_{rz}^1 = \tau_{rz}^2, z = 0 \tag{11}$$

Another boundary condition can be established by satisfying the free surface.

$$\sigma_z^1 = 0, \tau_{rz}^1 = 0, z = h \tag{12}$$

Substituting Equation (10) into Equation (9) using boundary condition Equation (11) and geometry condition $z = 0$, $z_i = s_i = -(2i - 1)h$, $R_i = \rho_i$ on the interface; recursive relations can be obtained:

$$\begin{cases} \eta_i^1(r, s_i) = \frac{\beta - \alpha}{1 - \beta} [2 \frac{\partial \phi_i^1(r, z_i)}{\partial z} - \kappa_1 \Omega_i^1(r, z_i)] \\ \delta_i^1(r, s_i) = -\frac{\beta - \alpha}{1 - \beta} \phi_i^1(r, z_i) - \frac{1 - \alpha}{2\Gamma} [\frac{\kappa_1}{1 - \beta} - \frac{\kappa_2}{1 + \beta}] \int \Omega_i^1(r, z_i) dz \\ z_i = -s_i, R_i = \rho_i \\ \phi_i^2(r, z_i) = \frac{1 - \alpha}{1 - \beta} \phi_i^1(r, z_i) - \frac{1 - \alpha}{2} [\frac{\kappa_1}{1 - \beta} - \frac{\kappa_2}{1 + \beta}] \int \Omega_i^1(r, z_i) dz \\ \Omega_i^2(r, z_i) = \frac{1 - \alpha}{1 + \beta} \Omega_i^1(r, z_i) \end{cases} \tag{13}$$

Substituting Equation (10) into Equation (9) using boundary condition Equation (12) and geometry condition $z = h$, $z_i = -(2i - 1)h$, $s_i = 2ih$, $R_{i+1} = \rho_i$ on the surface, another recursive relation can be obtained:

$$\begin{cases} \Omega_{i+1}^1(r, z_{i+1}) = -2 \frac{\partial \delta_{i-1}^1(r, s_i)}{\partial z} + (3 - 4v_1) \eta_i^1(r, s_i) - 2h \frac{\partial \eta_i^1(r, s_i)}{\partial z} \\ \phi_{i+1}^1(r, z_{i+1}) = (3 - 4v_1) \delta_{i-1}^1(r, s_i) - 4(1 - 2v_1)(1 - v_1) \int \eta_i^1(r, s_i) dz + h[2 \frac{\partial \delta_{i-1}^1(r, s_i)}{\partial z} + 2h \frac{\partial \eta_i^1(r, s_i)}{\partial z}] \\ s_i = -z_{i+1}, \rho_i = R_{i+1} \end{cases} \tag{14}$$

where κ is the Kappa parameter; $\kappa_1 = 3 - 4v_1$ and $\kappa_2 = 3 - 4v_2$ are Kappa parameters for the coating material and substrate material. α and β are Dundurs parameters [24] for the coating–substrate system.

Thus, $\varphi^1{}_1(r, z_1)$ and $\Omega^1{}_1(r, z_1)$ are the only factors needed to finish the cyclic recursion. Obviously, the Midlin solutions for a point normal load, acting on the surface of a semi-infinite space, satisfy the requirements.

$$\begin{cases} \phi_1^1(r, z_1) = -(1 - 2v_1) A_n \ln(R_1 - z_1) - \frac{h A_n}{R_1} \\ \Omega_1^1(r, z_1) = \frac{A_n}{R_1} \\ A_n = -\frac{p}{2\pi} \end{cases} \tag{15}$$

Now, φ and Ψ can be obtained using Equations (13)–(15). Green's functions can be obtained by substituting Φ and Ψ into Equation (9), and then separating variable p_z.

2.3. Green's Function for a Point Tangential Load Acting on the Surface of a Coated Isotropic Thermoelastic Material

As shown in Figure 4, we can see that coating I, with thickness h, is used to cover substrate II, and is connected only by the interface. Above the coating, there is a free surface, and a tangential force, marked p_x, is applied at point O_1 on said surface and is parallel to the x-axis. A Cartesian coordinate (x, y, z) is chosen with the x-axis and y-axis being along the interface, and the z-axis being perpendicular to the interface and passing through O_1. The x-axis and y-axis intersect at O, namely the global point of origin. Coordinate plane xOy coincides with the interface. The image points in the coating space are marked as O_k, and those in the substrate space are marked as C_k. Green's function for a point tangential load, acting on the surface of a coated isotropic thermoelastic material, can be obtained in the similar way (shown in Section 2.2).

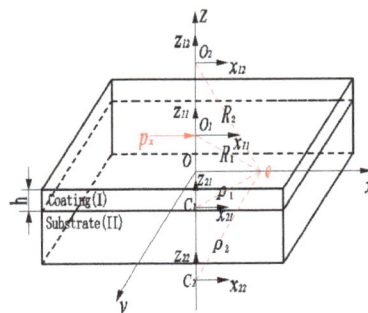

Figure 4. A point tangential force p_x acting on the surface of coated material.

Papokovitch's stress and displacement functions can be written as:

$$
\begin{cases}
u_x = \frac{3-4v}{4(1-v)}B_1 - \frac{1}{4(1-v)}\left[x\frac{\partial B_1}{\partial x} + y\frac{\partial B_2}{\partial x} + z\frac{\partial B_3}{\partial x} + \frac{\partial \delta}{\partial x}\right] \\
u_y = \frac{3-4v}{4(1-v)}B_2 - \frac{1}{4(1-v)}\left[x\frac{\partial B_1}{\partial y} + y\frac{\partial B_2}{\partial y} + z\frac{\partial B_3}{\partial y} + \frac{\partial \delta}{\partial y}\right] \\
u_z = \frac{3-4v}{4(1-v)}B_3 - \frac{1}{4(1-v)}\left[x\frac{\partial B_1}{\partial z} + y\frac{\partial B_2}{\partial z} + z\frac{\partial B_3}{\partial z} + \frac{\partial \delta}{\partial z}\right] \\
\sigma_x = \frac{v\mu}{1-v}\left(\frac{\partial B_1}{\partial x} + \frac{\partial B_2}{\partial y} + \frac{\partial B_3}{\partial z}\right) + \frac{\mu(1-2v)}{1-v}\frac{\partial B_1}{\partial x} - \frac{\mu}{2(1-v)}\left(x\frac{\partial^2 B_1}{\partial x^2} + y\frac{\partial^2 B_2}{\partial x^2} + z\frac{\partial^2 B_3}{\partial x^2} + \frac{\partial^2 \delta}{\partial x^2}\right) \\
\sigma_y = \frac{v\mu}{1-v}\left(\frac{\partial B_1}{\partial x} + \frac{\partial B_2}{\partial y} + \frac{\partial B_3}{\partial z}\right) + \frac{\mu(1-2v)}{1-v}\frac{\partial B_2}{\partial y} - \frac{\mu}{2(1-v)}\left(x\frac{\partial^2 B_1}{\partial y^2} + y\frac{\partial^2 B_2}{\partial y^2} + z\frac{\partial^2 B_3}{\partial y^2} + \frac{\partial^2 \delta}{\partial y^2}\right) \\
\sigma_z = \frac{v\mu}{1-v}\left(\frac{\partial B_1}{\partial x} + \frac{\partial B_2}{\partial y} + \frac{\partial B_3}{\partial z}\right) + \frac{\mu(1-2v)}{1-v}\frac{\partial B_3}{\partial z} - \frac{\mu}{2(1-v)}\left(x\frac{\partial^2 B_1}{\partial z^2} + y\frac{\partial^2 B_2}{\partial z^2} + z\frac{\partial^2 B_3}{\partial z^2} + \frac{\partial^2 \delta}{\partial z^2}\right) \\
\tau_{xy} = \frac{(1-2v)\mu}{2(1-v)}\left(\frac{\partial B_1}{\partial y} + \frac{\partial B_2}{\partial x}\right) - \frac{\mu}{2(1-v)}\left(x\frac{\partial^2 B_1}{\partial x\partial y} + y\frac{\partial^2 B_2}{\partial x\partial y} + z\frac{\partial^2 B_3}{\partial x\partial y} + \frac{\partial^2 \delta}{\partial x\partial y}\right) \\
\tau_{yz} = \frac{(1-2v)\mu}{2(1-v)}\left(\frac{\partial B_2}{\partial z} + \frac{\partial B_3}{\partial y}\right) - \frac{\mu}{2(1-v)}\left(x\frac{\partial^2 B_1}{\partial z\partial y} + y\frac{\partial^2 B_2}{\partial z\partial y} + z\frac{\partial^2 B_3}{\partial z\partial y} + \frac{\partial^2 \delta}{\partial z\partial y}\right) \\
\tau_{xz} = \frac{(1-2v)\mu}{2(1-v)}\left(\frac{\partial B_3}{\partial x} + \frac{\partial B_1}{\partial z}\right) - \frac{\mu}{2(1-v)}\left(x\frac{\partial^2 B_1}{\partial x\partial z} + y\frac{\partial^2 B_2}{\partial x\partial z} + z\frac{\partial^2 B_3}{\partial x\partial z} + \frac{\partial^2 \delta}{\partial x\partial z}\right)
\end{cases} \quad (16)
$$

One of the recursive relations can be expressed as:

$$
\begin{cases}
\phi_{\rho i}^1 = \frac{1-\Gamma}{1+\Gamma}\phi_{Ri}^1 \\
\Omega_{\rho i}^1 = \frac{\kappa_1(1-\Gamma)}{\kappa_1\Gamma+1}\Omega_{Ri}^1 + \frac{(1-\kappa_1)(1-\Gamma)}{(\kappa_1\Gamma+1)(\Gamma+1)}\int\frac{\partial\phi_{Ri}^1}{\partial x}dz - \frac{2(1-\Gamma)}{\kappa_1\Gamma+1}\left[x\frac{\partial\phi_{Ri}^1}{\partial z} + \frac{\partial\beta_{Ri}^1}{\partial z} - (z_i+(2i-1)h)\frac{\partial\phi_{Ri}^1}{\partial x}\right] \\
\beta_{\rho i}^1 = -\frac{\kappa_1+1}{2}\left(\frac{\kappa_1}{\kappa_1\Gamma+1} - \frac{\kappa_2}{\Gamma+\kappa_2}\right)\int\Omega_{Ri}^1 dz + \frac{\kappa_1(1-\Gamma)}{\kappa_1\Gamma+1}\beta_{Ri}^1 + \frac{(\kappa_1-1)(1-\Gamma)}{(\kappa_1\Gamma+1)(1+\Gamma)}\left[x\phi_{Ri}^1 - (z_i+(2i-1)h)\int\frac{\partial\phi_{Ri}^1}{\partial x}dz\right] + \\
\quad \frac{1}{2(1+\Gamma)}\left(\frac{\kappa_1^2-1+2\Gamma+2\kappa_1}{\kappa_1\Gamma+1} + 1 - \kappa_1 - 2m + \frac{(1-\Gamma)m(1-\kappa_2)\kappa_2}{\Gamma+\kappa_2}\right)\int\int\frac{\partial\phi_{Ri}^1}{\partial x}dzdz \\
\phi_{Ri}^2 = \frac{2}{1+\Gamma}\phi_{Ri}^1 \\
\Omega_{Ri}^2 = \frac{\kappa_2+1}{\Gamma+\kappa_2}\Omega_{Ri}^1 + \frac{(1-\kappa_2)(1-\Gamma)}{(\kappa_2\Gamma+1)(\Gamma+1)}\int\frac{\partial\phi_{Ri}^1}{\partial x}dz \\
\beta_{Ri}^2 = -\frac{\kappa_2+1}{2}\left(\frac{\kappa_1}{\kappa_1\Gamma+1} - \frac{\kappa_2}{\Gamma+\kappa_2}\right)\int\Omega_{Ri}^1 dz + \frac{\kappa_2+1}{\kappa_1\Gamma+1}\beta_{Ri}^1 + \left(\frac{(\kappa_2+1)}{(\kappa_1\Gamma+1)} - \frac{2}{1+\Gamma}\right)\left(x\phi_{Ri}^1 - z\int\frac{\partial\phi_{Ri}^1}{\partial x}dz\right) + \\
\quad \frac{1}{2m(1+\Gamma)}\left(\frac{\kappa_1^2-1+2\Gamma+2\kappa_1}{\kappa_1\Gamma+1} + 1 - \kappa_1 - 2m + \frac{(1-\Gamma)m(1-\kappa_2)\kappa_2}{\Gamma+\kappa_2}\right)\int\int\frac{\partial\phi_{Ri}^1}{\partial x}dzdz
\end{cases} \quad (17)
$$

Another recursive relation can be expressed as:

$$
\begin{cases}
\phi_{Ri+1}^1 = \phi_{\rho i}^1 \\
\Omega_{Ri+1}^1 = -2(1-2v_1)\int\frac{\partial\phi_{\rho i}^1}{\partial x}dz + \kappa_1\Omega_{\rho i}^1 - 2\left(x\frac{\partial\phi_{\rho i}^1}{\partial z} - (s_i-2ih)\frac{\partial\phi_{\rho i}^1}{\partial x} + h\frac{\partial\Omega_{\rho i}^1}{\partial z} + \frac{\partial\beta_{\rho i}^1}{\partial z}\right) \\
\beta_{Ri+1}^1 = \frac{\kappa_1^2-1}{2}\left(\int\int\frac{\partial\phi_{\rho i}^1}{\partial x}dzdz - \int\Omega_{\rho i}^1 dz\right) + \kappa_1\beta_{\rho i}^1 + (\kappa_1-1)\left[x\phi_{\rho i}^1 - (s_i-(2i+1)h)\int\frac{\partial\phi_{\rho i}^1}{\partial x}dz\right) + \\
\quad 2h\left(x\frac{\partial\phi_{\rho i}^1}{\partial z} - (s_i-2ih)\frac{\partial\phi_{\rho i}^1}{\partial x} + h\frac{\partial\Omega_{\rho i}^1}{\partial z} + \frac{\partial\beta_{\rho i}^1}{\partial z}\right)
\end{cases} \quad (18)
$$

where μ_2/μ_1 is noted as Γ and $(1-v_1)/(1-v_2)$ is noted as m.

The Midlin solutions for a point tangential load, acting on the surface of a semi-infinite space, can be written as:

$$
\begin{cases}
\phi_{R1}^1(x,y,z_1) = \frac{A_f}{R_1} \\
\Omega_{R1}^1(x,y,z_1) = -\frac{(1-2v_1)A_f x}{R_1(R_1-z_1)} \\
\beta_{R1}^1(x,y,z_1) = -\frac{(1-2v_1)A_f x}{R_1-z_1}\left(1-2v_1-\frac{h}{R_1}\right) \\
A_f = -\frac{p_x}{2\pi\mu_1}
\end{cases} \quad (19)
$$

where p_x is a tangential load. Now, B_1, B_2, B_3 and δ can be obtained by using Equation (17)–(19). Green's functions can be obtained by substituting B_1, B_2, B_3, and δ into Equation (16) and then separating variable p_x.

2.4. Green's Function for a Moving Point Heat Resource Acting on the Surface of a Coated Isotropic Thermoelastic Material

As shown in Figure 5, we can see that coating I, with thickness h, is used to cover substrate II, and is connected only by the interface. Above the coating, there is a free surface, and a moving point heat resource, marked H, is applied at point O_1 on said surface. A cylindrical coordinate (r,θ,z) is chosen with the r-axis being along the interface, and the z-axis being perpendicular to the r-axis and passing through O_1. The r-axis and z-axis intersect at O, namely, the global point of origin. Coordinate plane $rO\theta$ coincides with the interface. The image points in the coating space are marked as O_k and those in the substrate space are marked as C_k. The cyclic recursions and stress and displacement functions for point heat resource are given here.

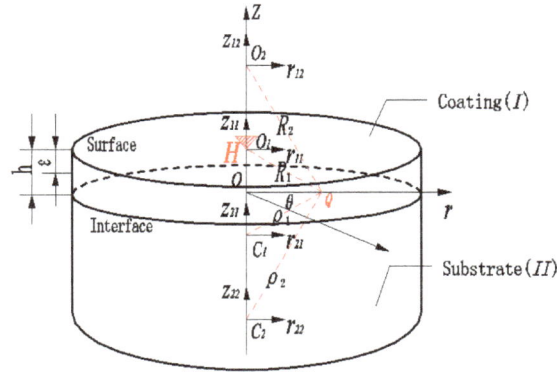

Figure 5. A moving point heat resource H acting on the surface of coated material.

The general solution for displacement and stress within a semi-infinite space for a thermal load can be written as:

$$\begin{cases} 2\mu u_r = \frac{\partial \psi_1}{\partial r} + z\frac{\partial \psi_2}{\partial r} \\ 2\mu u_z = \frac{\partial \psi_1}{\partial z} - (3-4v)\psi_2 + z\frac{\partial \psi_2}{\partial r} + 4(1-v)\psi_3 \\ \frac{2\mu t}{E_t} = \frac{\partial \psi_3}{\partial z} \\ \sigma_r = \frac{\partial^2 \psi_1}{\partial r^2} - 2v\frac{\partial \psi_2}{\partial z} + z\frac{\partial^2 \psi_2}{\partial r^2} - 2(1-v)\frac{\partial \psi_3}{\partial z} \\ \sigma_\theta = \frac{1}{r}\frac{\partial \psi_1}{\partial r} - 2v\frac{\partial \psi_2}{\partial z} + \frac{r}{z}\frac{\partial \psi_2}{\partial r} - 2(1-v)\frac{\partial \psi_3}{\partial z} \\ \sigma_z = \frac{\partial^2 \psi_1}{\partial z^2} - 2(1-v)\frac{\partial \psi_2}{\partial z} + z\frac{\partial^2 \psi_2}{\partial r^2} + 2(1-v)\frac{\partial \psi_3}{\partial z} \\ \tau_{rz} = \frac{\partial^2 \psi_1}{\partial z\partial r} - (1-2v)\frac{\partial \psi_2}{\partial r} + z\frac{\partial^2 \psi_2}{\partial r\partial z} + 2(1-v)\frac{\partial \psi_3}{\partial r} \end{cases} \quad (20)$$

where α is thermal expansion coefficient. λ is the Lame constant [26], its relationships with E and v are in the form of $\lambda = vE/(1+v)(1-2v)$. E_t is the thermal modulus, its relationships with E, v, and α are in the form of $E_t = 2(\lambda + 2\mu)/\alpha$.

Harmonic functions in the coating material can be written as:

$$\begin{cases} \psi_j^1 = \sum\limits_{i=1}^{\infty} (\psi_{ij}^1 + \psi_{ij}^2), (j = 1,2,3) \\ \psi_{ij}^1 = \sum\limits_{k=1}^{i} \psi_{ikj}^1, \psi_{ij}^2 = \sum\limits_{k=1}^{i} \psi_{ikj}^2, (j = 1,2,3) \\ \psi_{i11}^1 = D_{i111}^1(z_t \ln R_i^* + R_i) + D_{i112}^1 \ln R_i^* \\ \psi_{i1j}^1 = D_{i1j}^1 \ln R_i^*, (j = 2,3) \\ \psi_{ikj}^1 = D_{ijk1}^1 \frac{\partial^{2(k-2)}}{\partial z^{2(k-2)}}(\frac{1}{R_i}) + D_{ijk2}^1 \frac{\partial^{(2k-3)}}{\partial z^{(2k-3)}}(\frac{1}{R_i}), (k \geq 2, j = 1,2,3) \\ R_i^* = R_i - z_i \\ \psi_{i11}^2 = D_{i111}^2(s_t \ln \rho_i^* + \rho_i) + D_{i112}^2 \ln \rho_i^* \\ \psi_{i1j}^2 = D_{i1j1}^2 \ln \rho_i^* + D_{i1j2}^2 \frac{1}{\rho_i}, (j = 2,3) \\ \psi_{ik1}^2 = D_{ik11}^2 \frac{\partial^{2(k-2)}}{\partial z^{2(k-2)}}(\frac{1}{\rho_i}) + D_{ik12}^2 \frac{\partial^{(2k-3)}}{\partial z^{(2k-3)}}(\frac{1}{\rho_i}), (k \geq 2, j = 1) \\ \psi_{ikj}^2 = D_{ikj1}^2 \frac{\partial^{(2k-3)}}{\partial z^{(2k-3)}}(\frac{1}{\rho_i}) + D_{ikj2}^2 \frac{\partial^{2(k-1)}}{\partial z^{2(k-1)}}(\frac{1}{\rho_i}), (k \geq 2, j = 2,3) \\ \rho_i^* = \rho_i - s_i \end{cases} \quad (21)$$

Harmonic functions in the substrate material can be written as

$$\Psi_j^1 = \sum_{i=1}^{\infty} \Psi_{ij}^1 == \sum_{i=1}^{\infty} \sum_{k=1}^{i} \Psi_{ikj}^1, (j = 1, 2, 3)$$

$$\begin{cases} \Psi_{i11}^1 = L_{i111}^1 (z_t \ln R_i^* + R_i) + L_{i112}^1 \ln R_i^* \\ \Psi_{i1j}^1 = L_{i1j1}^1 \ln R_i^* + L_{i1j2}^1 \frac{1}{R_i}, (j = 2, 3) \\ \Psi_{ik1}^1 = L_{ik11}^1 \frac{\partial^{2(k-2)}}{\partial z^{2(k-2)}} \left(\frac{1}{R_i}\right) + L_{ik12}^1 \frac{\partial^{(2k-3)}}{\partial z^{(2k-3)}} \left(\frac{1}{R_i}\right), (k \geq 2, j = 1) \\ \Psi_{ikj}^1 = L_{ikj1}^1 \frac{\partial^{(2k-3)}}{\partial z^{(2k-3)}} \left(\frac{1}{R_i}\right) + L_{ikj2}^1 \frac{\partial^{2(k-1)}}{\partial z^{2(k-1)}} \left(\frac{1}{R_i}\right), (k \geq 2, j = 2, 3) \end{cases} \quad (22)$$

One of the boundary conditions can be established by satisfying the continuous conditions of displacement, stress, temperature, and thermal flux on the interface.

$$u_r^1 = u_r^2, u_z^1 = u_z^2, t^1 = t^2, \sigma_z^1 = \sigma_z^2, \tau_{rz}^1 = \tau_{rz}^2, \zeta_1 \frac{\partial t^1}{\partial z} = \zeta_2 \frac{\partial t^2}{\partial z} \quad (23)$$

Another boundary conditions can be established by satisfying the stress free and thermally insult conditions on the surface.

$$\sigma_z^1 = 0, \tau_{rz}^1 = 0, \zeta_1 \frac{\partial t^1}{\partial z} = 0 \quad (24)$$

Additionally, mechanical equilibrium and thermal equilibrium of an infinite plane, $\varepsilon < z < h$, should be satisfied.

$$\begin{cases} -2\pi\zeta_1 \int_0^{+\infty} \frac{\partial t^1}{\partial z} dz = \frac{qfPe_2\lambda_2}{a_h} \\ -2\pi\sigma_z \int_0^{+\infty} \sigma_z r dr = 0 \\ Pe_2 = \frac{Va_h}{\lambda_2} \end{cases} \quad (25)$$

where H is the heat flux. As in Grylitsky and Paul [27], we also introduce the Pe number into the above-mentioned equation for the thermoelastic problem. λ_2 is the thermal diffusion coefficient for the substrate material.

Substituting Equation (21) into Equation (20), using boundary condition (24) and equilibrium condition (25), one can obtain:

$$\begin{cases} -D_{1111}^1 + 2(1-v)D_{112}^1 - 2(1-v)D_{113}^1 = 0 \\ D_{1111}^1 - 2(1-v)D_{112}^1 + 2(1-v)D_{113}^1 = 0 \\ D_{1112}^1 + hD_{112}^1 = 0 \\ D_{113}^1 = -\frac{H\mu_1}{\pi\zeta_1 E_{t1}} \end{cases} \quad (26)$$

Thus, $D^1{}_{1111}$, $D^1{}_{112}$, $D^1{}_{1112}$, and $D^1{}_{113}$ can be obtained from Equation (26). Substituting Equation (21) into Equation (20) using boundary condition (23), one can obtain

$$\begin{cases} D_{1111}^1 - D_{1111}^2 = L_{1111}^1 \left(\frac{\mu_1}{\mu_2}\right) \\ D_{1112}^1 + D_{1112}^2 = L_{1112}^1 \left(\frac{\mu_1}{\mu_2}\right) \\ D_{1111}^1 - (3 - 4v_1)D_{112}^1 + 4(1-v_1)D_{113}^1 + D_{1111}^2 - (3 - 4v_1)D_{1121}^2 + 4(1-v_1)D_{1131}^2 = \\ [L_{1111}^1 - (3 - 4v_2)L_{1121}^1 + 4(1-v_2)D_{1131}^1]\left(\frac{\mu_1}{\mu_2}\right) \\ -D_{1112}^1 + D_{1112}^2 - (3 - 4v_1)D_{1122}^2 + 4(1-v_1)D_{1132}^2 = [-L_{1112}^1 - (3 - 4v_2)L_{1122}^1 + 4(1-v_2)D_{1132}^1]\left(\frac{\mu_1}{\mu_2}\right) \\ -D_{1111}^1 + 2(1-v_1)D_{112}^1 - 2(1-v_1)D_{113}^1 + D_{1111}^2 - 2(1-v_1)D_{1121}^2 + 2(1-v_1)D_{1131}^2 = \\ -L_{1111}^1 + 2(1-v_2)L_{1121}^1 - 2(1-v_2)L_{1131}^1 \\ -D_{1112}^1 - D_{1112}^2 + 2(1-v_1)D_{1122}^2 - 2(1-v_1)D_{1132}^2 = -L_{1112}^1 - 2(1-v_2)L_{1122}^1 + 2(1-v_2)L_{1132}^1 \\ D_{1111}^1 - (1 - 2v_1)D_{112}^1 + 2(1-v_1)D_{113}^1 + D_{1111}^2 - (1 - 2v_1)D_{1121}^2 + 2(1-v_1)D_{1131}^2 = \\ L_{1111}^1 - (1 - 2v_2)L_{1121}^1 + 2(1-v_2)L_{1131}^1 \\ D_{1112}^1 - D_{1112}^2 + (1 - 2v_1)D_{1122}^2 - 2(1-v_1)D_{1132}^2 = L_{1112}^1 + 2(1-v_2)L_{1122}^1 - 2(1-v_2)L_{1132}^1 \\ -D_{113}^1 + D_{1131}^2 = -L_{1131}^1 \left(\frac{\mu_1 E_{t2}}{\mu_2 E_{t1}}\right) \\ D_{1132}^2 = L_{1132}^1 \left(\frac{\mu_1 E_{t2}}{\mu_2 E_{t1}}\right) \\ -hD_{113}^1 - hD_{1131}^2 - D_{1132}^2 = (hL_{1131}^1 - L_{1132}^1)\left(\frac{\mu_1 E_{t2}\zeta_2}{\mu_2 E_{t1}\zeta_1}\right) \\ D_{1132}^2 = L_{1132}^1 \left(\frac{\mu_1 E_{t2}\zeta_2}{\mu_2 E_{t1}\zeta_1}\right) \end{cases} \quad . \quad (27)$$

Substituting D^1_{1111}, D^1_{112}, D^1_{1112}, and D^1_{113} into Equation (27), twelve coefficients D^2_{11jl} and L^1_{11jl} ($j = 1, 2, 3; l = 1, 2$) can be obtained. Using boundary conditions Equations (23) and (24), successively, equations such as Equations (26) and (27) can be established and each coefficient of the harmonic functions can be obtained.

Green's functions for the thermal load can be obtained by substituting ψ_1, ψ_2, ψ_3 and Ψ_1, Ψ_2, Ψ_3 into Equation (20) and separating variable H.

2.5. Solution Procedure

Figure 6 is a flowchart for solving thermoelastic contact problem of coated solids. The solving procedure can be conducted as follows:

- Initialize material parameters and structural parameters. Calculate Hertz contact radius a_h, Hertz contact pressure c_h, Hertz contact approach δ^0, and initial gap g^0.
- According to the material parameters and the coating thickness, construct a matrix of Green's functions for normal load G^p_{AA} using Equation (9) and Equations (13)–(15); construct matrix of Green's function for thermal load G^t_{AA} using Equations (20)–(27); construct matrix of Green's function for tangential load G^v_{AA} using Equations (16)–(19).
- Employ G^p_{AA} and the quadratic programming method, setting $g^{iter0} = g^0$, calculate Hertzian contact pressure distribution p^c using Equations (2)–(6). Set $p^{iter0} = p^c$.
- Obtain surface heat flux H^{iter}, according to Equation (1) and p^{iter}. Employing G^t_{AA} and H^{iter}, calculate thermally induced surface displacement u^t_{iter}. Update gap $g^{iter} = g^{iter} + u^t_{iter}$.
- Save p^{iter} as p_0, then update surface contact pressure p^{iter} by employing g^{iter}, G^p_{AA}, and the quadratic programming method.
- Calculate $\varepsilon^{iter} = \Delta_x\Delta_y \mid p^{iter} - p_0 \mid$ to judge convergence. If $\varepsilon^{iter} < \varepsilon_0$, go to step 7, otherwise go to step 4 for the next iteration.
- Obtain contact pressure p and heat flux H. Calculate σ_{AA} using Equation (7) by employing $G_{AA}{}^p$, $G_{AA}{}^v$, and $G_{AA}{}^t$.

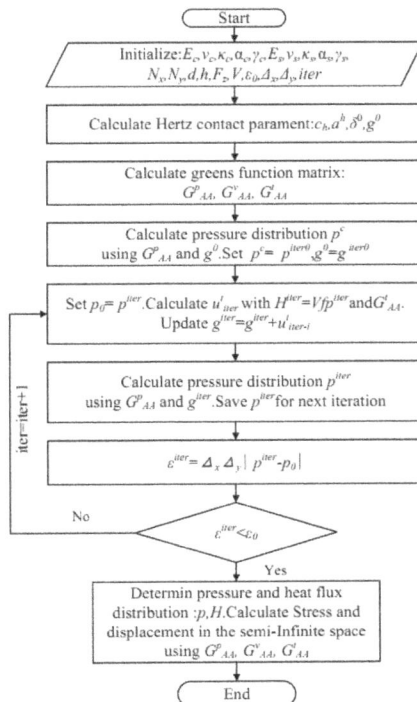

Figure 6. Flowchart for solving the thermoelastic contact problem of a coated solid.

3. Results and Discussion

The ratio of shear modulus between the coating and substrate is $\mu = \mu_c/\mu_s$, the ratio of thermal expansion coefficient between the coating and substrate is $\alpha = \alpha_c/\alpha_s$, and the ratio of the thermal conductive coefficient between the coating and substrate is $\zeta = \zeta_c/\zeta_s$. The following non-dimensional temperature rise, stress components, and coordinates are used:

$$Pe_2 = Va_h/\lambda_s, p_n = p/c_h, t_n = t\zeta_s/fc_hV, \sigma_{n-AA} = \sigma_{AA}/c_h, x_n = x/a_h, y_n = y/a_h, z_n = z/a_h \quad (28)$$

where Pe_2 is the Pecelet Number, λ_s is the thermal diffusion coefficient for the substrate material, and p_n is the non-dimensional contact pressure. t_n is the non-dimensional temperature rise. σ_{n-AA} is the non-dimensional stress components in an arbitrary direction. a_h and c_h are the Hertz contact radius and max contact pressure. Coating and substrate material's Poisson ratios are set to be 0.3. In this study, the substrate material is 52100 steel with $E_s = 210$ GPa, $\alpha_s = 1.17 \times 10^{-5}$ K^{-1}, $\zeta_s = 50.2$ W·m^{-1}·K^{-1}, and $\lambda_s = 1.0 \times 10^{-5}$ m^2·s^{-1}. Max Hertz contact pressure c_h is 1.09 GPa and the Hertz contact radius, a_h, is 100 μm. Sliding velocity, V, is 2.4 m·s^{-1}; thus, Pe_2 is 2.4.

3.1. Model Verification

In order to verify the validity of the present model, the numerical results, obtained by setting $\mu = 1$, $\alpha = 1$, and $\zeta = 1$, are compared with the exact analytical solution given by Johnson and Midlin [21] for the thermoelastic contact problem of solo material semi-space at $f = 0.2$ and $Pe_2 = 2.4$. The contour plot of the Von Mises stress and temperature on the xz plane are shown in Figure 7a,b, respectively, with numerical results being implied by the dotted line and the analytical results being implied by the solid line. All numerical simulations agree with those from the analytical solutions.

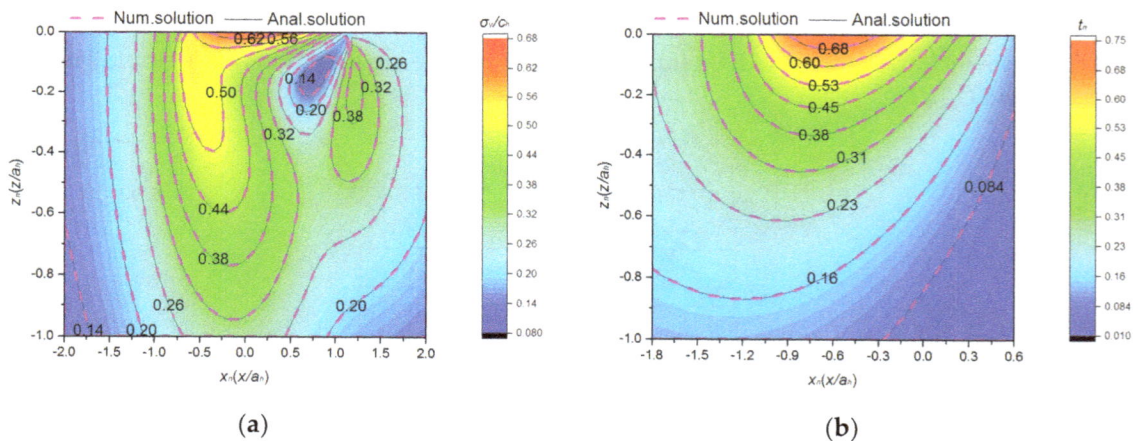

(a)   (b)

Figure 7. Comparison between the numerical solution and the analytical solution at $\mu = 1$, $\alpha = 1$, $\zeta = 1$ and $f = 0.2$; $Pe_2 = 2.4$. (**a**) Contour plot of σ_{n-v} obtained using the present model and analytical solution. (**b**) Contour plot of t_n obtained by the present model and analytical solution.

Note that, for verifying the validity of the present model further, the work by Wang [23] for the contact problem of a coated solids without thermal effect can be recovered by assuming $V = 0$ in the present model. Here, we choose the same material parameters as in Reference [9] and calculate the contact pressure distribution. Figure 8 shows the dimensionless contact stress at $h/a_h = 0.5$, $f = 0.2$, $V = 0$ with μ_c/μ_s varying from 2 to 4. It is seen that the present results agree very well with Liu's results.

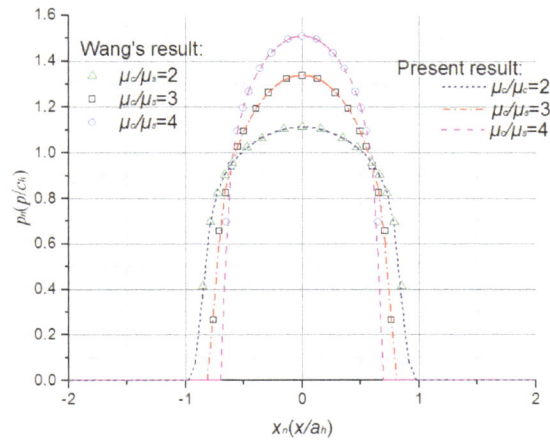

Figure 8. Comparison of contact pressure obtained from Wang's result and the present model at $h/a_h = 0.5$, $f = 0.2$, $V = 0$ with μ_c/μ_s varying from 2 to 4.

3.2. Effects of Friction Coefficient and Coating Thickness on Contact Pressure and Interface Stress Distribution

The effects of the friction coefficient on contact pressure and surface temperature are shown in Figure 9. With the increase of f, more frictional heats are generated within the contact zone, and the thermal expansion of the contact surface in the z-direction is enlarged. As shown in Figure 9b, this leads to a decrease in the contact area by 23% and an increase in maximum contact pressure by 307% when f increases from 0.1 to 0.5. A comment to be made regarding Figure 9a is that the magnitude of the surface temperature in proportion to f is offset by greater concentrations in the tail of the contact zone due to the moving of heat resources. Because of the higher thermal expansion in the tail of the contact zone, contact pressure is more concentrated in this area with the increase of f, as shown in Figure 9b.

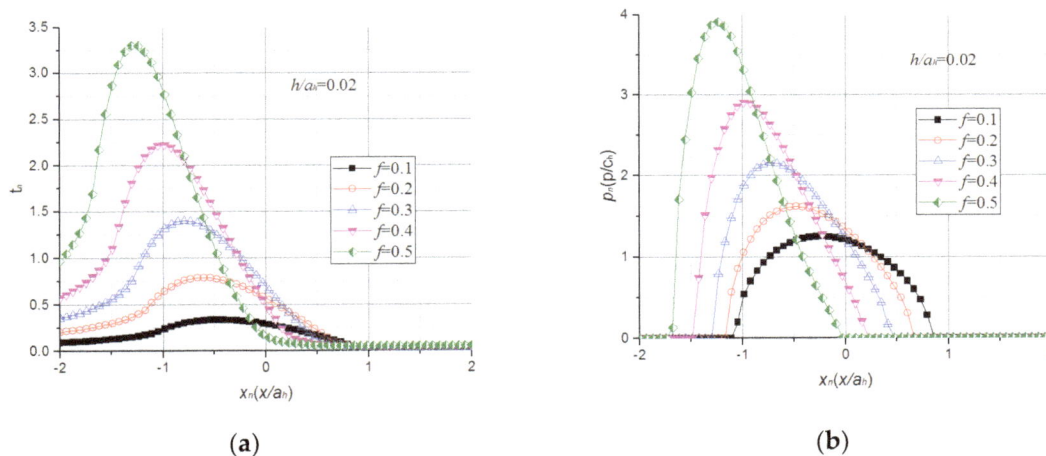

(a)	(b)

Figure 9. Effect of f on contact characteristics. (**a**) Effect of f on contact pressure distribution; (**b**) Effect of f on surface temperature distribution.

The effect of the friction coefficient on shear stress τ_{n-xz} along the intersecting line of plane xoz and the interface is shown in Figure 10, with variations of h/a_h at $\mu = 2$, $\alpha = 0.3$, $\zeta = 0.5$. With the increase of f, the maximum interfacial shear stress increases sharply for coatings with $h/a_h = 0.02$, as shown in Figure 10a, and moderately for coatings with $h/a_h = 0.25$ and $h/a_h = 0.5$, as shown in Figure 10c,d. For coatings with $h/a_h = 0.1$, the maximum interfacial shear stress decreases slightly, and then increases, as shown in Figure 10b. The maximum interfacial shear stress of coated solids with thin coatings (e.g., $h/a_h = 0.02$) is more sensitive to the change of friction coefficient than those that have thick coatings. Traditionally, employing thin coatings is preferable, as lower interfacial shear stress can be obtained as soon as $f < 0.3$. However, due to being affected by frictional heat, thin coatings (e.g., $h/a_h = 0.02$) are only favored when f is below 0.1.

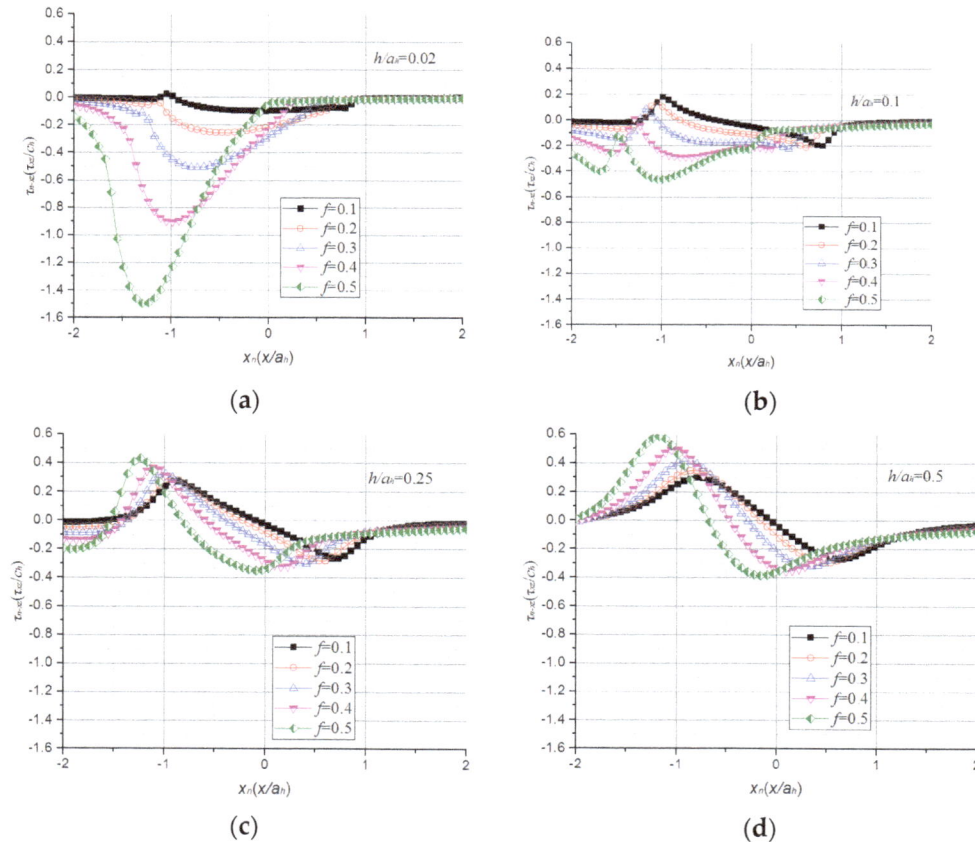

Figure 10. τ_{n-xz} along the intersecting line of plane xoz and interface with various f and h/a_h when $\mu = 2$, $\alpha = 0.3$, $\zeta = 0.5$. (**a**) With various f at $h/a_h = 0.02$. (**b**) With various f at $h/a_h = 0.1$. (**c**) With various f at $h/a_h = 0.25$. (**d**) With various f at $h/a_h = 0.5$.

The effect of the friction coefficient on transverse stress σ_{n-xx} along the intersecting line of plane xoz and the coating bottom can be inferred from Figure 11, which shows various h/a_h at $\mu = 2$, $\alpha = 0.3$, $\zeta = 0.5$. As shown in Figure 11c,d, coatings with a thickness of $h/a_h > 0.25$ may render interfacial transverse stress σ_{n-xx} compressive within the contact zone for various f, counteracting the brittle failure of the coating in general. Meanwhile, thin coatings with a thickness of $h/a_h = 0.02$ may encounter brittle failure at the end of the tail of the contact zone because σ_{n-xx} is always tensile in that zone, in all cases of f. For coatings with $h/a_h = 0.1$, brittle failure at the end of the tail of the contact zone may happen in the case where $f > 0.2$.

By compromising between interfacial shear stress and tensile transverse stress, coatings with thicknesses of about $h/a_h = 0.02$ are preferred in cases where $f < 0.1$; coatings with a thickness of about $h/a_h = 0.1$ are preferred in the case of $0.1 < f < 0.3$, and coatings with a thickness of about $h/a_h = 0.25$ are proposed in cases where $f > 0.3$.

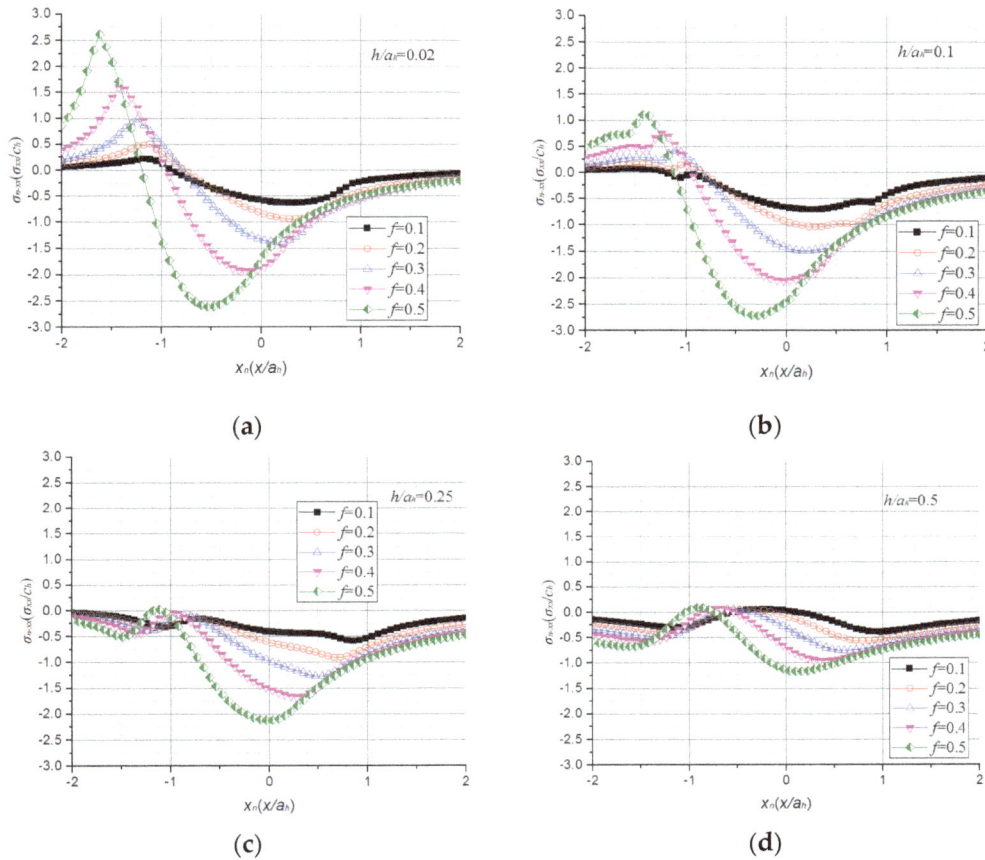

Figure 11. σ_{n-xx} along the intersecting line of plane xoz and interface with various f and h/a_h when $\mu = 2$, $\alpha = 0.3$, $\zeta = 0.5$. (**a**) with various f at $h/a_h = 0.02$; (**b**) with various f at $h/a_h = 0.1$; (**c**) with various f at $h/a_h = 0.25$; (**d**) with various f at $h/a_h = 0.5$.

3.3. Effect of a Coating's Elastic Modulus and Thermal Expansion Coefficient on Interface Stress Distribution

The effects of a coatings' elastic modulus ratio, μ, on shear stress τ_{n-xz} along the intersecting line of plane xoz and the interface are shown in Figure 12, with variations of α at $h/a_h = 0.02$, $\zeta = 0.5$, $f = 0.1$. As can be inferred from Figure 12c,d, the increase in stiffness of the coating results in a higher interfacial shear stress when α is greater than 1. However, when α is less than 1, the interfacial shear stress decreases dramatically with the increase in the stiffness of the coatings, as can be inferred from Figure 12a,b. With the increase of μ, thermally induced interfacial shear stress τ_{t-xz} always increases, as shown in Figure 13. However, the direction of the thermally induced interfacial shear stress component is opposite to that caused by tangential traction when $\alpha < 1$, as is shown in Figure 13a. Thus, the increase in the thermally induced interfacial shear stress results in the decrease of total interfacial shear stress. While $\alpha > 1$, the direction of thermally induced interfacial shear stress is the same as that caused by tangential traction, as shown in Figure 13b. Thus, the increase in the interfacial shear stress, which is induced by fractional heat, results in the increase of total interfacial shear stress under the conditions of $\alpha > 1$.

Another revealed feature is the effects of a coatings' elastic modulus and thermal expansion coefficient on the tensile transverse stress along the intersecting line of plane xoz and the interface, as shown in Figure 14, with variations of α and μ at $h/a_h = 0.02$, $\zeta = 0.5$, and $f = 0.1$. Figure 14b–d indicate that coatings with a thermal expansion coefficient of $\alpha > 0.6$ may render the interfacial transverse stress σ_{n-xx} compressive within the contact area on the interface for variations of μ, counteracting, in general, the brittle failure of the coating.

Thus, in order to obtain a lower interfacial shear stress and compressive transverse stress at the same time, $0.6 < \alpha_c/\alpha_s < 1$ can be recommended as the goal in order to optimize the thermoelastic properties of thin hard coatings.

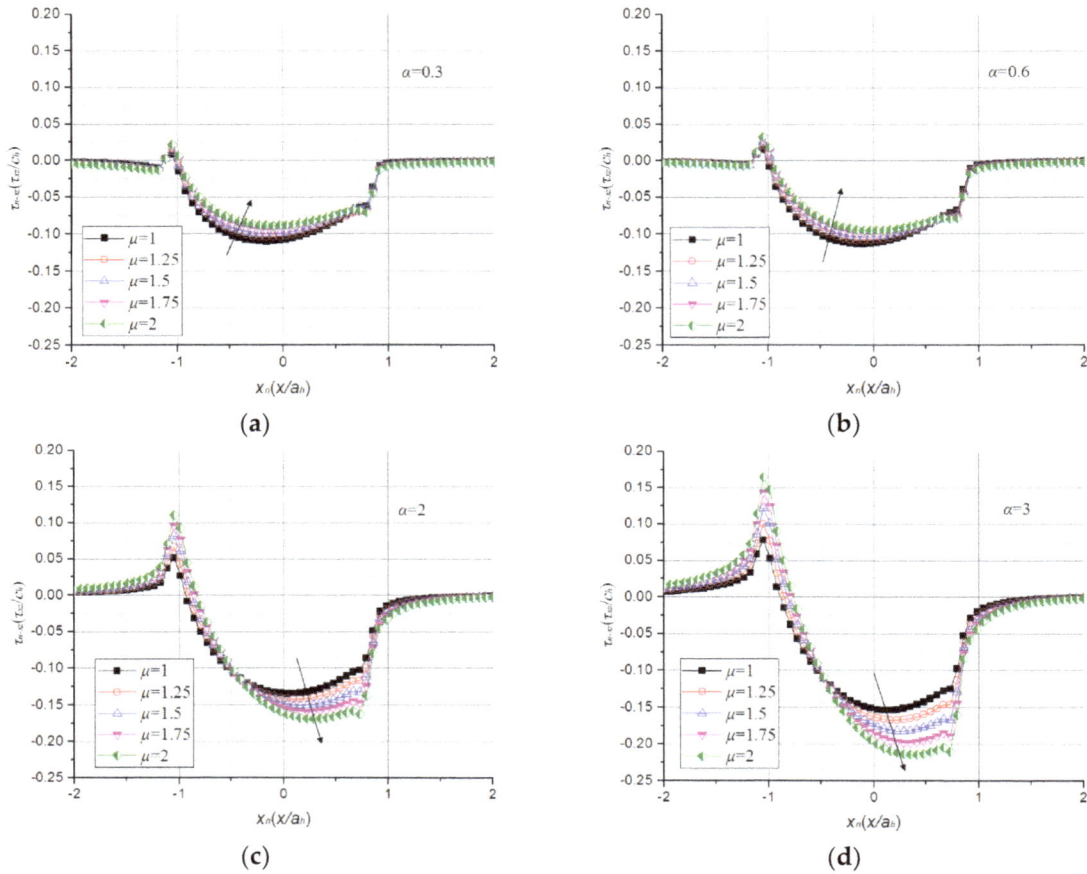

Figure 12. τ_{n-xz} along the intersecting line of plane xoz and interface with various α and μ when $h/a_h = 0.02, \zeta = 0.5, f = 0.1$. (**a**) With various μ at $\alpha = 0.3$; (**b**) with various μ at $\alpha = 0.6$; (**c**) with various μ at $\alpha = 2$; (**d**) with various μ at $\alpha = 3$.

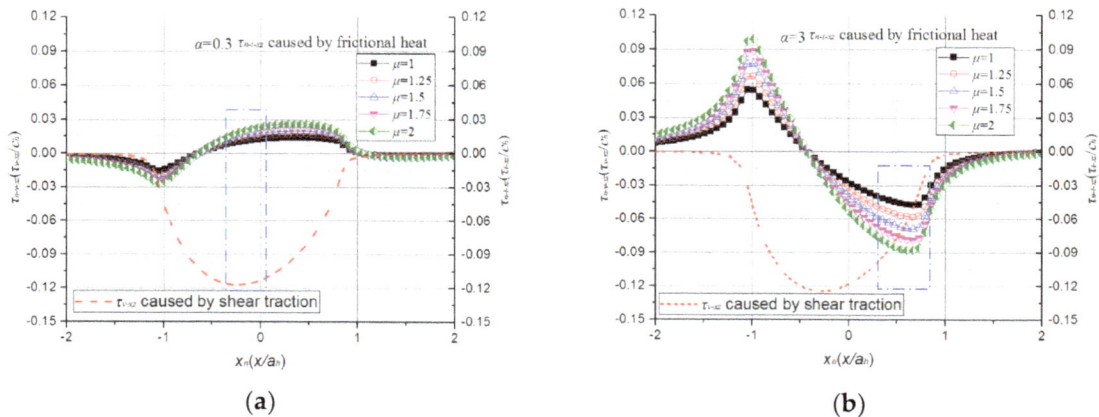

Figure 13. Distributions of thermally induced and tangential traction caused interfacial shear stresses for different α. (**a**) Direction of thermally induced interfacial shear stress is opposite to that caused by tangential traction when $\alpha < 1$; (**b**) Direction of thermally induced interfacial shear stress is same as that caused by tangential traction when $\alpha > 1$.

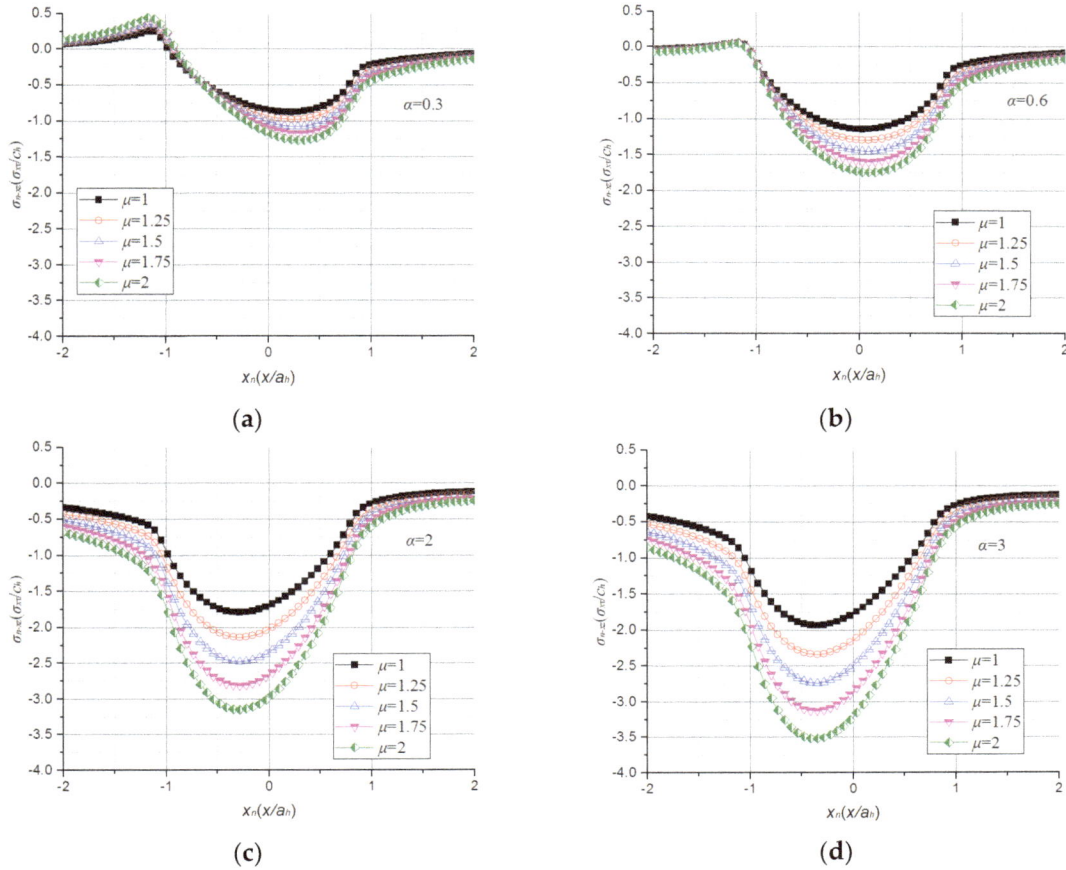

Figure 14. σ_{n-xx} along the intersecting line of plane xoz and interface with various α and μ when $h/a_h = 0.02$, $\zeta = 0.5$, $f = 0.1$. (**a**) With various μ at $\alpha = 0.3$; (**b**) with various μ at $\alpha = 0.6$; (**c**) with various μ at $\alpha = 2$; (**d**) with various μ at $\alpha = 3$.

3.4. Effect of a Coating's Thermal Conductiveness and Thermal Expansion Properties on Interface Stress Distribution

The effects of a coatings' thermal conductive ratio, ζ, on shear stress τ_{n-xz} along the intersecting line of plane xoz and the interface are shown in Figure 15, with variations of α at $h/a_h = 0.02$, $\mu = 2$, and $f = 0.1$. The results shown in Figure 15 confirm that the increase in the thermal conductive coefficient of a coating is always beneficial in reducing the interfacial shear stress. As mentioned above, contact pressure distribution is co-determined by contact load and surface temperature rise in thermoelastic contact conditions. Figure 16 illustrates that the increase in ζ results in decreases in contact pressure and interfacial shear stress. With respect to Figure 16a, the magnitude of the contact pressure grows in an inverse proportion to ζ, and the magnitude of the contact area is proportional to ζ. The fact that coatings with a higher thermal conductivity can facilitate heat transfer within a coated system is understood to be accountable for contact pressure relaxation. With the dispersing of surface shear traction, interfacial shear stress decreases substantially, as shown in Figure 16b.

The effects of a coatings' thermal conductive coefficient and thermal expansion coefficient on the tensile transverse stress along the intersecting line of plane xoz and the interface are provided in Figure 17, with variations of α and ζ at $h/a_h = 0.02$, $\mu = 2$, and $f = 0.1$. Figure 17a–d indicate that the dependence of interfacial tensile transverse stress on the variations of ζ is negligible.

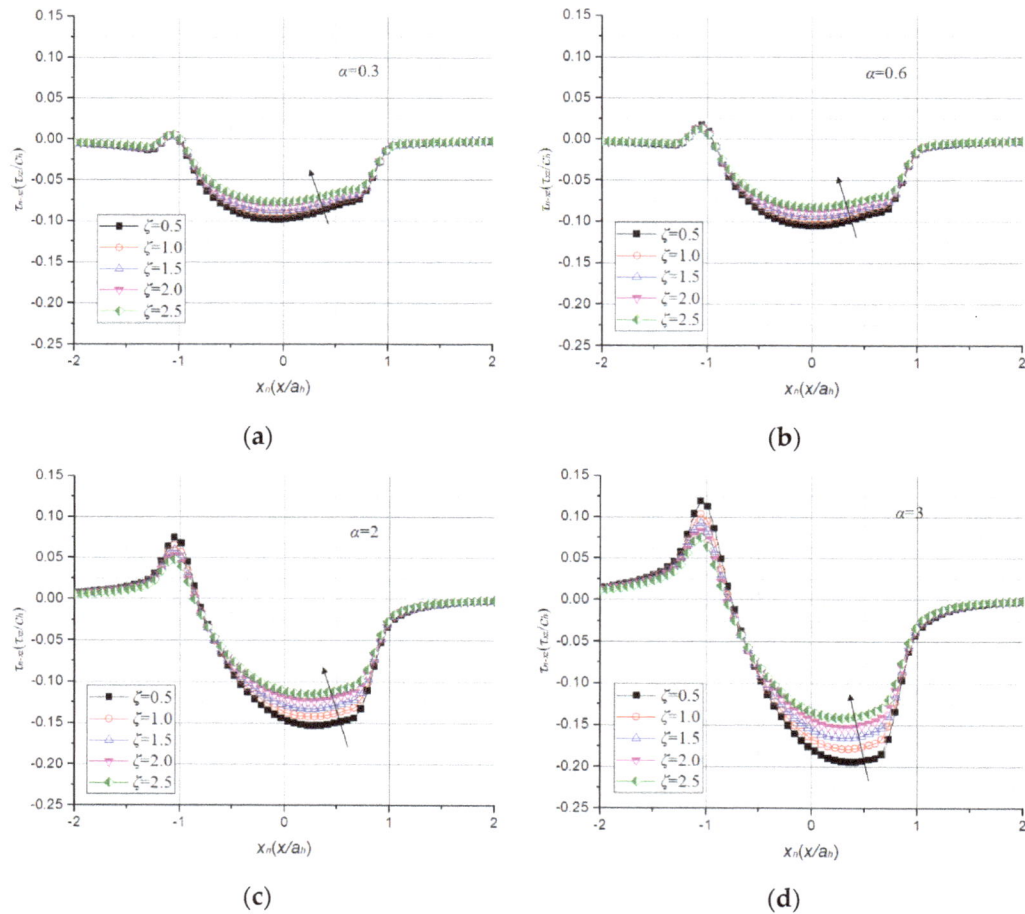

Figure 15. τ_{n-xz} along the intersecting line of plane xoz and interface with various α and ζ when $h/a_\mathrm{h} = 0.02$, $\mu = 2$, and $f = 0.1$. (**a**) with various ζ at $\alpha = 0.3$; (**b**) with various ζ at $\alpha = 0.6$; (**c**) with various ζ at $\alpha = 2$; (**d**) with various ζ at $\alpha = 3$.

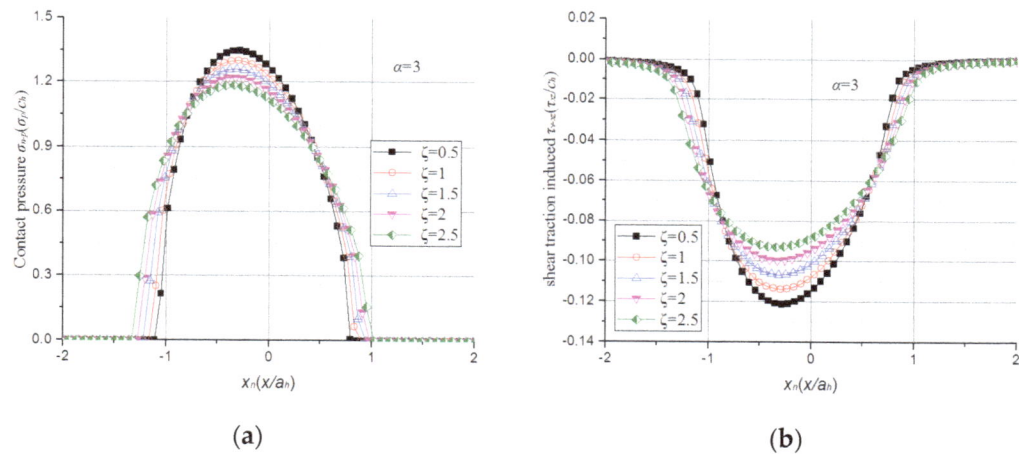

Figure 16. Effect of ζ on the contact pressure and interfacial shear stress component, caused by friction force. (**a**) Increase in ζ results in decreases in contact pressure; (**b**) Increase in ζ results in decreases in interfacial shear stress component, caused by friction force.

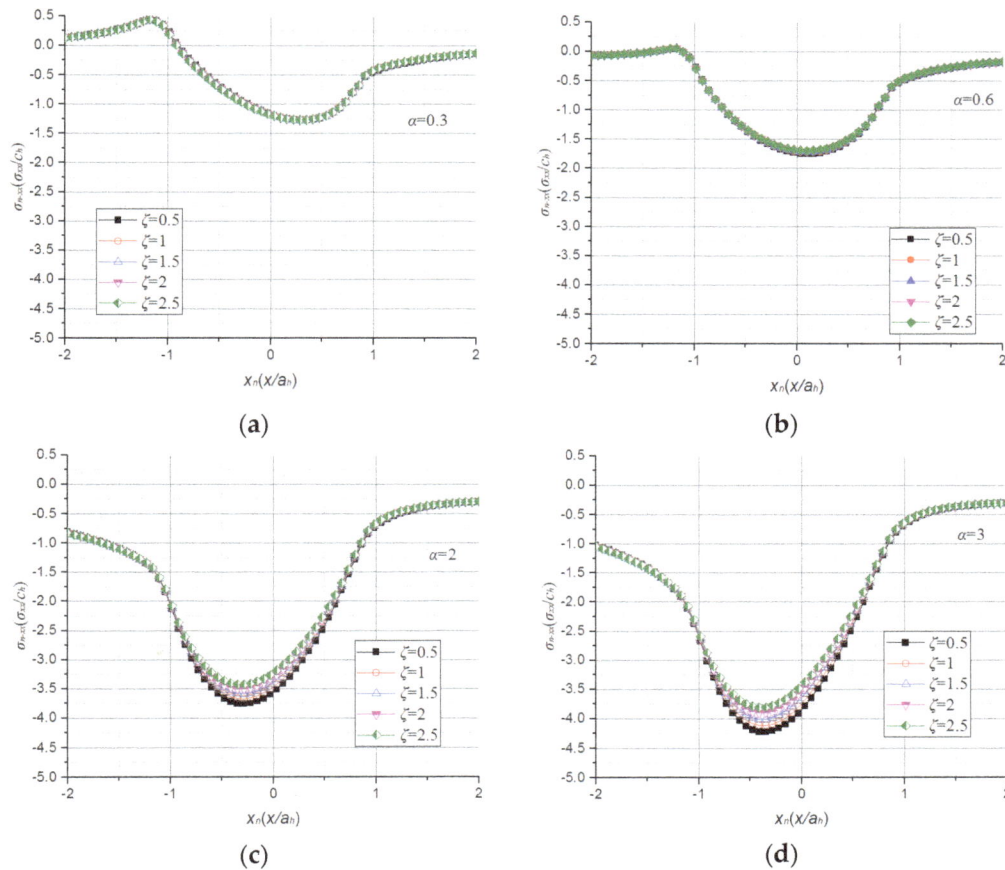

Figure 17. σ_{n-xx} along the intersecting line of plane xoz and interface with various α and ζ when $h/a_h = 0.02$, $\mu = 2$, and $f = 0.1$. (a) With various ζ at $\alpha = 0.3$; (b) with various ζ at $\alpha = 0.6$; (c) with various ζ at $\alpha = 2$; (d) with various ζ at $\alpha = 3$.

Thus, in order to obtain a lower interfacial shear stress and compressive transverse stress at the same time, a higher value of ζ_c / ζ_s can be recommended in order to improve the resistance to interfacial delamination of the coating material, when subjected to sliding contact with friction heat generation.

The quasi-static thermoelastic contact problem between a ball and a coated solid is discussed in the present study. The contact mode between the two bodies is assumed to be conformal, with coating and substrate all being isotropic and without defects on the interface. In considering those assumptions, the present model can be applied to the brittle coating–ductile substrate system, with the coating's elastic modulus being greater than the elastic modulus of the substrate material. The diameter of the ball should be 20 times greater than the width of the contact area, and the velocity of the ball should be less than 100 m·s^{-1}. The work present herein offers room for further viable and promising extensions. For instance, the assumptions made above could be relaxed to cover a broader range of applications. Moreover, in the sense that the frictional contact on the surface of a brittle coating may eventually give rise to damage patterns in the form of interface cracking, the feasible coupled crack/contact analysis in the brittle coating–ductile substrate system incorporating the contact-induced frictional heating effect would pose another interesting research topic of technological significance. Such issues need to be addressed and will be reported in forthcoming papers. The analytical work in this paper can help in the design of the coating–substrate system with a proper coating thickness to obtain a relative low interface stress. In addition, the analytical work in this paper can propose targets on the modification of the coating material in elastic and thermoelastic properties. In the sense that the thermoelastic properties of the coating material have effects on the loading bearing capacity of the coating–substrate system. Tribology tests, such as the ball-on-disk test, with relative velocity being similar to real working conditions, could be applied to study the load bearing capacity of the coating–substrate system due to its reflecting the real thermal load condition and contact condition well.

4. Conclusions

This paper investigated the interfacial mechanical characteristics of a hard coating/substrate system, which was involved in a thermoelastic contact. The image point method and interface mechanics were employed in order to obtain Green's functions for a normal load, tangential load, and thermal load. Based on Green's functions and the contact mechanics, the temperature field and stress field were obtained with a high degree of accuracy. The effects of coating thickness, friction coefficient, and thermoelastic properties on the interfacial stress characteristics were discussed. It was found that:

- Adequate design of coating thickness for different friction coefficients can be help in obtaining relatively low interfacial shear stress and tensile transverse stress at the same time. For brittle coatings ($E_c > E_s$), in the case where the friction coefficient is less than 0.1, a smaller coating thickness, with $h/a_h < 0.02$, is recommended. With a friction coefficient between 0.1 and 0.3, a coating with a thickness h/a_h of about 0.1 is proposed. When the friction coefficient is greater than 0.3, coating with a thickness h/a_h above 0.25 is preferable.

- Interfacial shear stress tends to be relieved by frictional heating in the case of $\alpha_c/\alpha_s < 1$ for brittle coatings. Meanwhile, tensile transverse stress on the interface tends to be eliminated in the case of $\alpha_c/\alpha_s > 0.6$. A thermal expansion coefficient ratio of $0.6 < \alpha_c/\alpha_s < 1$ can be recommended as a goal for the optimization of thin hard coatings-substrate systems.

- Coating materials with enhanced thermal conductivity are shown to lower surface contact pressure concentration, and, thus, decrease interfacial shear stress. Although the interfacial tensile transverse stress is affected, to a negligible extent, by the variations of the thermal conductive coefficient of coating materials, it is predictable that the increased thermal conductivity of a coating could improve the resistance to interfacial delamination of the coating material.

Finally, we believe that this paper provide an understanding of the mechanism of modifying interfacial failure for hard coatings by employing adequate analysis.

Acknowledgments: This work was supported by the National Natural Science Foundation of China (Grant No. 51275125) and the National Basic Research Program of China (Grant No. 2013CB632305).

Author Contributions: Chongyang Nie contributed the analysis software. Le Gu, Liqin Wang, and Chuanwei Zhang contributed to the helpful discussion of interfacial mechanics. Chi Zhang analyzed the calculation results and wrote the paper.

Conflicts of Interest: The authors declare no conflicts of interest.

References

1. Ebert, F.J. Fundamentals of design and technology of rolling element bearings. *Chin. J. Aeronaut.* **2010**, *23*, 123–136. [CrossRef]
2. Tong, Y.X.; Wang, L.Q.; Gu, L. Friction and wear behaviors of Si_3N_4 sliding against M50 bearing steel in vacuum. *Adv. Mater. Res.* **2010**, *97*, 1681–1684. [CrossRef]
3. Wang, L.Q.; Li, C.; Zheng, D.Z.; Gu, L. Nonlinear Dynamics Behaviors of a Rotor Roller Bearing System with Radial Clearances and Waviness Considered. *Chin. J. Aeronaut.* **2008**, *21*, 86–96.
4. Daisuke, Y.; Richard, J.C.; Peter, A.D. Wear mechanisms of steel roller bearings protected by thin, hard and low friction coatings. *Wear* **2005**, *259*, 779–788.
5. Bobzin, K.; Lugscheider, E.; Maes, M.; Gold, P.W.; Loos, J.; Kuhn, M. High-performance chromium aluminium nitride PVD coatings on roller bearings. *Surf. Coat. Technol.* **2004**, *188*, 649–654. [CrossRef]
6. Nie, C.Y.; Zheng, D.Z.; Gu, L.; Zhao, X.L.; Wang, L.Q. Comparison of interface mechanics characteristics of DLC coating deposited on bearing steel and ceramics. *Appl. Surf. Sci.* **2014**, *317*, 188–197. [CrossRef]
7. O'Sullivan, T.C.; King, R.B. Sliding contact stress-field due to a spherical indenter on a layered elastic half-space. *ASME. J. Tribol.* **1988**, *110*, 235–240. [CrossRef]

8. Wang, Z.J.; Wang, W.Z.; Wang, H.; Dong, Z.; Hu, Y.Z. Partial slip contact analysis on three-dimensional elastic layered half Space. *ASME. J. Tribol.* **2010**, *132*, 280–290. [CrossRef]

9. Liu, S.B.; Wang, Q.; Liu, G. A versatile method of discrete convolution and FFT (DC-FFT) for contact analyses. *Wear* **2000**, *243*, 101–111. [CrossRef]

10. Cai, S.; Bhushan, B. A numerical three-dimensional contact model for rough, multilayered elastic/plastic solid surfaces. *Wear* **2005**, *259*, 1408–1423. [CrossRef]

11. Kot, M. Contact mechanics of coating–substrate systems: Monolayer and multilayer coatings. *Arch. Civ. Mech. Eng.* **2012**, *12*, 464–470. [CrossRef]

12. Komvopoulos, K. Finite element analysis of a layered elastic solid in normal contact with a rigid surface. *ASME. J. Tribol.* **1988**, *110*, 477–485. [CrossRef]

13. Kral, E.R.; Komvopoulos, K.; Bogy, D.B. Finite Element Analysis of Repeated Indentation of an Elastic-Plastic Layered Medium by a Rigid Sphere, Part I: Surface Results. *ASME. J. Appl. Mech.* **1995**, *62*, 20–28. [CrossRef]

14. Ye, N.; Komvopoulos, K. Three-dimensional finite element analysis of elastic–plastic layered media under thermomechanical surface loading. *ASME. J. Tribol.* **2003**, *125*, 52–59. [CrossRef]

15. Yang, J.; Komvopoulos, K. Dynamic indentation of an elastic–plastic multilayered medium by a rigid cylinder. *ASME. J. Tribol.* **2004**, *126*, 18–27. [CrossRef]

16. Liu, J.; Ke, L.L.; Wang, Y.S. Two-dimensional thermoelastic contact problem of functionally graded materials involving frictional heating. *Int. J. Solids Struct.* **2011**, *48*, 2536–2548. [CrossRef]

17. Choi, H.J.; Paulino, G.H. Thermoelastic contact mechanics for a flat punch sliding over a graded coating/substrate system with frictional heat generation. *J. Mech. Phys.* **2008**, *56*, 1673–1692. [CrossRef]

18. Neha, S.; Kumar, N.; Dash, S.; Das, C.R.; Subba, R.R.V.; Tyagi, A.K. Scratch resistance and tribological properties of DLC coatings under dry and lubrication conditions. *Tribol. Int.* **2012**, *56*, 129–140.

19. Singh, R.K.; Tilbrook, M.T.; Xie, Z.H. Contact damage evolution in diamond-like carbon coatings on ductile substrates. *J. Mater. Res.* **2008**, *23*, 27–36. [CrossRef]

20. Moorthy, V.; Shaw, B.A. An observation on the initiation of micro-pitting damage in as-ground and coated gears during contact fatigue. *Wear* **2013**, *297*, 878–884. [CrossRef]

21. Shahsavari, S.; Desouza, A.; Bahrami, M.; Erik, K.E. Thermal analysis of air-cooled PEM fuel cells. *Int. J. Hydrogen Energy* **2012**, *37*, 18261–18271. [CrossRef]

22. Johnson, K.L. *Contact Mechanics*; Cambridge University Press: Cambridge, UK, 1985; pp. 171–193.

23. Wang, T.J.; Wang, L.Q.; Gu, L.; Zheng, D.Z. Stress analysis of elastic coated solids in point contact. *Tribol. Int.* **2015**, *86*, 52–61. [CrossRef]

24. Xu, J.Q.; Mutoh, Y. Analytical solution for interface stresses due to concentrated surface force. *Int. J. Mech. Sci.* **2003**, *45*, 1877–1892. [CrossRef]

25. Hou, P.F.; Jiang, H.Y.; Tong, J.; Xiong, S.M. Study on the coated isotropic thermoelastic material based on the three-dimensional Green's function for a point heat source. *Int. J. Mech. Sci.* **2014**, *83*, 155–162. [CrossRef]

26. Xu, J.Q. *Interface Mechanics*; Science Press: Beijing, China, 2006; pp. 70–80.

27. Grylitsky, D.V.; Pauk, V.J. Some quasistationary contact problem for half-space involving heat generation and radiation. *Int. J. Eng. Sci.* **1995**, *33*, 1773–1781. [CrossRef]

7

Field Evaluation of Red-Coloured Hot Mix Asphalt Pavements for Bus Rapid Transit Lanes in Ontario, Canada

Qingfan Liu [1,*], Sina Varamini [2] and Susan Tighe [1]

[1] Centre for Pavement and Transportation Technology, University of Waterloo, Waterloo, ON N2L 3G1, Canada; sltighe@uwaterloo.ca

[2] McAsphalt Industries Limited, Toronto, ON M1B 5R4, Canada; svaramini@mcasphalt.com

* Correspondence: qingfan.liu@uwaterloo.ca

Academic Editors: Andrea Simone and Claudio Lantieri

Abstract: Coloured pavements have been implemented by metropolitan areas to denote dedicated lanes for bus rapid transit to maintain a high level of safety. Transit benefits of these installations are well documented. However, field performance of various types of coloured pavement has not been investigated systematically, with questions not being answered. In collaboration with the Regional Municipality of York (ON, Canada) where red pavement sections have been in operation for years for its bus rapid transit lanes, the Centre for Pavement and Transportation Technology at the University of Waterloo (Waterloo, ON, Canada) assessed the performance of various types of red pavements including epoxy paint and red asphalt mixes. It was found that, with significant lower texture depth, epoxy paint surface has disadvantages to red asphalt pavement from a pavement texture and safety perspective. The red asphalt sections in this study were observed as lower yet compatible frictional levels to conventional black pavement. Various types of contamination onto the red pavement were observed during field survey. In addition, the ultraviolet radiation degraded the colour of red asphalt pavement over time and may make it less effective for lane designation. Long-term monitoring is recommended to evaluate the functional and structural performance of red asphalt pavement.

Keywords: red asphalt pavement; bus rapid transit; pavement texture; friction; field evaluation

1. Introduction

Many metropolitan areas around the world have implemented coloured pavements in their infrastructure to denote dedicated lanes for bus rapid transit (BRT) [1]. This concept moves away from car dependency around active modes of public transit and pedestrian facilities. However, developing a BRT system that is easily understood by right-of-way users is a necessity in order to maintain a high level of safety. This is traditionally accomplished through signage and lane markings, but the most effective solution is to have a different surface colour for designated lanes [1].

Lane colouring can be achieved by painting, applying a coloured thermoplastic, or laying a thin wearing course of coloured asphalt mixture. One of the major concerns is their durability under significant volumes of vehicle traffic and winter maintenance operations [2]. In collaboration with the University of North Carolina Highway Safety Research Center (Chapel Hill, NC, USA), the City of Portland investigated colour options for bike lane identification in the 1990s, and their analysis provides a broad analysis of material durability. It was reported that the most durable solution would be a dyed asphalt wearing course; however, this was not tested due to the high cost of implementation. Portland installed test sections of painted and thermoplastic colours and found that while the painted material wore away after the first winter, the thermoplastic proved to still be in good condition after one year [2].

A solution to the durability issue is to colour the entire surface by using a coloured asphalt mixture. This can be accomplished through a number of methods, depending on the desired colour of the pavement, including using coloured aggregates, adding pigments to conventional binders, adding pigments and using a clear synthetic binder, or a combination of the above methods. The most vibrant colours can be obtained by using a clear synthetic binder. However, these technologies have not been studied for their long-term performance characteristics, nor been considered for the long-life pavement designs. Additionally, the cost of using a clear binder is estimated to be five to eight times more than a conventional binder; depending on the asphalt grade.

Carry et al. [3] found that epoxy street paints produced the most durable solution, while asphaltic-based mixtures including Hot Mix Asphalt (HMA) and slurry surface treatment were promising and required further investigation. Moreover, the study concluded that, regardless of age and condition of the asphalt road surface, treatments experience intense wear at bus stops. This wearing was suggested to be due to factors including friction caused by buses' stopping and starting, and prolonged heat exposure from bus engines.

Lee and Kim [4] investigated HMA overlays incorporating coloured synthetic binders for use in bus lanes in Seoul. They performed Marshall stability test, indirect tensile strength, and modified Lottman moisture sensitivity tests to evaluate the strength and moisture susceptibility of the overlays. The results showed that the designed overlay was of higher strength and lower moisture susceptibility than conventional asphalt mixtures. Thermal properties have also been extensively studied as one of the major benefits of coloured asphalt is reduced heat absorption, reducing the impact that the pavements have on urban heat island (UHI) generation [5]. Coloured asphalt reduces the amount of thermal energy absorbed compared to black asphalt. This can reduce the urban heat island effect which has major benefits for cities. Other parameters that affect the thermal and ultraviolet (UV) absorption of asphalt are permeability, thermal conductivity, convection and heat capacity. A challenge with coloured asphalt might be an increased solar reflectance, causing glare problems for drivers. Since light-coloured layers reflect some light that is in the visible part of the solar spectrum, some of the reflected light could be reflected at lower angles and travel into the sight of drivers and pedestrians. Glare is not often reported as a major cause of accidents, but might play a significant role in the case of rain events, early morning and late afternoon.

Friction is of high importance for safe vehicle operation. Adding paints and polymers to the roadway surface may increase the risk of making a driving surface too smooth [3]. Researchers [6,7] have reported positive relationship between pavement friction and various texture indices.

Transit benefits of coloured pavement installations are well documented in terms of vehicle violation of the lanes. However, structural and functional performance has not been investigated systematically and data are scarcer [8]. This paper presents a field evaluation conducted by the Centre for Pavement and Transportation Technology (CPATT). The field evaluation includes amplitude of pavement surface texture, frictional, functional, and environmental characteristics of the coloured asphalt pavements for BRT lanes in York Region, Canada.

2. Test Site and Mixtures Design

Located in north of Toronto in the Province of Ontario in Canada, the regional Municipality of York has adopted a combination of coloured aggregate and red pigment for its BRT lanes as shown in Figure 1. The coloured asphalt pavement was to improve the level of safety through enhanced visibility.

Figure 1. A section of coloured bus rapid transit (BRT) lane at York Region, ON, Canada.

The test site consists of three sections as shown in Table 1. Initial mixture of red HMA consisted of a pink granite aggregate blend, red proprietary pigment and polymer-modified Performance Graded (PG) 70-28 asphalt binder. The aggregate blend consisted of 12.5 mm coarse aggregate, and crusher fines (washed, and unwashed) to meet the physical requirements of Superpave (Superior Performing Asphalt Pavements) 12.5 FC2 mixture type for use in Traffic Category "D" as per Ontario Provincial Standard Specification [9]. This type of mixture is intended to provide superior rutting resistance and skid resistance for a 20-year Equivalent Single Axle Load (ESAL) level of 10–30 million.

Table 1. Information of tested sections.

Section (As-Built Material)	Age at Test Date (Year)	Field Assessment
Epoxy paint	3	(1) surface texture
Red HMA -initial mix	3	(2) frictional property
Red HMA -new mix	1	(3) distress survey

HMA = Hot Mix Asphalt.

In brief, Superpave is a standard procedure of designing asphalt mixtures in the Canadian pavement industry. Initiated by the United States Department of Transportation Federal Highway Administration (FHWA) (Washington, DC, USA) during the late 1980s, Superpave was adopted in Canada as an improved mixture design procedure over the Hveem and Marshall methods. The Superpave is a system of mixture design for asphalt mixtures based upon mechanistic concepts, which includes: (1) an asphalt-grading system called Performance Grading (PG) with the intention of matching the physical binder properties to the desired level of resistance to rutting, fatigue and low-temperature cracking, subjected to local climate and environmental conditions, and (2) an approach to help design the aggregate structure based on volumetric analysis and requirements.

It should be noted that the pink aggregate used in the initial mixture is the same source as the aggregate Type B [9] with aggregate mineralogy and physical properties. After three years in service, sections of BRT lanes paved with initial red mixture exhibited cracks that raised concern about the integrity and performance of the initial red mixture. To address these concerns, a new red HMA mix was introduced by modifying the initial red mixture. The physical properties of the mixtures are shown in the Table 2.

Table 2. Red hot-mix asphalt and binder course asphalt properties.

Property	Sieve Size (mm)	OPSS [1] Requirement	Red HMA-Initial Mix	Red HMA-New Mix	Binder Course
	19.01	90–100	–	–	95.9
	16.0	–	100	100	88.4
	12.5	90–100	98.2	94.7	79.5
	9.5	45–90	83.4	79.1	70.8
	6.7	–	65.2	64.0	60.3
Gradation	4.75	45–55	56.2	55.0	53.2
(% passing)	2.36	28–58	48.0	43.0	41.1
	1.18	–	36.9	33.0	29.3
	0.600	–	27.9	24.3	22.1
	0.300	–	17.5	13.7	15.6
	0.150	–	10.2	6.4	8.20
	0.075	2–10	5.8	3.3	4.20
N_{des} (%G_{mm}) [3]		96.0	96.1	96	96
N_{ini} (%G_{mm}) [3]		≤ 89.0	89.1	89	88.5
N_{max} (%G_{mm}) [3]		≤ 98.0	97.8	97	97.3
Air Voids (%) at N_{design}		4.0	3.9	4.0	4.0
Voids in mineral aggregate (% minimum)		14.0	14.1	14.3	13.0
Asphalt binder performance grade		–	PG 70-28 P [2]	PG 64-34P [2]	PG 64-28 P [2]
Voids filled with asphalt (%)		65–75	72.1	72.2	72.0
Dust proportion (%)		0.6–1.2	1.33	0.7	1.0
Tensile strength ratio (%)		80	97.6	91.3	83.7
Asphalt film thickness (μm)		–	6.8	9.0	7.9
Asphalt cement content (%)		–	4.9	5.0	4.65

Note: [1] Ontario Provincial Standard Specification; [2] P stands for polymer-modified asphalt binder; [3] N_{des}, N_{ini}, N_{max} are number of gyrations at different compaction levels (design, initial, and maximum); G_{mm} is theoretical maximum specific gravity; PG, Performance Graded/Performance Grading.

3. Field Quantitative Evaluation

3.1. Three-Dimensional (3D) Pavement Surface Texture

The pavement surface texture was assessed by using a line-laser scanner, 3D non-contact line-laser pavement texture measurement system (Version 1002.05.20, Measurement Instrument Technology, Austin, TX, USA) [10] and based on the laser triangulation principle for a 3D texture measurement. The width of laser line for the line-laser scanner is 0.1 mm. A centre of gravity algorithm is employed to refine the laser line position for precise measurement to achieve the horizontal sampling interval of <0.05 mm. The line-laser scanner produces 3D surface texture measurements with a horizontal sample interval finer than 0.05 mm, and covering partially both microtexture and macrotexture ranges of the pavement surface. The accuracy of the line-laser scanner vertical texture height measurement is better than 0.05 mm. The recovered 3D texture height map was decomposed by using discrete wavelet transform to classify macrotexture and microtexture at various scales. Figure 2 shows a line-laser pavement texture scanner and a recovered 3D texture height map.

The digitally simulated 3D mean texture depth (MTD3) was calculated to represent macrotexture amplitude as shown in Equation (1). Detailed information about MTD3 can be found elsewhere [6,11].

$$\text{MTD3} = h_{2\%} - \frac{1}{A} \sum_{i=1}^{M} \sum_{k=1}^{N} \frac{1}{3} a h_{ik} \qquad (1)$$

where MTD3 = Digitally simulated 3D mean macrotexture depth from the line-laser scanner, $h_{2\%}$ = Texture height corresponding to 2% threshold of the bearing area curve of the macrotexture heights, A = The area of surface texture measurement (100×100 mm^2), a = The area of each data point of the macrotexture heights, h_{ik} = The elevation of any data point of the macrotexture heights, M and N = The number of data points in each direction of the macrotexture heights.

(a)

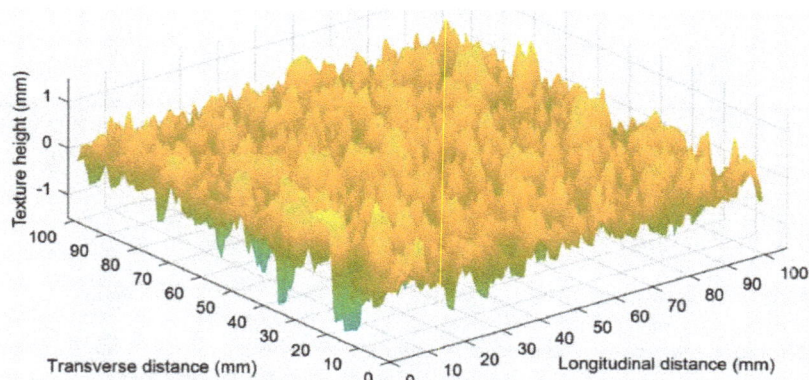

(b)

Figure 2. Pavement texture measurement: (**a**) a line-laser scanner; (**b**) a recovered 3D texture height map.

3.2. Pavement Frictional Property

Pavement frictional property was assessed by using a British pendulum tester (BPT) (Munro Instruments, Harlow, UK) following ASTM E303 standard [12]. BPT can provide an indirect measure of relative micro-texture and the well accepted pavement friction index: the British pendulum number (BPN) [13]. Figure 3 shows field test by using a BPT. The BPT tests were conducted in outer wheelpath at a surface temperature of $21 \pm 1\,°C$. Other temperatures were not tested.

The calculated 3D mean macrotexture depth and measured pavement frictional values are summarized as shown in Table 3.

Figure 3. Pavement friction assessment by using a British Pendulum tester (BPT).

Table 3. Field test results.

Section	Age at Test (year)	British Pendulum Number (BPN)	3D Mean Texture Depth (MTD3, mm)
Epoxy paint	3	–*	0.16
		–	0.27
		–	0.26
		–	0.30
		–	0.22
Red HMA–initial mix	3	61	0.63
		58	0.57
		54	0.62
		59	0.59
		59	0.46
Red HMA–new mix	1	70	0.53
		77	0.39
		76	0.32
		67	0.44
		66	0.44
Conventional HMA	1	79	0.60
		79	0.47
		77	–
		78	–
		84	–

* sections that are not tested.

As it can be seen from Figure 4, the averaged frictional values, BPN, varies from section to section with the conventional black asphalt section being the highest. In general, similar levels of frictional values were observed at the time of the test. Traffic conditions and ages of the tested sections were not considered for this test.

The calculated 3D mean macrotexture depth was plotted as shown in Figure 5. As expected, the epoxy paint section was observed as the lowest mean texture depth. This observation is in agreement with the findings from another study [3], that adding paints to the roadway surface may increase the risk of making a driving surface too smooth.

To test if the means of the calculated 3D texture of the three sections (conventional HMA being excluded because of too few measurements being conducted) are statistically different, the analysis of variance (ANOVA) was carried out for the MTD3. Table 4 presents the results of one-way ANOVA analysis for the MTD3 which has been previously found a reliable macrotexture index [11]. The purpose of one-way ANOVA is to test if the three sections have a common mean, and to determine whether MTD3 values are statistically different. The null hypothesis, there is no significant difference of the MTD3 among the three sections, is rejected given that p-value, 2.095×10^{-5}, is much smaller than the significance level, which is 5% in this study. It can be concluded that the mean values of MTD3 for epoxy paint, red HMA-initial mix, and red HMA-new mix are significantly different. Multiple comparisons between all pair-wise means of MTD3 were conducted to determine how they differ from one another. As it can be seen from Figure 6, which shows the mean values of MTD3 and their standard deviations for each of the three mixes, these comparisons suggested that epoxy paint surface has disadvantages to red HMA pavement from a surface texture point of view.

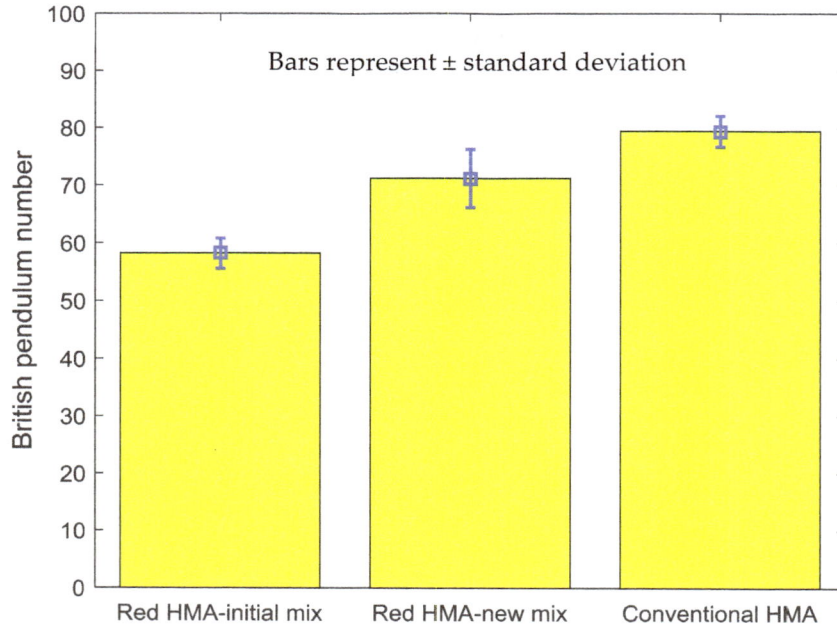

Figure 4. Pavement frictional values of tested sections.

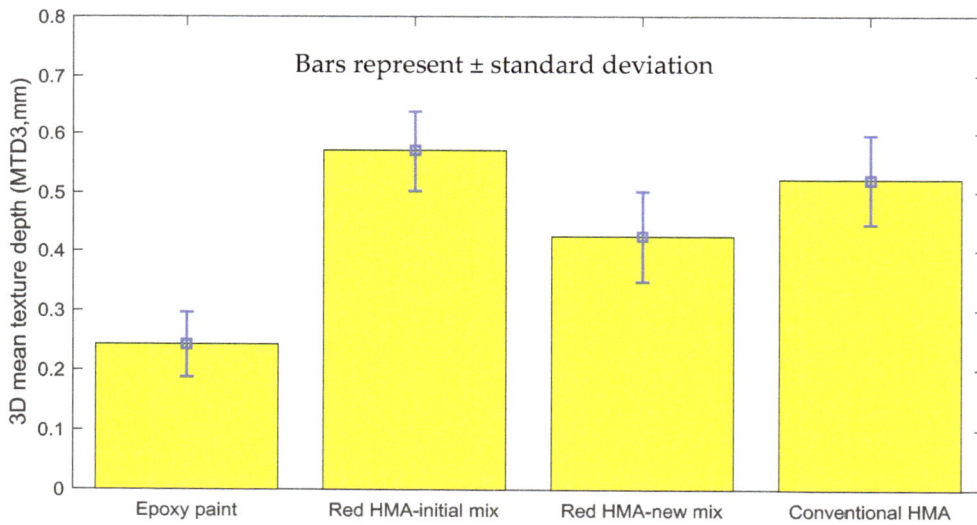

Figure 5. Pavement 3D mean macrotexture depth of tested sections.

Table 4. One-way analysis of variance (ANOVA) results for 3D mean texture depth (MTD3).

Source of Variability	Sum of Squares (SS)	Degrees of Freedom (DF)	Mean Squares (SS/DF)	F-Statistic	p-Value
Among sections	0.271	2	0.135	30.14	2.095×10^{-5}
Within section	0.054	12	0.005		
Total	0.325	14	–	(F-critical = 3.885)	

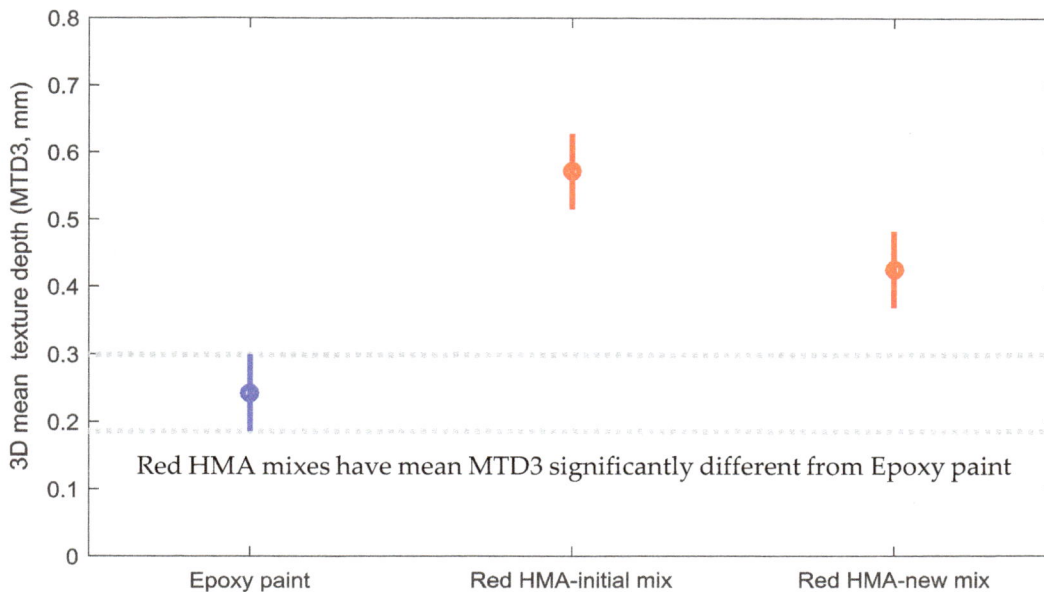

Figure 6. Pavement 3D mean macrotexture depths of tested sections.

The significant higher 3D mean texture depth of the two red HMA mixes than that of epoxy paint indicated a better friction for the red HMA sections based on the positive relationship between 3D texture amplitude and the pavement friction values [6,7]. These observations suggested that the red HMA pavement has significant advantages to the epoxy paint section of the BRT in York Region from pavement texture, friction, and traffic safety perspectives.

4. Field Pavement Survey

4.1. Pavement Surface Visualization

During the paving of red HMA-initial mixture, a site visit was conducted to assess the paved sections which exhibited surficial tire scuff marks as shown in Figure 7a. Although the appearance of scuffing and tire marks was found to be aesthetically unpleasant, they did not seem to affect the overall performance of the pavement sections, nor indicate a sign of poor workmanship or improper materials. After reviewing the aggregate properties as shown in Table 2, it was noted that more than 50 percent of the aggregate blend was consisted of a fine aggregate (passed sieve size of 4.75 mm) combined with a proprietary red pigment. This resulted in promoting a tighter surface texture and more aesthetically pleasing finish, which may cause the surface texture to be sensitive to tire scuffing. This sensitivity is even expected to be higher during warm periods (i.e., summer times). It seems that the tire scuffing disappeared, Figure 7b, in time under normal traffic conditions, becoming less visible. Tire scuffing was mitigated by using the New Red asphalt mixture after using coarser aggregate blend in the design.

Another visible aesthetic problem was found to be the appearance of oil, grease, fuel, or other automotive fluids dripped onto the pavement. These oil spots were found to be dominating at bus stops, and behind the intersections with buses at the full stop. It should be noted that the appearance of so called "oil spots" can be spotted at road sections that carry a higher rate of heavy commercial truck traffic. Accumulation of these drips over time can create a continuous streak that causes discoloration of the pavement as shown in Figure 8. Besides being aesthetically unpleasant, oil drips/streaks were suspected of containing minerals that dissolve or soften the asphalt binder and can result in surface deterioration and defects over a longer period of time. A preventive maintenance activity was suggested for inclusion in the life cycle to clean oil spots on a regular basis.

Figure 7. Tire marks at red asphalt sections paved by the Initial Red mixture: (**a**) newly paved section; (**b**) two-year-old section.

Figure 8. Monitored discoloration caused by oil drips and formation of oil streaks.

The UV properties of a coloured surface are important as UV radiation will degrade the colour over time and may make it less effective for lane designation. Figure 9 compares the colours of the tested sections in this research. It should be noted that these sections were compared under same levels of traffic loading and climatic conditions.

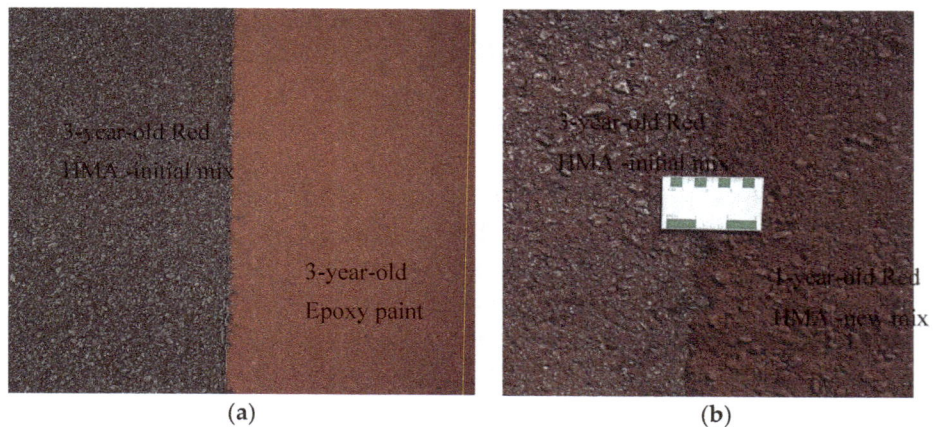

Figure 9. Visualization of tested surfaces: (**a**) Red HMA-initial mix and epoxy paint; (**b**) Red HMA-initial mix and Red HMA-new mix.

4.2. Pavement Distress Assessment

It was observed that sections paved with the initial red HMA mixture exhibited early signs of premature transverse (thermal cracking) and longitudinal (fatigue cracking) as shown in Figure 10. This field observation was evidenced by laboratory flexural beam fatigue test [14], which suggests that adding pigment may have caused the average fatigue life to decrease significantly for the red HMA-initial mixture in comparison to a conventional black asphalt mixture [15]. Same laboratory study suggests improved level of fatigue and thermal cracking resistance for the red HMA-new mix.

A site visit conducted after one year of paving confirmed no premature thermal cracking and fatigue cracking for sections with the red HMA-new mixture.

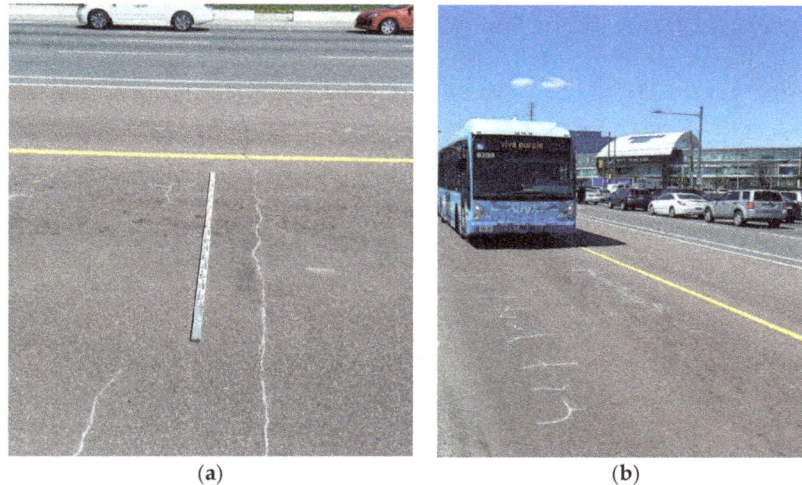

(a)   (b)

Figure 10. Early signs of cracks observed in sections paved with the Initial Red mixture after few years of in-service: (**a**) premature thermal cracking; (**b**) wheelpath fatigue cracking. The white colour of cracks is because of a chalk used for mapping.

5. Conclusions

The functional and structural performance of three sections of red coloured pavement including epoxy paint section and two types of red asphalt mixes were evaluated in the field by CPATT at the University of Waterloo. Epoxy paint surface was observed as the lowest texture depth, which may raise the concern that colouring the pavement by using epoxy paint may make a driving surface too smooth. The two types of red asphalt sections were observed as lower yet compatible frictional levels to conventional black asphalt pavement.

Surficial tire scuff marks were observed for newly paved red asphalt pavement sections. These tire scuffings were becoming less noticeable in time under normal traffic conditions. Another visible aesthetic problem was found to be the appearance of oil, grease, fuel, or other automotive fluids dripped onto the pavement. Besides being aesthetically unpleasant, oil drips/streaks were suspected of containing minerals that dissolve or soften the asphalt binder and can result in surface deterioration and defects over a longer period of time. The ultraviolet radiation degraded the colour of red asphalt pavement over time and may make it less effective for lane designation. Early signs of premature thermal cracking and fatigue cracking were observed for the section of red asphalt with initial mix in this research. This field observation was evidenced by laboratory flexural beam fatigue test which found that adding pigment decreased the average fatigue life significantly for red asphalt initial mix compared to conventional back asphalt mixtures. After one year in service, no premature thermal cracking and fatigue cracking occurred for the red asphalt section with new mix design.

This paper highlighted the texture, frictional, and structural performance of the red asphalt pavements which were three or one year old at time of field assessment. Long-term evaluation is recommended to monitor the performance of the red asphalt pavement for bus rapid transit dedicated lanes.

Acknowledgments: The authors of this paper gratefully acknowledge the financial support from the Regional Municipality of York and Metrolinx. Appreciation is also extended to the Norman W. McLeod in Sustainable Engineering at the University of Waterloo.

Author Contributions: As research associates, Qingfan Liu and Sina Varamini conducted field tests, field survey, and data analysis. Susan Tighe supervised and led this research.

Conflicts of Interest: The authors declare no conflict of interest.

References

1.   Assoc, N.N.C. *Seattle Transit Master Plan Briefing Book*; Seattle Department of Transportation: Washington DC, USA, 2011. Available online: http://www.seattle.gov/transportation/docs/tmp/briefingbook/SEATTLE%20TMP%200%20COVER%20TOC.pdf (accessed on 22 April 2017).

2.   Birk, M.; Burchfield, R.; Flecker, J. *Portland's Blue Bike Lanes: Improved Safety through Enhanced Visibility*; UNC Highway Safety Research Center, City of Portland Office of Transportation: Portland, OR, USA, 1999.

3.   Carry, M.; Donnell, E.; Rado, Z.; Hartman, M.; Steven, S.S. Red Bus Lane Treatment Evaluation. Available online: http://nacto.org/docs/usdg/red_bus_lane_evaluation_carry.pdf (accessed on 22 April 2017).

4.   Lee, H.; Kim, Y. Laboratory Evaluation of Color Polymer Concrete Pavement with Synthetic Resin Binder for Exclusive Bus Lanes. *Transp. Res. Rec. J. Transp. Res. Board* **2007**, *1991*, 124–132. [CrossRef]

5.   Synnefa, A. Measurement of Optical Properties and Thermal Performance of Coloured Thin Layer Asphalt Samples and Evaluation of Their Impact on the Urban Environment. In Proceedings of the Second International Conference on Countermeasures to Urban Heat Islands, Berkeley, CA, USA, 19–23 September 2009.

6.   Liu, Q.; Shalaby, A. Relating concrete pavement noise and friction to three-dimensional texture parameters. *Int. J. Pavement Eng.* **2015**. [CrossRef]

7.   Liu, Q.; Tighe, S.; Shalaby, A. Assessment of Airfield Runway Macrotexture and Friction Using Three-Dimensional Laser-Based Measurement. Available online: http://docs.trb.org/prp/17-01393.pdf (accessed on 22 April 2017).

8.   Varamini, S.; Farashah, M.K.; El-Hakim, M.; Tighe, S.L. Coloured Asphalt Bus Rapid Transit Lanes in the Regional Municipality of York: Integrating Laboratory Performance Testing into Sustainable Pavement Asset Management. Available online: http://www.tac-atc.ca/sites/tac-atc.ca/files/conf_papers/varamini_0.pdf (accessed on 22 April 2017).

9.   Ontario provincial standards specification (OPSS). Available online: http://www.raqsb.mto.gov.on.ca/techpubs/OPS.nsf/OPSHomepage (accessed on 22 April 2017).

10.   *Operation Manual, LS-40 Pavement Surface Analyzer*; Version 1002.05.20; HyMIT LLC: Austin, TX, USA, 2013.

11.   Liu, Q.; Gonzalez, M.; Tighe, S.L.; Shalaby, A. Three-dimensional surface texture of Portland cement concrete pavements containing nanosilica. *Int. J. Pavement Eng.* **2016**. [CrossRef]

12.   ASTM E303. (American Society for Testing and Materials). *Standard Test Method for Measuring Surface Frictional Properties Using the British Pendulum Tester*; ASTM International: West Conshohocken, PA, USA, 1993.

13.   Hall, J.W.; Smith, K.L.; Titus-Glover, L.; Wambold, J.C. Guide for Pavement Friction. Available online: http://redlightrobber.com/red/links_pdf/Guide-for-Pavement-Friction-NCHRP-108.pdf (accessed on 22 April 2017).

14.   *ASTM D7460–10–Standard Test Method for Determining Fatigue Failure of Compacted Asphalt Concrete Subjected to Repeated Flexural Bending*; ASTM International: West Conshohocken, PA, USA, 2010.

15.   Effect of Colouring pigment on Asphalt Mixture Performance: Case for use in Ontario. Available online: https://trid.trb.org/view.aspx?id=1393441 (accessed on 22 April 2017).

8

Statistical Determination of a Fretting-Induced Failure of an Electro-Deposited Coating

Kyungmok Kim

School of Aerospace and Mechanical Engineering, Korea Aerospace University, 76 Hanggongdaehang-ro, Deogyang-gu, Goyang-si, Gyeonggi-do 412-791, Korea; kkim@kau.ac.kr

Academic Editors: Eva Pellicer and Jordi Sort

Abstract: This paper describes statistical determination of fretting-induced failure of an electro-deposited coating. A fretting test is conducted using a ball-on-flat plate configuration. During a test, a frictional force is measured, along with the relative displacement between an AISI52100 ball and a coated flat specimen. Measured data are analyzed with statistical process control tools; a frictional force versus number of fretting cycles is plotted on a control chart. On the control chart, critical number of cycles to coating failure is statistically determined. Fretted surfaces are observed after interrupting a series of fretting tests. Worn surface images and wear profiles provide that the increase on the kinetic friction coefficient after a steady-state sliding is attributed to the substrate enlarged at a contact surface. There is a good agreement between observation of worn surfaces and statistical determination for fretting-induced coating failure.

Keywords: electro-deposited coating; coating failure; fretting; statistical process control

1. Introduction

In the design of tribo-components, it is essential to evaluate the tribological properties of the components, since they crucially affect the performance of an entire system. In order to minimize friction between two mating components, solid coatings are often applied into contact surfaces. Evaluation of a solid coating is carried out with a simplified tribotest, prior to a field test [1]. In a tribotest, a coating is evaluated with the kinetic friction coefficient or the wear rate under the conditions similar to those found in actual contact situations. The wear rate of a coating was typically calculated by measuring volumetric loss of materials after a series of wear tests [2,3]. The kinetic friction coefficient is a useful indicator because it is possible to measure the coefficient in the course of a friction test. Some studies evaluated tribological performances of solid coatings on the basis of a kinetic friction coefficient evolution [4–6]. Fretting tests with thermally sprayed solid coatings were terminated when the kinetic friction coefficient reached the critical value similar to those observed at the substrate-to-substrate contact. Then, the number of cycles to the critical value was taken into account as the fretting lifetime of a solid coating [5,6]. Meanwhile, the endurance life of a coating was proposed as the number of fretting cycles when the friction coefficient came to be three times greater than the initial value [4]. The dissipated energy approach was used to establish the durability and the safe running time of a coating under fretting conditions [7]; an energy-Wohler wear chart was introduced, in which the critical dissipated energy density corresponded to the time when a coating wore off.

Electro-deposited coatings are widely used for automotive components because they maintain high corrosion resistance. Electro-deposited coatings are often applied to frictional contacts such as automotive seat sliding rails. In the previous studies with epoxy-based electro-deposited coatings, there were difficulties in determining the durability of the coating with the criteria described above; the coating was found to fail earlier than the critical values proposed by the criteria [8]. Thus, an adequate method is needed for determining the fretting lifetime of electro-deposited coatings.

Statistical process control (SPC) is a well-established technique for ensuring process quality, thereby being used in manufacturing and production industries. Particularly, SPC tools were successfully used for monitoring a manufacturing process subjected to tool wear [9,10]. For monitoring tool wear, SPC tools allowed distinguishing two sources of variation; one was natural variability, and the other was special causes such as improperly adjusted tools, operator errors, and inferior raw materials. In SPC, measured output data (e.g., surface roughness) were analyzed with control charts such as the average x-bar chart and a range chart. Measured data from a process without special causes were placed between the upper and the lower control limits on the average x-bar chart. If some data were found beside the control limits, a process was considered to be instable, said to be out of control. Thus, it was possible to detect a process that became out of control.

For monitoring the state of a cutting tool, acoustic emission signals were measured and analyzed with SPC tools [11]. SPC tools were used for predicting the formation of voids and defects in friction stir welding [12]. Variance of process forces was found to be correlated to the formation of defects in friction stir welds. SPC was used for detecting faults in a bearing [13–15]. Parameters indicating the state of a bearing were measured and analyzed with SPC tools. The parameters included vibration, stator current noise, and bearing housing temperature. A set of vibration data measured in a rolling element ball bearing was evaluated with SPC control tools [13]. Instead of vibration, stator current noise was monitored for detecting in situ bearing faults [14]. Housing temperature of a sliding bearing was determined in dry friction conditions [15]. Change of housing temperature in the course of a test was used for identifying the state of bearing condition. However, it is necessary to exclude the variance of housing temperature that arises from the change of room temperature.

In this study, a fretting test was conducted with an electro-deposited coating used for automotive seat tracks. In order to develop a criterion to identify the failure of a fretted electro-deposited coating, statistical analysis was employed with measured frictional force data. Control charts in statistical process control technique were used to identify the transition of the frictional force. Additionally, fretted surfaces were captured and wear profiles were measured to observe surface damages at various numbers of fretting cycles. Finally, comparison between observation of the wear profile and statistical determination was employed.

2. Materials and Methods

Figure 1 shows an in-house developed fretting machine used for this study. The fretting machine enables the linear reciprocating motion of a flat specimen to a fixed counterpart. The reciprocating motion was imposed in a sine wave. The relative displacement between a specimen and a counterpart (ball) was continuously measured with a laser displacement sensor (Model LK-081, a resolution of 0.003 mm and a linearity of $\pm 0.1\%$, Keyence Corp., Itasca, IL, USA). A cylindrical ball holder was permitted to vertical movement in a rigid arm. Dead weights were placed on the holder for the purpose of inducing normal force at the contact between a flat specimen and a ball. A load-cell (Interface, Inc., Atlanta, GA, USA) was connected to a rigid arm for measuring frictional force. In this study, a commercial AISI 52100 steel ball (a diameter of 5 mm and R_a of about 0.025 μm) was used as a counterpart.

In this study, a normal force of 49 N, a displacement amplitude of 0.2 mm, and a frequency of 1 Hz were applied to produce fretting damage on a flat specimen. The magnitude of normal force was similar to those found in the automotive seat slide tracks.

An epoxy-based cathodic electro-deposited coating was applied onto the substrate of a flat specimen (high-strength steel, material designation: SPFC 440); in order for the coating layer to be deposited on the substrate, the substrate was immersed in a tank filled with the electrocoat and connected to a rectifier as the corresponding electrode. The counter electrode was immersed at the same time and a direct charge was provided under the conditions described in Table 1. The coating contained an epoxy resin with a metal catalyst and a crosslinker of blocked aromatic isocyanates. Figure 2 shows the microstructure image of an electro-deposited coating. The thickness of a deposited coating was dependent upon parameters such as applied voltage, electro-deposited (ED) time, and pH. In this study, the coating thickness on the substrate was measured with a microscope; the initial coating thickness was 22 ± 2 μm and the arithmetic average roughness (R_a) of the coating was 0.518 ± 0.051 μm. Pencil hardness of the coating was 5H (corresponding to about 34.9 in Knoop hardness). Fretting tests were conducted with the specimen having a coating thickness of 22 μm for minimizing the effect of coating thickness on a frictional force evolution. All tests were conducted under room temperature (about 25 °C) and ambient humidity (about 55% RH).

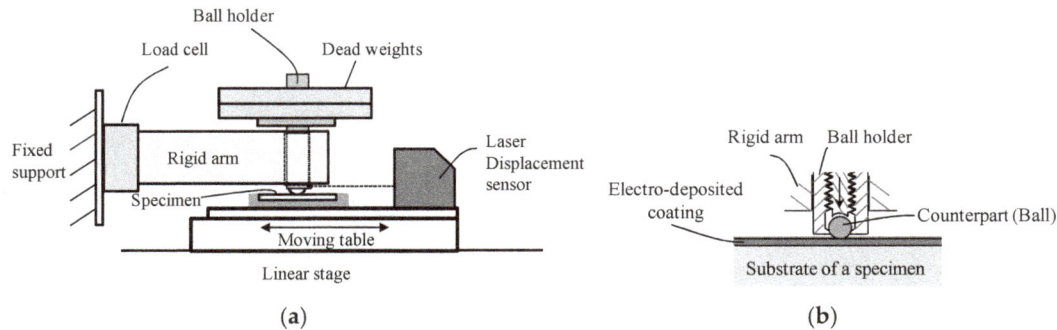

Figure 1. A schematic diagram of a fretting testing machine: (a) test rig; (b) ball-on-flat plate contact.

Table 1. Condition of cathodic electro-deposition.

Specification	Value
Solid content	15%–22%
Pigment binder ratio	0.15
pH	5.9–6.3
Conductivity	120–180 μS/mm
Voltage	230–290 V
Electro-deposition (ED) time	About 180 s

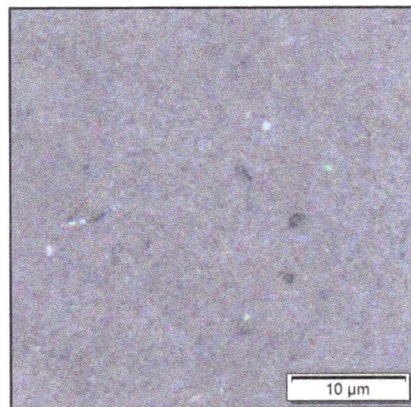

Figure 2. Microstructure image of an electro-deposited coating.

3. Results and Discussion

A fretting test was conducted at the displacement amplitude of 0.2 mm, a normal force of 49 N, and a frequency of 1 Hz. The test was terminated at the friction of 25 N; it was identified from the previous studies that the kinetic friction between an AISI 52100 steel ball and a high strength steel plate was about 25 N at a normal force of 49 N [8]. During a test, a frictional force and the relative displacement between a ball and a plate were recorded.

Figure 3 shows a fretting loop with respect to number of cycles. All fretting loops were quasi-rectangular shaped. Frictional force peaks were observed in the loops found at the 750th and 800th cycles due to ploughing effect. Figure 4 shows the evolution of a slip ratio for an electro-deposited coating at the displacement amplitude of 0.2 mm. A slip ratio was defined as the ratio of an actual sliding distance to a total displacement. The actual sliding distance was determined when a frictional force was zero on a fretting loop. It was identified that the transition between a reciprocal sliding regime and a fretting regime (gross slip) was observed at a slip ratio of 0.95 [16]. As shown in Figure 4, all slip ratios remained lower than 0.95, ensuring that a test was completed within a gross slip regime.

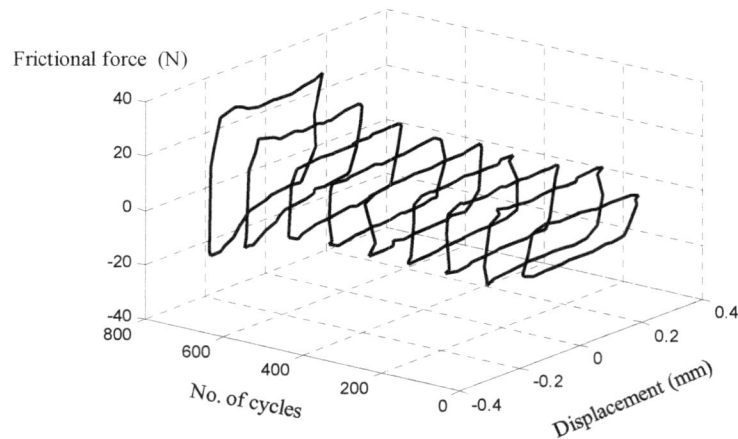

Figure 3. Fretting loops of an electro-deposited coating at a normal force of 49 N, an imposed displacement amplitude of 0.2 mm, and a frequency of 1 Hz.

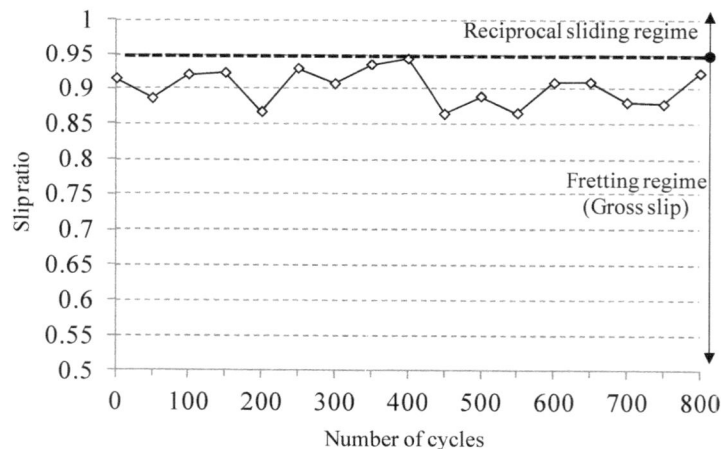

Figure 4. Slip ratios of an electro-deposited coating at a normal force of 49 N, an imposed displacement amplitude of 0.2 mm, and a frequency of 1 Hz.

For statistical analysis, two frictional forces were selected on a fretting loop: The maximum value and the force at zero displacement (movement to the left side). Note that the forces could be identified apparently in a fretting loop and the maximum value is used for determine the coefficient of kinetic friction. The two force values were then averaged. Figure 5 shows the average frictional force with respect to number of cycles. The initial value was 8.63 N. After the initial increase, the value remained almost steady (so-called steady-state sliding stage). During steady-state sliding, variance of the frictional force was attributed to surface roughening of an electro-deposited coating layer. In addition, wear debris formed might remain within the contact zone. After 600 cycles, the average frictional force was determined to increase rapidly; wear debris might be removed from the contact zone.

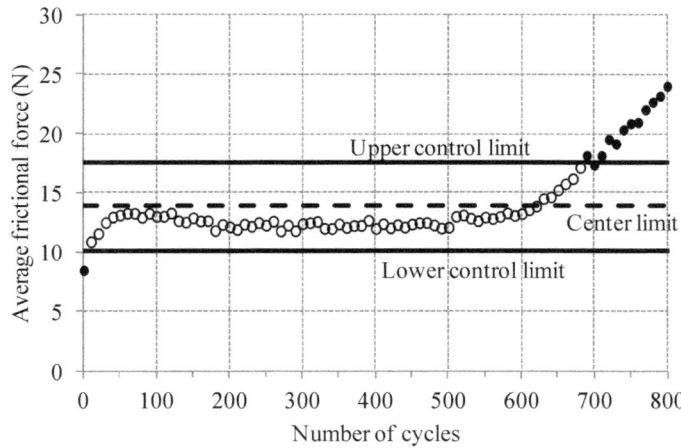

Figure 5. Evolution of the average frictional force and control limits.

Analysis of statistical process control (SPC) was employed with the average frictional forces. The purpose of the statistical analysis was to distinguish two sources of variation; one was surface roughening of a coating layer, and the other was the partial contact between the ball and the substrate. In order to determine the cycle when the partial contact occurs between the ball and the substrate, the following procedure was employed:

- The average frictional force versus the number of cycles plot was obtained. In addition, the difference between two selected frictional values was plotted over the number of cycles. The difference was called Range (R).
- Control limits for the average and the range were calculated according to Tables 2 and 3.
- Control limits were then plotted on the average frictional force evolution (the average x-bar chart) and R-chart as shown in Figures 5 and 6.
- After excluding the data that were out of control limits, control limits were revised.
- When all data remained between the revised control limits, the revision was terminated, and the final cycle was proposed as the one that the partial contact started to occur between the ball and the substrate.

Table 2. Definitions of control limits for the average frictional force chart. $\overline{Q}$ and N_s denote the average frictional force and the number of subgroups, respectively. A_2 is 1.88 for two sample sizes. R is the difference between selected frictional forces.

Control Limit	Definition
Center limit (CL)	$\frac{\sum \overline{Q}}{N_s}$
The upper control limit (UCL)	$CL + A_2 \times \frac{\sum R}{N_s}$
The lower control limit (LCL)	$CL - A_2 \times \frac{\sum R}{N_s}$

Table 3. Definitions for R-chart. R is the range between selected frictional forces. N_s denotes number of subgroups. D_3 and D_4 denote the constants, and they were zero and 3.267, respectively.

Control Limit	Definition
Center limit (CL_R), $\overline{R}$	$\frac{\sum R}{N_s}$
The upper control limit (UCL_R)	$D_4 \times \overline{R}$
The lower control limit (LCL_R)	$D_3 \times \overline{R}$

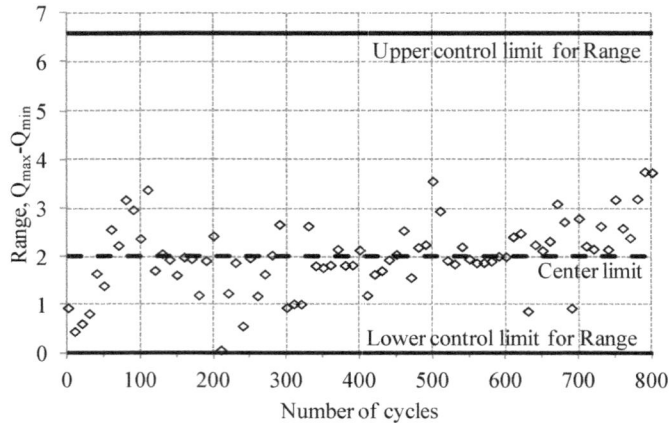

Figure 6. *R*-chart for the difference between selected frictional forces.

Figure 5 shows the evolution of the average frictional force ($\overline{Q}$) with three control limits. Control limits for the average frictional force evolution were determined with the parameters shown in Table 2. $\overline{Q}$ and N_s denote the average frictional force and the number of subgroups, respectively. In this study, a subgroup is associated with a cycle. Each subgroup contained two frictional force values. The constant, A_2, is dependent upon the sample size ($A_2 = 1.88$ for two samples in one subgroup [17]). *R* is the difference between selected frictional forces. In this study, the total number of cycles was 800, and the cycle interval was 10. Thus, the number of subgroups was 80. If all of the average frictional forces were placed between the upper control limit and the lower control limit in the chart, a process was considered as in control. That is, the "in-control process" can be defined as the state that a ball slid on a coating layer. Small variance of the frictional force was attributed to the surface roughening of a coating layer. If the average value exceeded the upper control limit, a process became statistically out of control. It means that a ball partially slid on the surface of the substrate. That is, an out-of-control process might correspond to the state that some parts of a coating layer wore off. In Figure 5, dark solid markers were placed outside the control limits; the first value remained below the lower control limit, while other values after 670 cycles were above the upper control limit. The first value was the average frictional forces measured during the initial running-in period. Thus, one could exclude the initial value in determining coating failure.

Figure 6 shows Range (*R*) chart for a fretted electro-deposited coating. The Range was defined as the difference between selected frictional forces found in a fretting loop. The upper and the lower control limits, and the center limit for R-chart were defined as Table 3. The constants, D_3 and D_4, are dependent upon selected sample size ($D_3 = 0$ and $D_4 = 3.267$ for two samples in each subgroup [17]). It was identified from Figure 6 that all range values were found to remain between the upper and the lower control limits.

For the purpose of additional analysis, the average frictional forces outside the upper and the lower control limits in Figure 5 were removed and then statistical analysis was re-employed; control limits were recalculated. On the modified frictional force chart, some data exceeded the upper control limit. Thus, the control limits were revised after excluding the data that exceeded the control limit. Figure 7 shows the revised chart for the average frictional force. It was identified from the plot that all values were placed between two control limits, indicating that a ball slid on a coating layer by the 670th cycle. In other words, it could be suggested that the durability of an electro-deposited coating was 670 fretting cycles at a given test condition.

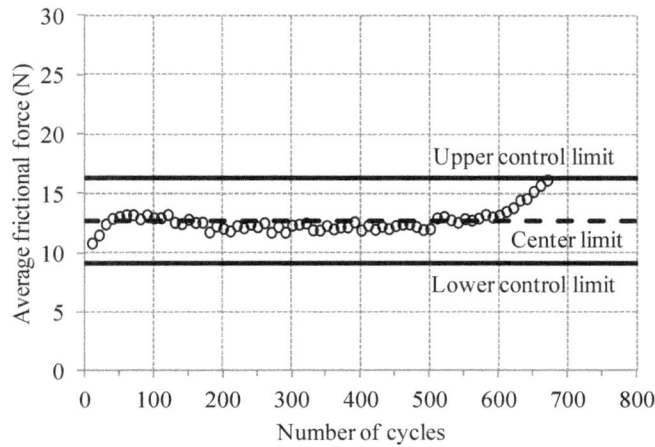

Figure 7. The revised average x-bar chart for the evolution of the average frictional force. Control limits were revised.

In order to observe surface damages under the fretting condition, fretting tests were interrupted at various numbers of cycles. The coefficient of the kinetic friction was determined during a fretting test. The kinetic friction coefficient was defined as the ratio of the maximum frictional force to the imposed normal force. Fretted surface images were captured by optical microscope (250×) (Shenzhen Supereyes Technology Co., Ltd., Shenzhen, Guangdong, China) and wear profiles were measured by surface profilometer (Mitutoyo Corp., Takatsuku, Kawasaki, Japan). Figure 8 shows the fretting images at various numbers of cycles. After the initial cycle, a crater was observed. At the 100th cycle, the crater became bigger. At the 500th cycle, a coating layer was observed to remain within the contact zone. Wear profile shows that the substrate did not appear in the contact. Note that the average coating thickness was 0.022 mm. Bight areas near contact edges were due to the reflection of light during image capture. Measured wear profile at the 700th cycle shows that the maximum wear depth exceeded the coating depth, indicating that the substrate appeared at the contact. It can be identified that the increase on the friction coefficient was attributed to the ball-on-substrate contact. At the 800th cycle, it was observed that the entire contact surface was severely damaged and the substrate was roughened along the direction parallel to the sliding direction.

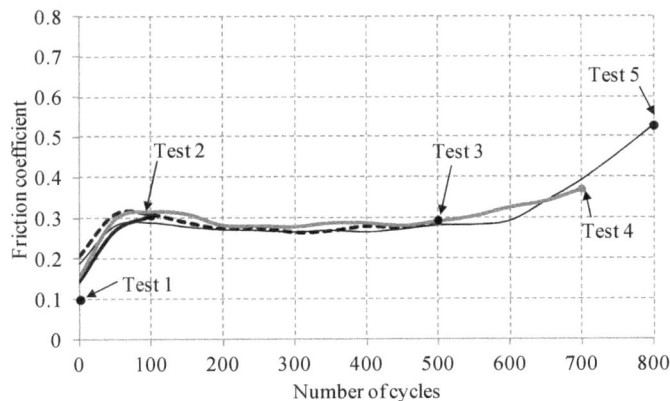

(a)

Figure 8. *Cont.*

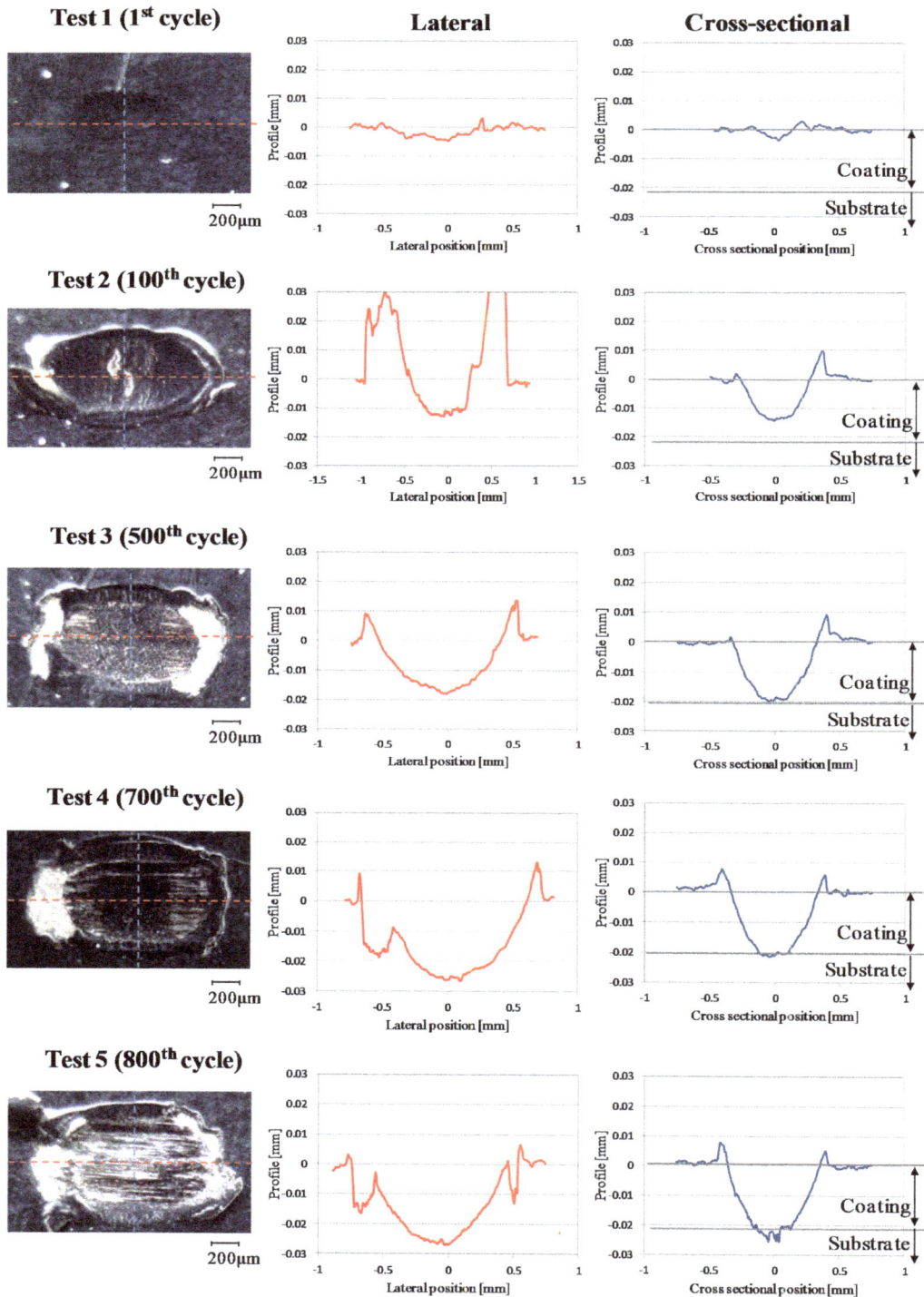

Figure 8. Measurement of the friction coefficient and observation of fretted surfaces: (**a**) the friction coefficient evolution; (**b**) worn surface images (250×) and surface profiles. Note that the initial coating thickness was 0.022 mm. Bright areas near contact edges were associated with reflection of light in a microscope.

Figure 9 shows the ratio of the maximum wear depth to initial coating thickness. The maximum wear depth ratio of greater than unity indicates that some parts of the coating in a contact zone wore off and the ball slid on the surface of the substrate. It was shown that the maximum wear depth exceeded the initial coating depth between the 500th and the 700th cycles. This is in good agreement with statistical analysis result; in the statistical analysis, the critical cycle for in-control process was the 670th cycle. Therefore, one might consider the 670th cycle as the durability of a fretted electro-deposited coating under the given test conditions.

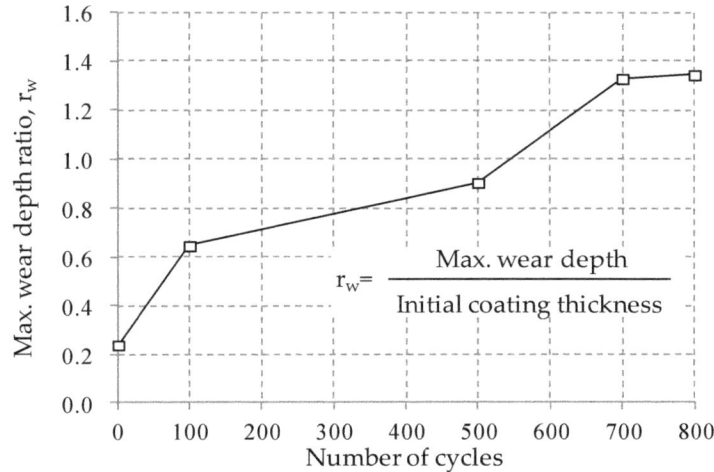

Figure 9. The maximum wear depth evolution according to number of cycles. The maximum wear depth ratio defined as the ratio of the maximum wear depth to initial coating thickness.

In this study, fretting was induced to the contact between an electro-deposited coating and an AISI 52100 steel ball. Thus, the displacement amplitude of greater than 0.2 mm needs to be taken into account for determining the failure of an electro-deposited coating under a reciprocal sliding condition. In addition, other solid lubricant coatings need to be evaluated with the proposed statistical method. In this statistical analysis, two frictional forces in a fretting loop were selected. Thus, future work needs to include additional force selection in a fretting loop for statistical analysis.

4. Conclusions

This paper investigated the fretting-induced failure of a cathodic electro-deposited coating for automotive components. A frictional force and the relative displacement between a specimen and a counterpart were measured during a fretting test. Statistical analysis was employed with measured frictional forces. The following conclusions were drawn:

- Control charts in statistical process control were found to be useful for identifying the quality of a fretted electro-deposited coating in terms of the kinetic friction coefficient. It was identified that control charts could detect the cycle when the kinetic friction became instable.
- Optical observation of worn surfaces and measurement of wear profiles showed that the coating was found to fail partially at the cycle when the kinetic friction coefficient increased.
- It could be suggested that the durability of the fretted electro-deposited coating can be defined as the cycle when the average value of frictional forces exceeded the upper control limit on the charts.

Future work needs to include reciprocal sliding tests with various solid lubricant coatings, and various counterparts such as stainless steel and ceramic balls. Additional force selection per cycle would be useful for accurate statistical analysis.

Acknowledgments: This work was supported by the National Research Foundation of Korea (NRF) grant funded by the Korean government (MSIP) (No. 2016R1C1B1008483).

Conflicts of Interest: The authors declare no conflict of interest.

References

1. Bhushan, B. *Introduction to Tribology*, 1st ed.; John Wiley & Sons, Inc: New York, NY, USA, 2002.
2. Fouvry, S.; Kapsa, P.; Vincent, L. Analysis of sliding behaviour for fretting loadings: Determination of transition criteria. *Wear* **1995**, *185*, 35–46. [CrossRef]
3. Nallasamy, P.; Saravanakumar, N.; Nagendran, S.; Suriya, E.M.; Yashwant, D. Tribological investigations on MoS₂-based nanolubricant for machine tool slideways. *Proc. Inst. Mech. Eng. Part J J. Eng. Tribol.* **2015**, *229*, 559–567. [CrossRef]
4. Langlade, C.; Vannes, B.; Taillandier, M.; Pierantoni, M. Fretting behavior of low-friction coatings: Contribution to industrial selection. *Tribol. Int.* **2001**, *34*, 49–56. [CrossRef]
5. Korsunsky, A.M.; Torosyan, A.T.; Kim, K. Development and characterization of low friction coatings for protection against fretting wear in aerospace components. *Thin Solid Films* **2008**, *516*, 5690–5699. [CrossRef]
6. Kim, K.; Korsunsky, A.M. Dissipated energy and fretting damage in CoCrAlY-MoS₂ coatings. *Tribol. Int.* **2010**, *43*, 676–684. [CrossRef]
7. Liskiewicz, T.; Fouvry, S.; Wendler, B. Development of a Wöhler-like approach to quantify the Ti(C_xN_y) coatings durability under oscillating sliding conditions. *Wear* **2005**, *259*, 835–841. [CrossRef]
8. Kim, K. A study of the frictional characteristics of metal and ceramic counterfaces against electro-deposited coatings for use on automotive seat rails. *Wear* **2014**, *320*, 62–67. [CrossRef]
9. Wu, Z. An adaptive acceptance control chart for tool wear. *Int. J. Prod. Res.* **1998**, *36*, 1571–1586. [CrossRef]
10. Motorcu, A.R.; Gullu, A. Statistical process control in machining, a case study for machine tool capability and process capability. *Mater. Des.* **2006**, *27*, 364–372. [CrossRef]
11. Houshmand, A.A.; Kanatey-Ahibu, E. Statistical process control of acoustic emission for cutting tool monitoring. *Mech. Syst. Signal. Process.* **1989**, *3*, 405–424. [CrossRef]
12. Jata, K.V.; Mahoney, M.W.; Mishra, R.S. *Friction Stir Welding and Processing III*, 1st ed.; Wiley: Hoboken, NJ, USA, 2005.
13. Wang, W.; Zhang, W. Early defect identification: application of statistical process control methods. *J. Qual. Mainten. Eng.* **2008**, *14*, 225–236. [CrossRef]
14. Zhou, W. Bearing fault detection via stator current noise cancellation and statistical control. *IEEE. Trans. Ind. Electron.* **2008**, *55*, 4260–4269. [CrossRef]
15. Lepiarczyk, D.; Gawedzki, W.; Tarnowski, J. Thermal analysis of the friction process of sliding bearings using a statistical approach. *Tribologia* **2016**, *4*, 157–165.
16. Varenberg, M.; Etsion, I.; Halperin, G. Slip index: A new unified approach to fretting. *Tribol. Lett.* **2004**, *17*, 569–573. [CrossRef]
17. Montgomery, D.C. *Introduction to Statistical Quality Control*, 6th ed.; John Wiley & Sons, Inc.: New York, NY, USA, 1997.

9

Ride Quality Due to Road Surface Irregularities: Comparison of Different Methods Applied on a Set of Real Road Profiles

Giuseppe Loprencipe * and Pablo Zoccali

Department of Civil, Constructional and Environmental Engineering, Sapienza University of Rome, Via Eudossiana 18, 00184 Rome, Italy; pablo.zoccali@uniroma1.it

* Correspondence: giuseppe.loprencipe@uniroma1.it

Academic Editor: Andrea Simone

Abstract: Road roughness evaluation can be carried out using different approaches. Among these, the assessment of ride quality level perceived by road users is one of the most-used. In this sense, different evaluation methods have been developed in order to link the level of irregularities present on road surface profiles with the induced detrimental effects in terms of discomfort. In particular, relationships between wavelength content of road profiles and consequent level of comfort perceived had been investigated by using, in general, a mean panel ratings approach. In this paper, four ride quality evaluation methods (Ride Number, Michigan Ride Quality Index (*RQI*), Minnesota Ride Quality Index and frequency-weighted vertical acceleration, a_{wz}, according to ISO 2631 were applied to a set of real road profiles. The obtained results were analyzed, investigating a possible relation between the different indices, comparing them also with the most-used road roughness method worldwide: the International Roughness Index (*IRI*). The analyses carried out in this work have highlighted how the various rating scales may lead to a different ride quality assessment of the same road pavements. Furthermore, comparing the a_{wz} with the values obtained for the other three methods, it was found that their rating scales are set for speeds within the range 80–100 km/h. For this reason, it is necessary to identify new thresholds to be applied for lower speeds, as in the case of urban roads. In this sense, the use of the ISO 2631 approach would seem to be a useful tool.

Keywords: ride quality; ride Number; Michigan *RQI*; Minnesota *RQI*; road surface irregularities; ISO 2631; *IRI*; real road profiles

1. Introduction

Road roughness is an important issue for the assessment of road pavement condition [1,2], and an important aspect to be included in any Pavement Management System (*PMS*) [3–6]. Road roughness evaluation can be carried out using a number of different approaches. Among these, the most common ones are based on the assessment of detrimental effects induced by irregularities on road surfaces, like the dynamic increment of loads transmitted to pavements [7], road users' comfort [8] and noise generated due to road traffic [9].

The International Roughness Index (*IRI*) is the most used method worldwide and it was developed to take into account general effects (both on pavements and users) induced by irregularities of road pavement surface [10]. Some authors, like Kropáč and Múčka [11] and Loizos and Plati [12], have highlighted some limits of the applicability of IRI as a method for the evaluation of road roughness. In particular, in [13], the inability of *IRI* to describe car body vertical vibration due to the presence of certain wavelengths (i.e., >20 m) of road profile was described. In recent decades, several indices have been developed as alternative methods to consider various effects [14]. Particular attention has been paid to the assessment of the influence of road unevenness on vehicles and passenger vibrations [15,16], and the evaluation of possible correlations between existing roughness indicators and vehicles vibration response [17].

Cantisani and Loprencipe [18] proposed calculating the whole body vibration induced on passengers inside road vehicles, by means of the process described in the ISO 2631 standard [19], to assess ride quality. In this way, it would be possible to reflect the comfort perceived by road users. Many authors have analyzed existing relations between IRI and vertical accelerations measured on driver and/or passenger seats, considering different types of vehicles and different velocities [17,20–22]. Most of these studies found a linear regression with R^2 values within the range from 0.76 to 0.99. Other indicators, like the Ride Number (RN) and the Michigan and Minnesota Ride Quality Index (RQI), have been developed through Mean Panel Rating (MPR) tests, in order to take into account road customers' opinions.

In the literature, to the best of our knowledge, no relationships or comparisons between the latter indices are present. In some studies, on the other hand, it is possible to find some relationships between each of the aforementioned ride quality indices and IRI. In particular, Sayers and Karamihas [2] found a relationship between IRI and Profile Index (PI), which is a parameter at the base of RN calculations, having an R^2 value of 0.82. In this case, the IRI range considered was from 0.5 to 7 m/km. A similar range was also considered in [23] where, in addition to the comparison between IRI and PI (R^2 variable 0.96–0.98), a direct relationship between IRI and RN (R^2 = 0.98) was provided, although related to a narrower range of IRI values (from 0.5 to 1.6 m/km).

In [24], relationships between Michigan RQI (RQI_{Mich}) and IRI for different types of pavement (i.e., flexible, rigid, composite) are depicted, although corresponding equations and R^2 values are not reported (but the result seems to be good). The calculation of the Minnesota RQI (RQI_{Mn}) is instead, as will be described in the following section, based on IRI values.

In this paper, a comparison of the RN, RQI_{Mich} and RQI_{Mn} was carried out, applying them to a set of 3905 samples of real road (asphalt pavement) profiles, having section lengths equal to 100 m. These indices were selected because they are based on MPR and the corresponding threshold values currently adopted in certain countries are available in literature. The final purpose is to compare the different ride quality thresholds defined for each method, also evaluating the existence of possible correlations between them. Furthermore, the capability of the ISO 2631 [19] approach as a road unevenness indicator was investigated, comparing the results with those obtained using the consolidated methods (RN, RQI_{Mich} and RQI_{Mn}). Finally, all the aforementioned approaches were compared with IRI, which is the most-used road roughness evaluation method worldwide, as stated in [25], where IRI specifications around the world are reported. In this way, this work intends to highlight the need of standardizing mean panel tests, adopting also homogeneous speed-related threshold values to be used for ride quality evaluation.

2. Ride Quality Evaluation Methods

Most of the ride quality indices taken into account in the present work are based on MPR. Each of them was developed by in-situ experiments, where different samples of drivers and road pavement sections were considered. The calculation of the above-mentioned indices for the road profile samples analyzed in the present work was performed as described in the following sections.

2.1. Ride Number (RN)—ASTM E 1489

The Ride Number (RN) is a mathematical processing of longitudinal profiles that allows the estimation of the subjective ride quality perceived by road users. The calculation is performed by means of the following Equations (1) and (2), reported in the ASTM E1489 standard [26] and developed by Karamihas and Sayers [27]:

$$RN = 5 \times e^{-160 \times (PI)} \tag{1}$$

where

$$PI = \sqrt{\frac{PI_L^2 + PI_R^2}{2}} \tag{2}$$

where PI_L and PI_R are the Profile Indices of the left and right wheel paths. They are the computed Root Mean Square (RMS) of the filtered slopes of the measured elevation profiles of the both wheel paths. The range of RN values is from 0 to 5.0, where an RN of 5.0 is considered to be a road inducing a perfect ride quality. With some exceptions, the wavelengths' range of interest for RN is similar to that of IRI, as reported in [2]. In particular, RN presents a higher sensitivity to low wavelengths than IRI, which has a greater sensitivity to wavelengths of 16 m or longer than RN.

2.2. Michigan Ride Quality Index (RQI_Mich)

RQI_{Mich} is a roughness evaluation method developed by the Michigan Department of State Highways (Lansing, MI, USA) in order to predict users' opinions from road profiles [28]. The calculation is based on a research study where users' opinions were linked to wavelength content of the profile elevation Power Spectral Density (PSD) functions. In particular, three significant wavebands were identified: 0.61–1.52 m for the short waveband; 1.52–7.62 m for the medium waveband; and from 7.62 to 15.24 m for the long waveband. The variance in each waveband is calculated using a filter process. Finally, the Michigan RQI is calculated using the following Equation (3):

$$RQI_{Mich} = 3.077 \times \ln(VAR_1 \times 10^8) + 6.154 \times \ln(VAR_2 \times 10^8) + 9.231 \times \ln(VAR_3 \times 10^8) - 141.85 \tag{3}$$

where VAR_1, VAR_2 and VAR_3 are, respectively, the variances of the profile in the long, medium and short wavebands. As reported by Lee et al. [24], an RQI_{Mich} value between 0 and 30 indicates excellent ride quality, a value from 31 to 54 it indicates good ride quality, while values from 55 to 70 indicate fair ride quality. Pavements having RQI_{Mich} values greater than 70 are considered to have poor ride quality.

2.3. Minnesota Ride Quality Index (RQI_Mn)

RQI_{Mn} is a roughness evaluation method developed by the Minnesota Department of Transportation (Mn/DOT, St. Paul, MN, USA) in order to take into account customer's opinion; correlating it with IRI values calculated for over 120 test sections as reported in the document "2015—Pavement Condition Annual Report" compiled by Mn/DOT [29]. As already stated, IRI is the roughness evaluation method that is most used worldwide, and the algorithm used for its calculation was developed by Sayers [30] and reported in ASTM E1926 [31]. Two different correlation equations were found, Equation (4) for bituminous pavements and Equation (5) for concrete pavements as specified in the document "An Overview of Mn/DOT's Pavement Condition Rating Procedures and Indices" [32]:

$$RQI_{Mn,flexible} = 5.697 - (2.104) \times \sqrt{IRI} \tag{4}$$

$$RQI_{Mn,rigid} = 6.634 - (2.813) \times \sqrt{IRI} \tag{5}$$

where IRI value is in (m/km). Considering the kind of road pavements analyzed in this work, only Equation (4), related to bituminous pavements, was taken into account. As for RN, the scaling rate range of RQI_{Mn} varies from 0 to 5.0, with the different ride quality categories specified in Table 1.

Table 1. Minnesota RQI categories and ranges.

Numerical Rating	Verbal Rating
4.1–5.0	Very Good
3.1–4.0	Good
2.1–3.0	Fair
1.1–2.0	Poor
0.0–1.0	Very Poor

2.4. Whole-body Vibration (a_{wz})—ISO 2631

An additional method that can be used for the evaluation of road customers' comfort is the process provided by ISO 2631 for comfort assessment in public transport. This method is based on the measurement of vertical acceleration inside road vehicles, which are used to determine *RMS* accelerations through the evaluation of *PSD* with regard to all 23 one-third-octave bands that represent the frequency range of interest for human response to vibrations (0.5–80 Hz) described in the ISO 2631 [19]. Once the *RMS* accelerations are known, it is possible to calculate the vertical weighted *RMS* acceleration (a_{wz}) using the following Equation (6):

$$a_{wz} = \sqrt{\sum_{i=1}^{23} \left(W_{k,i} \times a_{iz}^{RMS} \right)^2} \tag{6}$$

where $W_{k,i}$ are the frequency weightings in one-third-octave bands for seated positions, provided by the standard; and a_{iz} is the vertical *RMS* acceleration for the *i*-th one-third-octave band. Then, the calculated values can be compared with the threshold values proposed by ISO 2631 for public transport (Table 2), in order to identify the comfort level perceived by users in all roads sections.

The current standard does not contain clearly-defined vibration exposure limits between adjacent comfort levels, because many factors (e.g., user age, acoustic noise, temperature, etc.) combine to determine the degree to which discomfort will possibly be noted or tolerated. For this reason, the ISO standard provides several comfort levels introducing an overlapping zone between two adjacent levels. To determine the frequency-weighted vertical acceleration on users due to road roughness, several simulations were performed using the 8 degree of freedom (d.o.f.) full-car model developed by Cantisani and Loprencipe [18] and calibrated in order to represent the behavior of a common passenger car. In particular, a speed range from 30 to 130 km/h was considered in order to evaluate correlation trends between the a_{wz} and the other three aforementioned methods (*RN*, *RQI*$_{Mich}$ and *RQI*$_{Mn}$) as a function of the traveling velocity.

Table 2. Comfort levels related to a_{wz} threshold values as proposed by ISO 2631 for public transport.

a_{wz} Values (m/s^2)	Comfort Level
<0.315	Not uncomfortable
0.315–0.63	Little uncomfortable
0.5–1	Fairly uncomfortable
0.8–1.6	Uncomfortable
1.25–2.5	Very uncomfortable
>2.5	Extremely uncomfortable

3. Data Set of Road Profiles and Performed Comparative Analyses

3.1. Data Set of Road Profiles

A set of about 200 km of real road profiles, belonging to the Italian road network, was sampled with a spatial increment of 2.5 cm. For each lane, two paths (right and left) at the main rutting alignments were measured using a high-speed laser/inertial profilometer. Each profile path was divided into profile sections of 100 m, which is the most common length reference for road roughness evaluation (with regards to *IRI*) used in several countries, as reported by Múčka [25]. Thus, 1987 sections were taken into account. In order to characterize and classify the profile sample available, a preliminary analysis based on the ISO 8608 standard [33] was carried out. In particular, to classify road surface profiles according to the aforementioned standard, the *PSD* of elevations was calculated using Fast Fourier Transform (*FFT*) and the Hanning window. Then, the smoothing and fitting processes described in Loprencipe and Zoccali [34] were performed. As reported in Table 3, the real road profiles considered in this study belong only to the following classes: A (very good), B (good), C (average) and D (poor); with a significant predominance of the second one (class B profiles).

Table 3. Percentage of real profiles belonging to a specific ISO 8608 class.

Class A (Very Good) (%)	Class B (Good) (%)	Class C (Average) (%)	Class D (Poor) (%)
18.1	57.9	22.6	1.4

3.2. Comparative Analyses and Thresholds Adopted

The comparison of the different approaches previously described, involved both the search for possible correlations between these indices and by evaluating their ability to assess each single profile in the same way. For this reason, a study about possible ride quality evaluation agreements provided by RN, RQI_{Mich} and RQI_{Mn} methods was also carried out. To perform this kind of analysis, it was decided to consider, for each index, four ride quality levels as defined in Table 4; comparing, then, the assessment provided by the various comfort indices for each available profile data.

The adopted thresholds described in Table 4 had already been reported in some documents for RQI_{Mich} limit values [24], RQI_{Mn} [31] and a_{wz} [18]. In contrast, for RN the division of the rating scale reported in [26] was performed in order to be consistent with the four ride quality levels considered.

Table 4. Ride quality thresholds considered for evaluation agreement analysis.

Ride Quality Level	RN	RQI_{Mich}	RQI_{Mn}	a_{wz} (m/s^2)
Very Good	4.1–5.0	0–30	4.1–5.0	<0.315
Good/Fair	3.1–4.0	31–54	3.1–4.0	0.315–0.565
Mediocre	2.1–3.0	55–70	2.1–3.0	0.565–0.9
Poor and Very Poor	0.0–2.0	>70	0.0–2.0	>0.9

The resulting percentage of agreement (PoA) between the two different methods was then calculated according to Equation (7):

$$PoA_{IJ} = \frac{N_{IJ,verygood} + N_{IJ,good} + N_{IJ,mediocre} + N_{IJ,poor}}{N_{tot}} \times 100 \qquad (7)$$

where $N_{IJ,verygood}$, $N_{IJ,good}$, $N_{IJ,mediocre}$, and $N_{IJ,poor}$ are the number of profiles evaluated as providing, respectively, a very good, good, mediocre and poor ride quality level by both ride quality evaluation methods I and J ($I \neq J$). N_{tot} is the total number of examined road profile samples.

It is useful to remember that the ride quality level thresholds for RN, RQI_{Mich} and RQI_{Mn} are not speed-related. Although the limits provided by the ISO 2631 are the same, the a_{wz} value strongly depends, by means of the mechanical model used, on the velocity considered for its calculation.

In order to provide a more detailed analysis of the examined ride quality assessment approaches, these indices were also compared to IRI, which is the world's most popular road roughness evaluation method.

3.3. International Roughness Index (IRI) Thresholds

The IRI was elaborated from a World Bank study in the 1980s [30]. It is based on a mathematical model called quarter-car and was developed in order to assess not only the ride quality on road pavements, but also other detrimental effects, such as dynamic load increment (on both vehicle and pavement) due the presence of irregularities on road surfaces. The calculation is performed using a model that calculates the simulated suspension motion on a profile and divides the sum by the distance traveled according to the Equation (8):

$$IRI = \frac{1}{l} \int_0^{l/v} |\dot{z}_s - \dot{z}_u| dt \tag{8}$$

where l is the length of the profile in km, v is the simulated speed equal to 80 km/h, $\dot{z}_s$ is the time derivative of vertical displacement of the sprung mass in m, and $\dot{z}_u$ is the time derivative of vertical displacement of the unsprung mass in m. The final value is expressed in slope units (e.g., m/km or mm/m). In the present work, the algorithm proposed by the ASTM E1926 standard [31] for *IRI* calculation was used.

As reported in [25], there is a high heterogeneity of *IRI* thresholds adopted around the world. In fact, *IRI* limit values mainly depends from several aspects: road surface type (i.e., asphalt or cement concrete pavements), road functional category, average annual daily traffic (*AADT*), legal speed limit and segment length considered for *IRI* calculation. In addition, they change depending on whether we talk about new, reconstructed or in-service roads. As already stated, the most common segment length indicated in non-US countries is equal to 100 m [25].

Between the different parameters affecting *IRI* specifications, surely the most important is the maximum traveling velocity allowed on the road, whose roughness level is meant to be assessed. For this reason, some authors [18,35] have proposed speed-related *IRI* thresholds to be used for the evaluation of ride quality. In particular, Yu et al. [35] defined five ride quality levels providing for each of them the corresponding limit values based on the jolt and jerk experienced by the raters within a speed range from 10 km/h to 120 km/h. They analyzed 102 longitudinal profiles with a length of 150 m collected in the Strategic Highway Research Program (*SHRP*) Long-Term Pavement Performance (*LTPP*). In Table 5, the suggested thresholds found in correspondence of some velocities are shown.

Table 5. IRI thresholds at different speeds suggested by Yu et al. [35].

Ride Quality Level	IRI Thresholds at Different Speeds (m/km)					
	20 km/h	40 km/h	60 km/h	80 km/h	100 km/h	120 km/h
Very Good	<5.72	<2.86	<1.90	<1.43	<1.14	<0.95
Good	5.72–8.99	2.86–4.49	1.90–2.99	1.43–2.24	1.14–1.79	0.95–1.49
Fair	9.00–11.39	4.50–5.69	3.00–3.79	2.25–2.84	1.80–2.27	1.50–1.89
Mediocre	11.40–16.16	5.70–8.08	3.80–5.40	2.85–4.05	2.28–3.24	1.90–2.70
Poor	>16.16	>8.08	>5.40	>4.05	>3.24	>2.70

Cantisani and Loprencipe [18] examined 124 *LTPP* profiles of 320 m, defining four ride quality levels and calculating the corresponding thresholds from the relation found between *IRI* and the vertical frequency-weighted acceleration (a_{wz}) at several speeds (from 30 to 90 km/h), by means of the 8 d.o.f. full-car model previously described in Section 2.4. As can be seen by comparing the *IRI* thresholds suggested in [18] (reported in Table 6) with the ones in Table 5, a generally good agreement between the two aforementioned studies can be found.

Table 6. IRI thresholds at different speeds suggested by Cantisani and Loprencipe [18].

Ride Quality Level	IRI Thresholds at Different Speeds (m/km)						
	30 km/h	40 km/h	50 km/h	60 km/h	70 km/h	80 km/h	90 km/h
Very Good	<4.17	<3.41	<2.98	<1.87	<1.60	<1.42	<1.15
Good/Fair	4.17–8.34	3.41–6.83	2.98–5.95	1.87–3.73	1.60–3.20	1.42–2.84	1.15–2.31
Mediocre	8.34–11.92	6.83–9.75	5.95–8.51	3.73–5.33	3.20–4.58	2.84–4.06	2.31–3.30
Poor	>11.92	>9.75	>8.51	>5.33	>4.58	>4.06	>3.30

The importance of considering appropriate speed-related *IRI* thresholds is also underlined by Múčka in [25], where a suggested relation for *IRI* threshold values as function of velocity limit is reported. In particular, Múčka did not suggest any specific *IRI* limit values, but he provided Equation (9) to be adopted:

$$IRI(v_2) = \left(\frac{v_1}{v_2}\right)^{0.5} \times IRI(v_1) \tag{9}$$

which is based on the assumption that the same level of vibration response can be achieved for two different speeds v_1 and v_2. Adopting the aforementioned equation would mean that the condition of the same quarter-car suspension relative velocity response for the two different speeds (v_1 and v_2) is met.

4. Results and Discussion

A preliminary study concerning the analysis of the *IRI* relation reported in Equation (9), which was introduced by Múčka [25], talking about international *IRI* specifications for new/reconstructed roads. Specifically, it was decided to compare the ratio between *IRI* thresholds at two adjacent velocities (e.g., *IRI*(30)/*IRI*(20) or *IRI*(80)/*IRI*(70)), calculated using Equation (9), and obtained by employing both the threshold values suggested by Yu et al. [35] and those suggested by Cantisani and Loprencipe [18] with regard to very good ride quality level. It can be supposed, in fact, that this ride quality level characterizes new or reconstructed roads. The results of the above-mentioned comparison are then depicted in Figure 1, where the results obtained by applying the procedure described by Cantisani and Loprencipe [18] to the real profile samples considered in the present work and taking into account a segment length of 100 m are also reported.

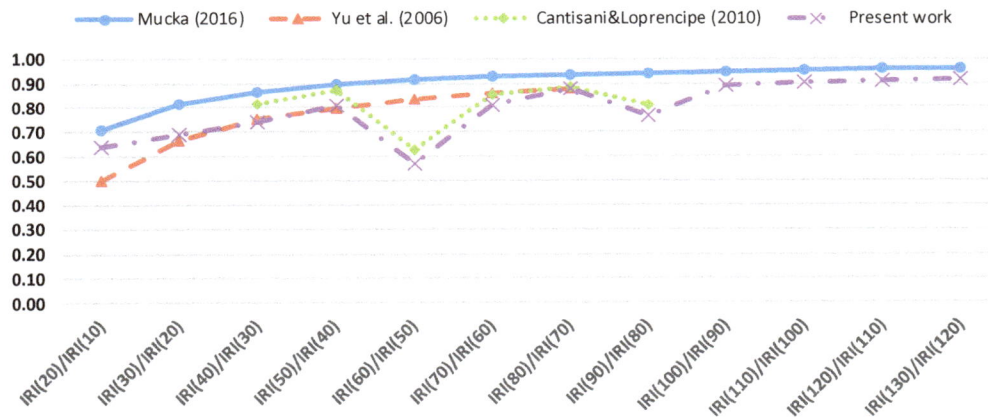

Figure 1. Ratio between *IRI* thresholds at different speeds provided by various studies.

As can be seen, similar trends are found for all of the examined studies and, in particular, values very close to the ones presented in [18] are found in the present work, where the same simulation model but different road profile samples were used. Furthermore, some anomalies in both of the latter approaches' trends can be noted to correspondend to the *IRI* thresholds ratio between speeds of 60 and 50 km/h, and between 90 and 80 km/h. This unexpected behavior was probably due to the mechanical parameters considered in the simulation model at these speeds (i.e., the mechanical properties were defined as a function of the traveling velocity).

Before proceeding with the analysis of the a_{wz} values at different speeds, a comparison between the results obtained for the other three users' comfort evaluation methods was performed. Because no indications were found about section length to be used for their calculation, it was decided to consider the same length of 100 m as commonly adopted for *IRI* calculation. In this way, all the wavelengths' contents of interest for the various examined ride quality methods are taken into account. In particular, pretty good correlations between them were found, as can be seen in Figures 2–4 where R^2 values within 0.78–0.95 are shown. The highest R^2 value (0.95) was obtained for RQI_{Mich}-RN regression equation (Figure 2).

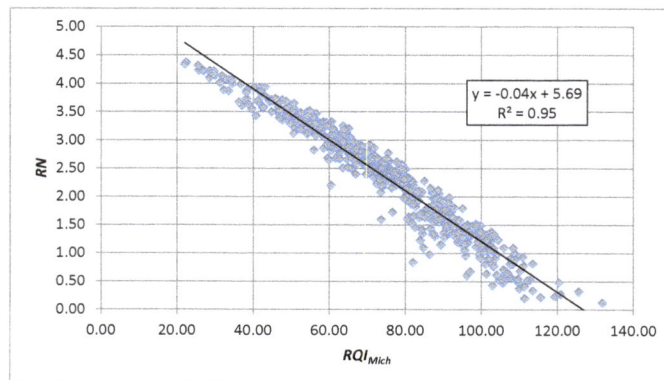

Figure 2. Correlation between RQI_{Mich} and RN.

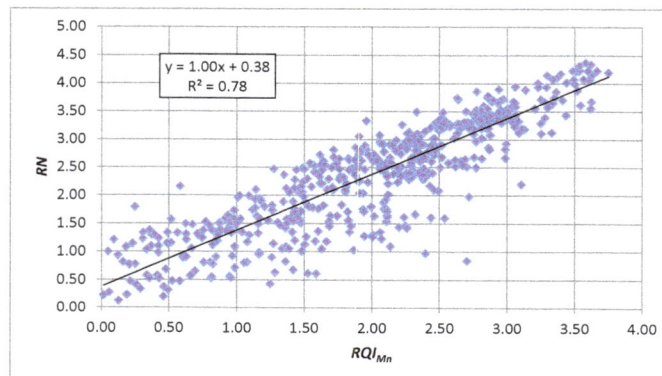

Figure 3. Correlation between RQI_{Mn} and RN.

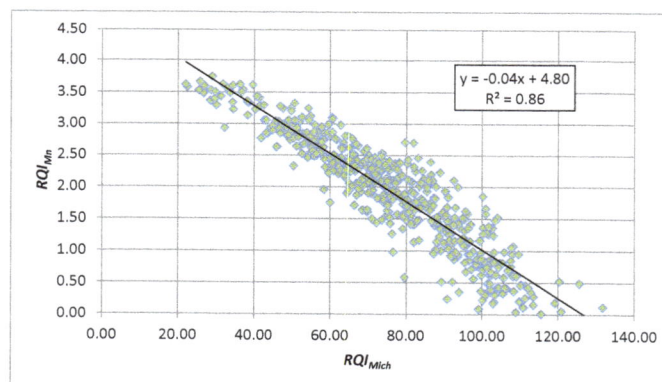

Figure 4. Correlation between RQI_{Mich} and RQI_{Mn}.

Although quite good correlations were found between these indices, each index presents a specific scale rating based on the performed panel rating tests, for which a standardization does not exist.

In fact, different distributions of the real profile samples among the four ride quality levels were obtained, based on the method that was being considered, as can be seen in Figure 5. Thus, the choice of road ride quality evaluation method meaningfully influences the maintenance actions planning.

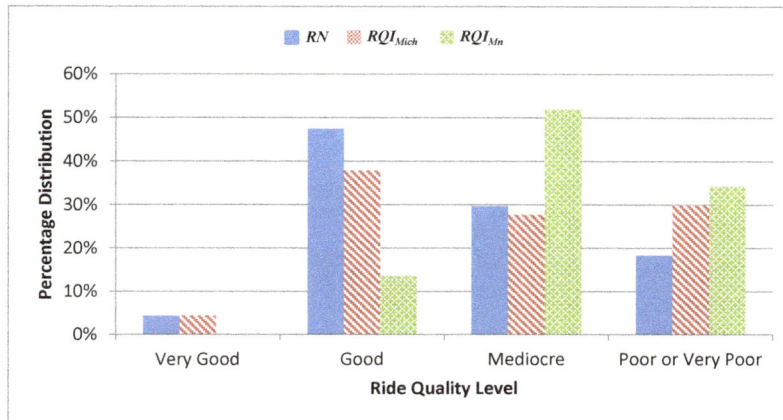

Figure 5. Ride Quality Level Percentage Distribution for the real profiles samples.

Counting the amount of profiles evaluated in the same way by the different indicators, a percentage of agreement (*PoA*, calculated according to Equation (7) for each pair of indices) greater than 60% was obtained just for RQI_{Mich} and RQI_{Mn} comparison (see Figure 6). The worst agreement, instead, was found between RN and RQI_{Mn}. This result could be expected by looking at Figure 3. In fact, both these indices have the same rating scale (from 0 to 5.0) and ride quality categories (see Table 4), but the linear regression equation found presents an intercept value equal to 0.38; which means that a switch between the rating scales of the two methods exists. In particular, the RQI_{Mn} provides a more severe evaluation of road profiles, as highlighted by the results shown in Figure 5.

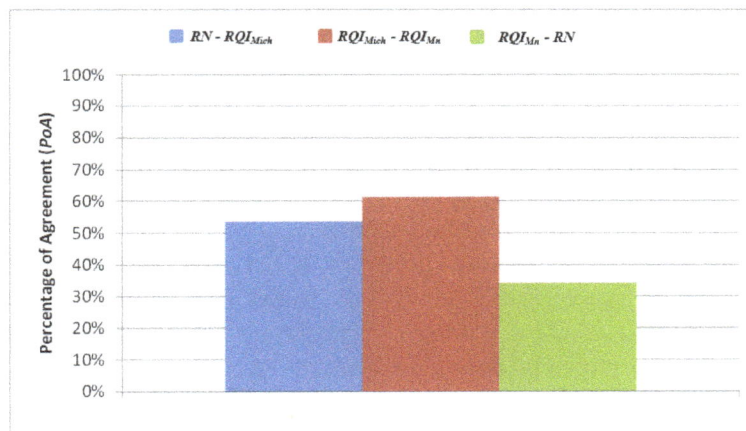

Figure 6. Percentage of agreement (*PoA*) between the three ride comfort evaluation methods.

As already stated in Section 2.4, the results obtained for RN, RQI_{Mich} and RQI_{Mn} were also compared with the vertical frequency-weighted acceleration a_{wz} calculated at several speeds (from 30 to 130 km/h). By analogy with the study of profile evaluation percentage agreement (*PoA*), calculated according to Equation (7), reported above, the same analysis was also carried out for the a_{wz} method, using the thresholds shown in Table 4, defined as the middle point of the overlapping zone provided by ISO 2631 for two adjacent comfort levels.

As can be seen in Figure 7, varying the traveling speed of the simulation vehicle, significant changes in the percentage agreement were found for all three methods with regards to a_{wz} index values.

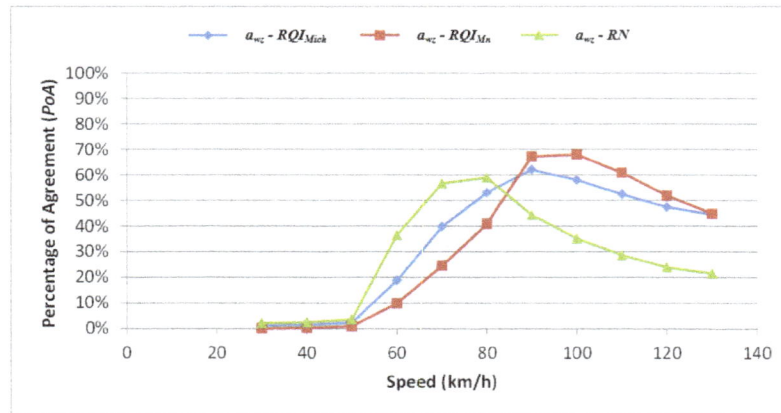

Figure 7. Percentage of agreement (*PoA*) between *RN*, RQI_{Mich} and RQI_{Mn} with ISO 2631 a_{wz}.

The highest values of the percentage agreement between a_{wz} and each of the other three ride quality evaluation methods were found to be within the range 55%–70%. Although these results are not too high, due to the fact that the panel rating tests have been performed using different vehicles, general considerations can be deduced. The best percentage agreement for the three ride quality evaluation methods, in fact, was found to correspond to different velocities. In particular, for *RN* it was found for a vehicle traveling at 80 km/h, while for RQI_{Mich} at 90 km/h. For RQI_{Mn}, the greatest percentage was found for simulation speeds equal to 100 km/h. These results could be an explanation of the different correlations found between the three indices (*RN*, RQI_{Mn} and RQI_{Mich}), which were in all cases lower than 65%. In fact, the ride quality thresholds of the aforementioned methods seem to be calibrated for different speeds of reference. A confirmation of the goodness of the results obtained can be found in the algorithm for the calculation of the *RN*, where a parameter similar to the velocity used in *IRI* calculation is set at 80 km/h, which is the value of the speed found for the maximum percentage agreement.

Looking at Figure 7, it can be noted that none of the three consolidated indices (*RN*, RQI_{Mich} and RQI_{Mn}) seem to be adequate for evaluating users' comfort on urban roads or, in general, on roads having maximum legal speed limits lower than 50 km/h.

In addition to the profile evaluation agreement study, the correlations between the three consolidated ride quality evaluation methods and a_{wz} were investigated. The R^2 values found in correspondence of the speeds at which the highest profile assessment agreements varies from 0.66 (RN-a_{wz}) to 0.75 (RQI_{Mn}-a_{wz}); as can be noted in Figures 8–10.

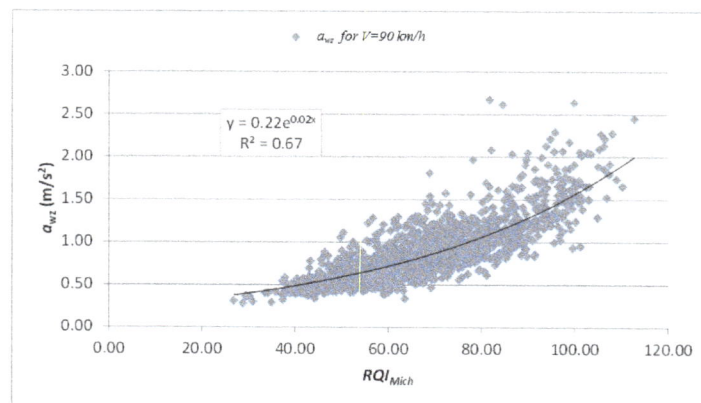

Figure 8. Correlation between RQI_{Mich} and a_{wz} at 90 km/h.

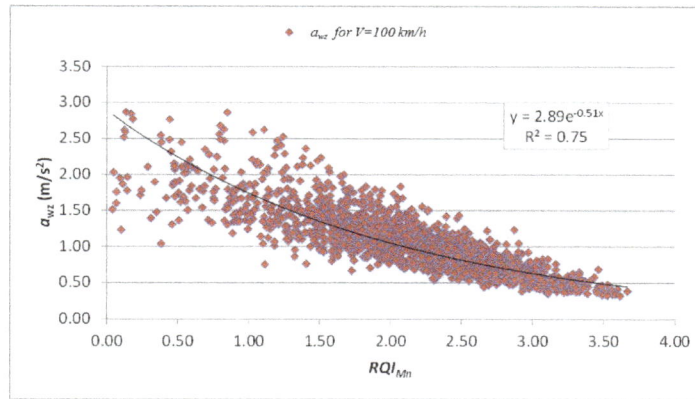

Figure 9. Correlation between RQI_{Mn} and a_{wz} at 100 km/h.

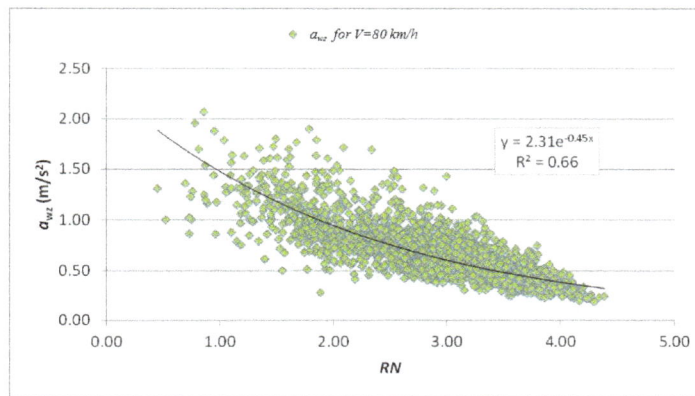

Figure 10. Correlation between RN and a_{wz} at 80 km/h.

Unlike for the other methods, the use of the a_{wz} allows the taking into account of the different users' perceptions of road conditions based on the traveling speed, as can be noted in Figure 11, where the distribution among the four ride quality levels of the real profiles set is reported for different velocities. To simplify the vision of the plot, not the whole range of speed considered (30–130 km/h), but just speeds within the range from 70 up to 110 km/h are represented. As can be seen, significant variation in ride quality judgement is obtained.

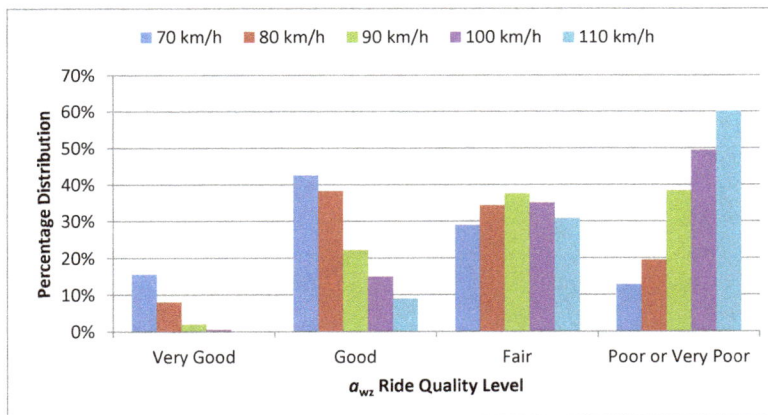

Figure 11. Ride quality level percentage distribution for the real profile samples calculating the a_{wz} at different speeds.

Evaluating the correlation between the various ride quality methods considered in this paper with IRI, R^2 values of 0.78 and 0.85 were found respectively, for RN (Figure 12) and RQI_{Mich} (Figure 13).

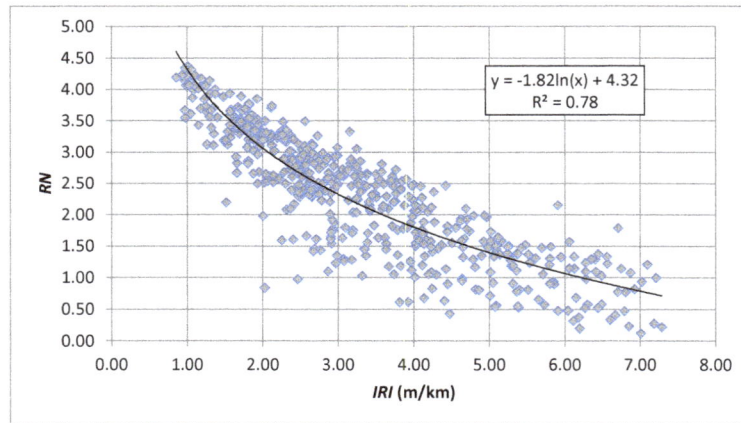

Figure 12. Correlation between RN and IRI.

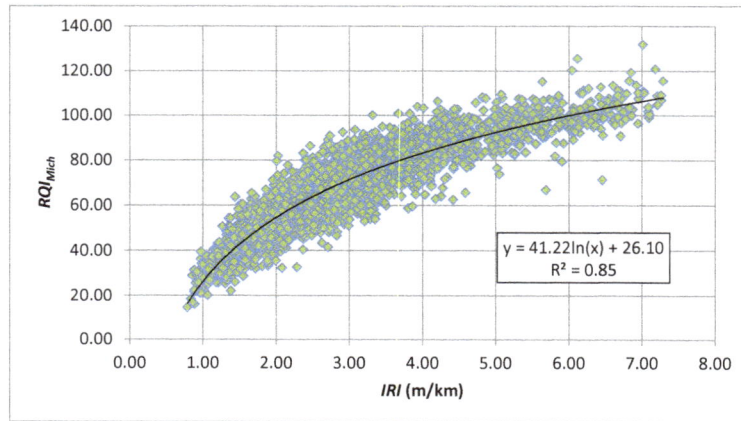

Figure 13. Correlation between RQI_{Mich} and IRI.

Obviously, a perfect correlation ($R^2 = 1$) was found between RQI_{Mn} and IRI (Figure 14), since the calculation of the first one is based on IRI values using Equation (4) previously described.

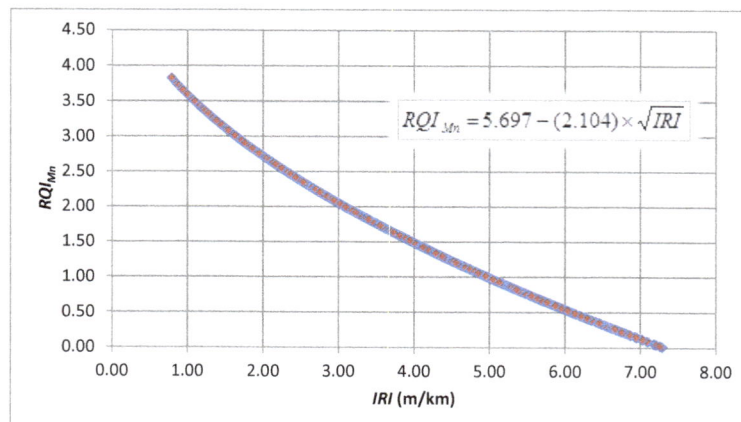

Figure 14. Correlation between RQI_{Mn} and IRI.

Starting from the regression equation found for each ride quality evaluation index and the thresholds reported in Table 4 for the different ride quality levels (very good, good, mediocre and poor), it is possible to determine the corresponding *IRI* limit values. The results are then reported in Table 7.

Table 7. *IRI* thresholds based on the different ride quality indices limit values in (m/km).

Ride Quality Level	IRI (RN)	IRI (RQI$_{Mich}$)	IRI (RQI$_{Mn}$)
Very Good	<1.13	<1.10	<0.58
Good/Fair	1.13–1.95	1.10–1.97	0.58–1.92
Mediocre	1.95–3.39	1.97–2.90	1.92–2.92
Poor and Very Poor	>3.39	>2.90	>2.92

As can be seen, looking at the thresholds related to the very good ride quality level the RQI_{Mn} was confirmed to be the more conservative approach, while the *RN* represents a less conservative approach. Although there is not perfect matching between the *IRI* thresholds calculated from each method and the limit values suggested by Yu et al. [35] and/or Cantisani and Loprencipe [18] corresponding to all ride quality levels, it is nevertheless possible to note that all aforementioned ride quality evaluation methods were mainly developed to be used on roads characterized by speed limits within the range from 80 to 120 km/h.

Considering, then, the *IRI* specifications suggested by the two aforementioned studies, already reported in Tables 5 and 6, the percentage agreement in the evaluation of the examined real profile samples between *IRI* and a_{wz} approaches at various speeds was evaluated. As can be seen in Figure 15, in this case for all the considered velocities percentages greater than 50% were always found.

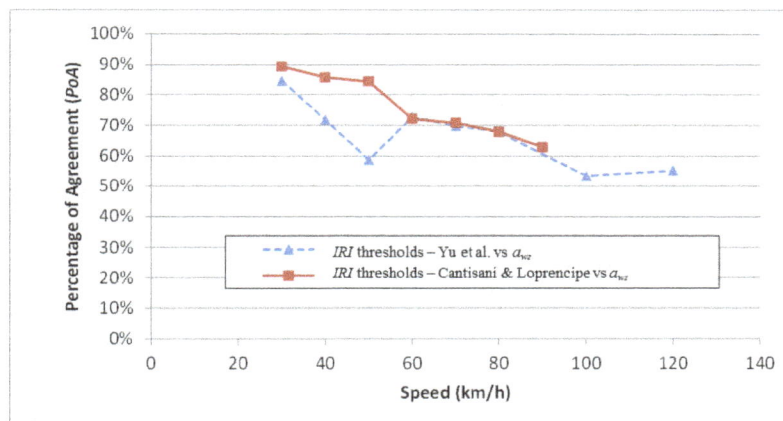

Figure 15. Percentage agreement between *IRI* thresholds suggested by Yu et al. [35] and Cantisani and Loprencipe [18] with ISO 2631 a_{wz}.

These results highlight the chance of using the a_{wz} approach for road ride quality evaluation for a wide range of traveling speeds as an alternative method to *IRI*, mainly for velocities lower than 50 km/h, where the percentage agreement is greater than 80%. Of course, the proposed approach will need calibration and validation phases in order to be correctly used; in fact, the a_{wz} is strongly affected by the type of vehicle considered for the measurement.

In addition, Kirbaş and Karaşahin [36] found a good correlation between a_{wz} and Pavement Condition Index (*PCI*). Therefore, using the a_{wz} approach at a preliminary step in order to locate priorities along road networks seems to be possible, although some distresses (e.g., crack distress) do not meaningfully affect this index. Once critical sections have been located, it would be then possible to plan adequate inspections and surveys in order to understand the causes of the distresses, and then select the most appropriate maintenance actions.

All of the analyses related to the *IRI* method have also underlined the need to homogenize its threshold values by defining appropriate speed-related limits to be adopted by road agencies.

5. Conclusions

In this paper, the analysis of a set of real road profiles was carried out according to different ride quality evaluation methods, comparing then the assessment provided by each of them.

In particular, the agreement in the assessment of road profiles and their correlation was investigated. The main results can be summarized as follows:

- Although pretty good correlations (R^2 = 0.78–0.95) exist between the three consolidated ride quality evaluation methods (RN, RQI_{Mich} and RQI_{Mn}) taken into account in the present work, the assessment of the same examined road profile can significantly differ according to the approach used. In this sense, the need to standardize *MPR* in order to obtain homogeneity in ride quality evaluation is clear.

- The RQI_{Mn} was found to be the most conservative indicator; in fact, its scale rating seems to be calibrated for a speed greater than 80 km/h (around 100 km/h), as shown by the results obtained using a simulation model representative of the behavior of a typical passenger car. Similar results were found evaluating the relations and correlations between RN, RQI_{Mich} and RQI_{Mn} with *IRI* (R^2 respectively equal to 0.78, 0.85 and 1). Using the regression equations found *IRI* thresholds corresponding to each ride quality index limit values were calculated. This analysis confirmed that the RQI_{Mn} is the most conservative approach among the three ride quality assessment methods. It is also important to highlight that the RQI_{Mn} calculation is based on *IRI* values, which is the most-used road roughness index worldwide.

- None of the consolidated methods (RN, RQI_{Mich} and RQI_{Mn}) present thresholds appropriate for use on urban roads that have legal speed limits lower than 50 km/h. By contrast, some *IRI* specifications intended for use on urban roads have been adopted in some countries and more specific speed-related threshold values have been suggested in literature. Although much attention is paid to this aspect of the use of *IRI* for road roughness evaluation, there is still the need to homogenize the criteria and the values to be adopted for the various road categories.

Considering the last point, the use of the a_{wz} method seems to be a promising and valid alternative for the evaluation of ride quality, with particular attention to urban road networks, where the use of the common profilometers (both contact and no-contact types) presents some application limits. Furthermore, it is a speed-related approach, which means that road sections having different legal speed limits can be properly assessed. However, deeper studies on the influence of different types of road vehicle and of irregularities at different velocities should be carried out, together with panel rating tests, in order to define appropriate a_{wz} thresholds to be used for ride quality evaluation.

Author Contributions: Giuseppe Loprencipe and Pablo Zoccali carried out the data and results analyses of the work. Pablo Zoccali wrote the manuscript and was in charge of the overall outline and editing of the manuscript. He was involved in the revision and completion of the work. Giuseppe Loprencipe contributed to the outline as well as to the revision, completion, and editing of the manuscript.

Conflicts of Interest: The authors declare no conflict of interest.

Abbreviations

The following abbreviations are used in this manuscript:

a_{wz}	Frequency-weighted vertical acceleration
IRI	International Roughness Index
LTPP	Long-Term Pavement Performance
MPR	Mean Panel Rating

PCI	Pavement Condition Index
PI	Profile Index
PoA	Percentage of Agreement
PMS	Pavement Management System
PSD	Power Spectral Density
RMS	Root Mean Square
RN	Ride Number
RQI	Ride Quality Index
RQI_{Mich}	Michigan Ride Quality Index
RQI_{Mn}	Minnesota Ride Quality Index
SHRP	Strategic Highway Research Program

References

1. Loprencipe, G.; Cantisani, G.; di Mascio, P. Global assessment method of road distresses. In *Life-Cycle of Structural Systems: Design, Assessment, Maintenance and Management—Proceedings of the 4th International Symposium on Life-Cycle Civil. Engineering, Proceedings of the International Association For Life-cycle Civil Engineering 2014, Tokyo, Japan, 16 November 2014*; CRC Press/Balkema: Leiden, The Netherlands, 2015; Available online: https://www.crcpress.com/Life-Cycle-of-Structural-Systems-Design-Assessment-Maintenance-and-Management/Furuta-Frangopol-Akiyama/p/book/9781138001206 (accessed on 28 February 2017).

2. Sayers, M.W.; Karamihas, S.M. *The Little Book of Profiling*; The Regent of the University of Michigan: Ann Arbor, MI, USA, 1998; Volume 2.

3. Bonin, G.; Folino, N.; Loprencipe, G.; Oliverio, R.C.; Polizzotti, S.; Teltayev, B. Development of a Road Asset Management System in Kazakhstan. In Proceedings of the TIS 2017 International Congress on Transport—Infrastructure and Systems, Rome, Italy, 10–12 April 2017. Available online: https://www.crcpress.com/Transport-Infrastructure-and-Systems-Proceedings-of-the-AIIT-International/DellAcqua-Wegman/p/book/9781138030091 (accessed on 28 February 2017).

4. Han, D.; Kobayashi, K. Criteria for the development and improvement of PMS models. *KSCE J. Civ. Eng.* **2013**, *17*, 1302–1316. [CrossRef]

5. Shahin, M.Y. *Pavement Management for Airports, Roads, and Parking Lots*, 2nd ed.; Springer: New York, NY, USA, 2005.

6. Loprencipe, G.; Pantuso, A.; di Mascio, P. Sustainable Pavement Management System in Urban Areas Considering the Vehicle Operating Costs. *Sustainability* **2017**, *9*, 453. [CrossRef]

7. Chatti, K.; Lee, D. Development of new profile-based truck dynamic load index. *Transp. Res. Rec. J. Transp. Res. Board* **2002**, *1806*, 149–159. [CrossRef]

8. Ahlin, K.; Granlund, J.; Lundström, R. Whole-Body Vibration When Riding on Rough Roads—A shocking Study. *Swed. Natl. Road Adm.* **2000**, *31*, 81.

9. Cantisani, G.; Fascinelli, G.; Loprencipe, G. Urban road noise: The contribution of pavement discontinuities. In *ICSDEC 2012: Developing the Frontier of Sustainable Design, Engineering, and Construction, Proceedings of the International Conference on Sustainable Design, Engineering, and Construction 2012, Fort Worth, TX, USA, 7–9 November 2012*; Elsevier: Amsterdam, The Netherlands, 2012. [CrossRef]

10. Gillespie, T.D. Everything You Always Wanted to Know about the IRI, but Were Afraid to Ask. In Proceedings of Road Profile Users Group Meeting, Lincoln, NE, USA, 22–24 September 1992.

11. Kropáč, O.; Múčka, P. Be careful when using the International Roughness Index as an indicator of road unevenness. *J. Sound Vib.* **2005**, *287*, 989–1003. [CrossRef]

12. Loizos, A.; Plati, C. An alternative approach to pavement roughness evaluation. *Int. J. Pavement Eng.* **2008**, *9*, 69–78. [CrossRef]

13. Múčka, P.; Granlund, J. Is the road quality still better? *J. Transp. Eng.* **2012**, *138*, 1520–1529. [CrossRef]

14. Múčka, P. Current approaches to quantify the longitudinal road roughness. *Int. J. Pavement Eng.* **2016**, *17*, 659–679. [CrossRef]

15. Kropáč, O.; Múčka, P. Effects of longitudinal road waviness on vehicle vibration response. *Veh. Syst. Dyn.* **2009**, *47*, 135–153. [CrossRef]

16. Ahlin, K.; Granlund, N.O.J. Relating Road Roughness and Vehicle Speeds to Human Whole Body Vibration and Exposure Limits. *Int. J. Pavement Eng.* **2002**, *3*, 207–216. [CrossRef]

17. Múčka, P. Correlation among road unevenness indicators and vehicle vibration response. *Int. J. Pavement Eng.* **2013**, *139*, 771–786. [CrossRef]

18. Cantisani, G.; Loprencipe, G. Road roughness and whole body vibration: Evaluation tools and comfort limits. *J. Transp. Eng.* **2010**, *136*, 818–826. [CrossRef]

19. *ISO 2631-1:1997—Mechanical Vibration and Shock—Evaluation of Human Exposure to Whole-Body Vibration—Part 1: General Requirements*; ISO: Geneve, Switzerland, 1997; Multiple, Distributed through American National Standards Institute (ANSI).

20. Fichera, G.; Scionti, M.; Garescì, F. Experimental correlation between the road roughness and the comfort perceived in bus cabins. *SAE Tech. Pap.* **2007**, *1*, 0352. [CrossRef]

21. Wang, S.; Zhang, J.; Yang, Z. Experiment on Asphalt Pavement Roughness Evaluation Based on Passengers' Physiological and Psychological Reaction. In *ICCTP 2010: Integrated Transportation Systems: Green, Intelligent, Reliable, Proceedings of the 10th International Conference of Chinese Transportation Professionals, Beijing, China, 4–8 August 2010*; Wei, H., Wang, Y., Rong, J., Weng, J. Eds.; American Society of Civil Engineers: Reston, VA, USA, 2010; pp. 3852–3863.

22. Zhang, J.; Du, Y.; Su, R. Investigating the relationship between pavement roughness and heart rate variability by road driving test. In Proceedings of the 3rd International Conference on Road Safety and Simulation, Washington, DC, USA, 14–16 September 2011.

23. Titi, H.H.; Rasoulian, M. Evaluation of Smoothness of Louisiana Pavements Based on International Roughness Index and Ride Number. *Jordan J. Civ. Eng.* **2008**, *2*, 238–249.

24. Lee, D.; Chatti, K.; Baladi, G. Use of Distress and Ride Quality Data to Determine Roughness Thresholds for Smoothing Pavements as a Preventive Maintenance Action. *Transp. Res. Rec. J. Transp. Res. Board* **2002**, *1816*, 43–55. [CrossRef]

25. Múčka, P. International Roughness Index specifications around the world. *Road Mat. Pavement Des.* **2017**, *18*, 929–965. [CrossRef]

26. American Society of Testing and Materials. *Standard Practice for Computing Ride Number of Roads from Longitudinal Profile Measurements Made by an Inertial Profile Measuring Device*; ASTM E1489-98 International: West Conshohocken, PA, USA, 2013.

27. Karamihas, S.M.; Sayers, M.W. *Interpretation of Road Roughness Profile Data*; FHWA/RD-96/101; Federal Highway Administration: McLean, VA, USA, 1996.

28. Karamihas, S.M. *Critical Profiler Accuracy Requirements*; UMTRI 2005-24; University of Michigan: Ann Arbor, MI, USA, 2005.

29. 2015 Pavement Condition Annual Report. Minnesota Department of Transportation. Available online: http://www.dot.state.mn.us/materials/pvmtmgmtdocs/AnnualReport_2015.pdf (accessed on 28 February 2017).

30. Sayers, M.W. On the calculation of international roughness index from longitudinal road profile. *Transp. Res. Rec.* **1995**, *1501*, 1–12.

31. American Society of Testing and Materials. *Standard Practice for Computing International Roughness Index of Roads from Longitudinal Profile Measurements*; ASTM E1926:08 International: West Conshohocken, PA, USA, 2015.

32. Janisch, D. An Overview of Mn/DOT's Pavement Condition Rating Procedures and Indices. Available online: http://www.dot.state.mn.us/materials/pvmtmgmtdocs/Rating_Overview_State.pdf (accessed on 28 February 2017).

33. *ISO 8608:1995—Mechanical Vibration—Road Surface Profiles—Reporting of Measured Data*; ISO: Geneve, Switzerland, 1995; Multiple, Distributed through American National Standards Institute (ANSI).

34. Loprencipe, G.; Zoccali, P. Use of generated artificial road profiles in road roughness evaluation. *J. Mod. Transp.* **2017**, *25*, 23–34. [CrossRef]

35. Yu, J.; Chou, E.; Yau, J.T. Development of speed-related ride quality thresholds using international roughness index. *Transp. Res. Rec. J. Transp. Res. Board.* **2006**, *1*, 47–53. [CrossRef]

36. Kirbaş, U.; Karaşahin, M. Investigation of ride comfort limits on urban asphalt concrete pavements. *Int. J. Pavement Eng.* **2016**, 1–7. [CrossRef]

10

Influence of the Electrolyte Concentration on the Smooth TiO$_2$ Anodic Coatings on Ti-6Al-4V

María Laura Vera [1,2,*], **Ángeles Colaccio** [2], **Mario Roberto Rosenberger** [1,2], **Carlos Enrique Schvezov** [1,2] and **Alicia Esther Ares** [1,2,*]

[1] Instituto de Materiales de Misiones (IMAM), CONICET-UNaM, Posadas 3300, Misiones, Argentina; rrmario@fceqyn.unam.edu.ar (M.R.R.); schvezov@fceqyn.unam.edu.ar (C.E.S.)

[2] Facultad de Ciencias Exactas, Químicas y Naturales (FCEQyN), UNaM, Posadas 3300, Misiones, Argentina; angelescolaccio@gmail.com

* Correspondence: lauravera@fceqyn.unam.edu.ar (M.L.V.); aares@fceqyn.unam.edu.ar (A.E.A.)

Academic Editor: S. D. Worley

Abstract: To obtain smooth TiO$_2$ coatings for building a new design of Ti-6Al-4V heart valve, the anodic oxidation technique in pre-spark conditions was evaluated. TiO$_2$ coating is necessary for its recognized biocompatibility and corrosion resistance. A required feature on surfaces in contact with blood is a low level of roughness ($R_a \leq 50$ nm) that does not favor the formation of blood clots. The present paper compares the coatings obtained by anodic oxidation of the Ti-6Al-4V alloy using H$_2$SO$_4$ at different concentrations (0.1–4 M) as electrolyte and applying different voltages (from 20 to 70 V). Color and morphological analysis of coatings are performed using optical and scanning microscopy. The crystalline phases were analyzed by glancing X-ray diffraction. By varying the applied voltage, different interference colors coatings were obtained. The differences in morphologies of the coatings caused by changes in acid concentration are more evident at high voltages, limiting the oxidation conditions for the desired application. Anatase phase was detected from 70 V for 1 M H$_2$SO$_4$. An increase in the concentration of H$_2$SO$_4$ decreases the voltage at which the transformation of amorphous to crystalline coatings occurs; i.e., with 4 M H$_2$SO$_4$, the anatase phase appears at 60 V.

Keywords: anodic oxidation; titanium dioxide; electrolyte concentration; sulfuric acid

1. Introduction

With the objective of building a new mechanical heart valve design [1], the Ti-6Al-4V alloy coated with titanium dioxide (TiO$_2$) has been selected as the building material [2]. TiO$_2$ coatings have demonstrated hemocompatibility properties appropriate for use in this type of prosthesis [3–5].

Ti-6Al-4V alloy (grade 5 in ASTM B367) is an α/β Ti alloy widely used in biomedical applications due to its corrosion resistance and high bio- and hemocompatibility [6,7]. In general, the recognized properties of Ti alloys are mainly due to the formation of a natural TiO$_2$ oxide at room temperature, which can reach a thickness of 2–10 nm [6,7]. This native thin film oxide often has a high density of defects (mainly cracks), which reduce mechanical properties such as wear resistance or chemical properties such as corrosion resistance [8]. Therefore, it becomes necessary to add thicker and more protective coatings than the TiO$_2$ natural oxide to improve the properties. Anodic oxidation is a viable technique due to its low cost, simplicity of application, and control of the coatings' characteristics. This electrochemical process allows coatings of oxides to be obtained that have greater thickness and density than those that are naturally grown [9,10]. Among the variables of this technique that most affect the characteristics of the oxide (thickness, color, homogeneity, roughness, crystalline structure, etc.) are current density, applied voltage, anodizing time, temperature, conductivity, and pH of the electrolyte [9–11].

Two characteristics required for coatings that will be in contact with blood are homogeneity and a low level of roughness ($R_a \leq 50$ nm) to avoid the promotion of blood clots (thrombosis) [2,12]. Another property that influences the bio- and hemocompatibility of TiO_2 coatings is their crystalline structure. At low pressure, TiO_2 can present three crystalline phases: Anatase, rutile, or brookite. According to the literature, both amorphous phases and crystalline phases such as anatase and rutile would be bio-compatible [12].

The anodic oxidation technique—with oxidation voltages below the production of spark discharge phenomenon—can be used to obtain homogeneous TiO_2 coatings with low roughness. Spark discharge produces porous and crystalline oxides and a rougher surface than desired due to the formation of sparks or electric arcs [9,13]. This phenomenon leads to a variation of the technique known as Anodic Spark Deposition [14]. The voltage at which the spark starts varies with the nature and concentration of the electrolyte employed [13,15–17]. This effect limited the oxidation voltage used in the present work, because (as previously mentioned) homogeneous and low roughness coatings are necessary for hemocompatible applications.

From the above, the primary objective of the present work was to define appropriate conditions (concentration of electrolyte and applied voltage) to obtain smooth and uniform coatings of TiO_2 by anodic oxidation of Ti-6Al-4V alloy in sulfuric acid as electrolyte, to be used in the construction of a cardiovascular device.

2. Materials and Methods

2.1. Synthesis of the Coatings

2.1.1. Preparation of the Substrates

The substrates used for oxidation were flat samples of Ti-6Al-4V alloy with a surface area of 1×2 cm^2 and thickness of 0.2 cm. They were polished with abrasive SiC papers with decreasing granulometry (from # 120 up to # 1500), with diamond paste of 1 µm (Prazis, Argentina) lubricated with ethylene glycol (Cicarelli, Santa Fe, Argentina), finishing with 4:1 mix of colloidal silica (Mastermet-Buehler, Lake Bluff, IL, USA) and hydrogen peroxide. The mirror surfaces were then cleaned with water and detergent, rinsed with alcohol, and hot air dried. One of the tested substrates was not coated (TiG5).

2.1.2. Anodic Oxidation

Oxidation of the samples was carried out at room temperature (25 °C) applying a DC electric current between the Pt cathode and Ti-6Al-4V anode, separated from each other by 5 cm in a beaker glass containing the electrolyte. The electrolytes were sulphuric acid (H_2SO_4) solutions, and the concentration was varied from 0.1 to 4 M (0.1 M, 0.5 M, 1 M, 2 M, and 4 M). The applied voltages were 20 V, 40 V, 60 V, and 70 V. Anodization time was 1 min. Immediately after oxidation, the oxidized samples were rinsed with demineralized water and dried with hot air. Evolution of voltage and current density was recorded during the oxidation. Anodization of the sample made with 4 M H_2SO_4 at 70 V had some problems due to drawbacks in cell contacts as a result of the severe anodization conditions, as evidenced by the occurrence of spark discharge and melting of the sample holder due to the high temperature achieved during spark.

The samples were labeled as follows: A letter S corresponding to the electrolyte (H_2SO_4), followed by the concentration (in M), the letter V and the corresponding voltage (in V)—e.g., S0.1-V20, S1-V40, etc.

2.2. Coatings Characterization

The surface of the oxides was observed by optical microscopy and scanning electron microscopy SEM, using and Arcano and Carl Zeiss Supra 40 equipment (Germany), respectively. The difference between the colors observed in the macrographs and in the micrographs is explained by the fact that in the first case it was illuminated with a white (fluorescent) light, and in the second with an incandescent light with a predominant yellow spectrum. All micrographs were taken under the same conditions of illumination, because it is extremely important to standardize the illumination to be able to compare the effects in the color, especially when the changes are very subtle.

The roughness of the coatings was measured using the Time Group (China) TR200 profilometer with a cut-off length of 0.8 mm, a sampling length of 0.8 mm, and a number of sampling of 5 in 4 mm. R_a is the average roughness and R_z is the distance between the highest peak and the lowest valley in each sampling length. Four measurements were performed on each sample, and the results were averaged, with relative error smaller than 10%.

The crystalline phases present in the coatings were analyzed by X-ray diffraction (XRD) using a Philips (Netherlands) PW 3710 diffractometer with a CuKα wavelength (λ CuKα = 1.5418 Å), using a Philips thin-film accessory allowing operation with a ground-beam geometry with incident angle of 1°. The crystallite size (L) of the crystalline anatase phase was estimated by Scherrer's formula, $L = K\lambda/(w\cos\theta)$, using the XRD profile. λ is the X-ray wavelength in nm and w (in radians) is the peak width of the diffraction peak profile at half maximum height. θ is the Bragg diffraction angle of the anatase (101) peak. K is a constant related to crystallite shape, normally taken as 0.9, assuming that the crystallites are spherical.

3. Results and Discussion

3.1. Influence of the Electrolyte Concentration (H_2SO_4) on the Characteristics of the Coatings

3.1.1. Variation of Voltage and Current Density with Oxidation Time

Figure 1a shows the current density (i) and voltage (V) characteristic behaviors of a potentiostatic anodization process at low voltages [17]. When V increases, i also increases up to a maximum, indicating the formation of the TiO_2 barrier layer; then, i rapidly decreases because the oxide increases its electrical resistance. When the voltage was established on 40 V, i became gradually stable due to the equilibrium between the oxide growth and dissolution rates [9,14,18]. The reduction of the oxide growth rate was probably due to titanium or oxygen diffusion through the coating becoming the controlling step. As this rate becomes small, for practical purposes, the oxide thickness reaches a limiting value, indicating that it depends on the applied voltage [17–20].

The temporal evolutions of current density in anodizations at 60 V with different electrolyte concentrations are shown in Figure 1b. The maximum and stabilization i values increased with the H_2SO_4 concentration due to the increase of conductivity from 47.2 mS/cm of 0.1 M to 778.5 mS/cm of 4 M [18,21] that should induce breakdown and spark at 60 V, less than the 70 V reported for 1 M on Ti-6Al-4V anodization [13,22]. Therefore, the stabilization of i values at 2 M and 4 M curves are ten times higher than the corresponding value for 1 M.

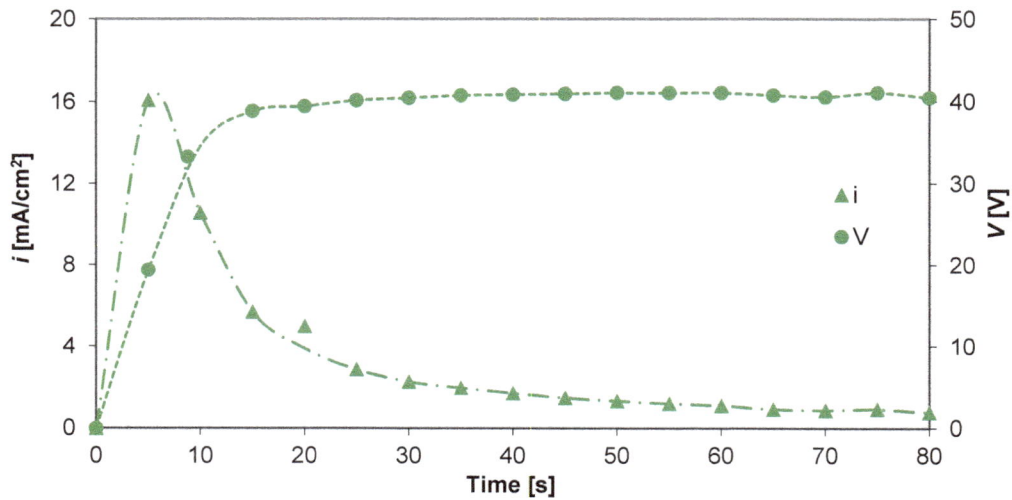

(a)

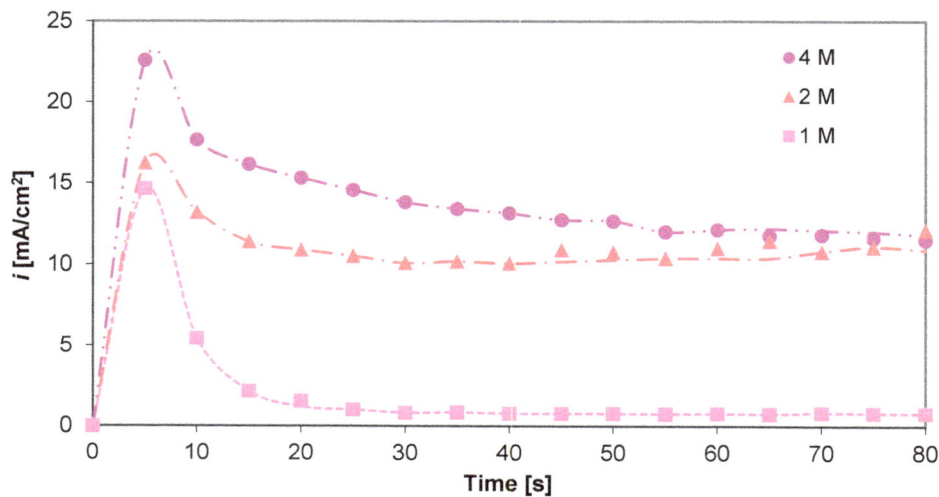

(b)

Figure 1. (a) Evolution of current density and voltage during the anodic oxidation of Ti-6Al-4V at 40 V; (b) Evolution of current density during the anodic oxidation performed with different sulfuric acid concentration (1 M, 2 M, and 4 M) at 60 V.

3.1.2. Color of the Coatings

In Figure 2, optical micrographs and macrographs (in the corner insets of each micrograph) of the samples are presented. In the macrograph, the color of the coatings obtained with different electrolyte concentration and voltage are observed. These colors are the same as observed with the naked eye.

The inset of Figure 2a shows a uniform blue color of sample S0.1-V20. When the acid concentration increased from 0.1 M to 4 M, a decrease in the tonality of this color was observed (see Figure 2a,e,i,m,q). On the other hand, the corresponding micrographs of 20 V samples show higher and clearer streaks with hue variations when the concentration increases.

Macrographs and micrographs of samples obtained at 40 V (Figure 2b,f,j,n,r) show uniform light green color and yellow color, respectively, independent of acid concentration.

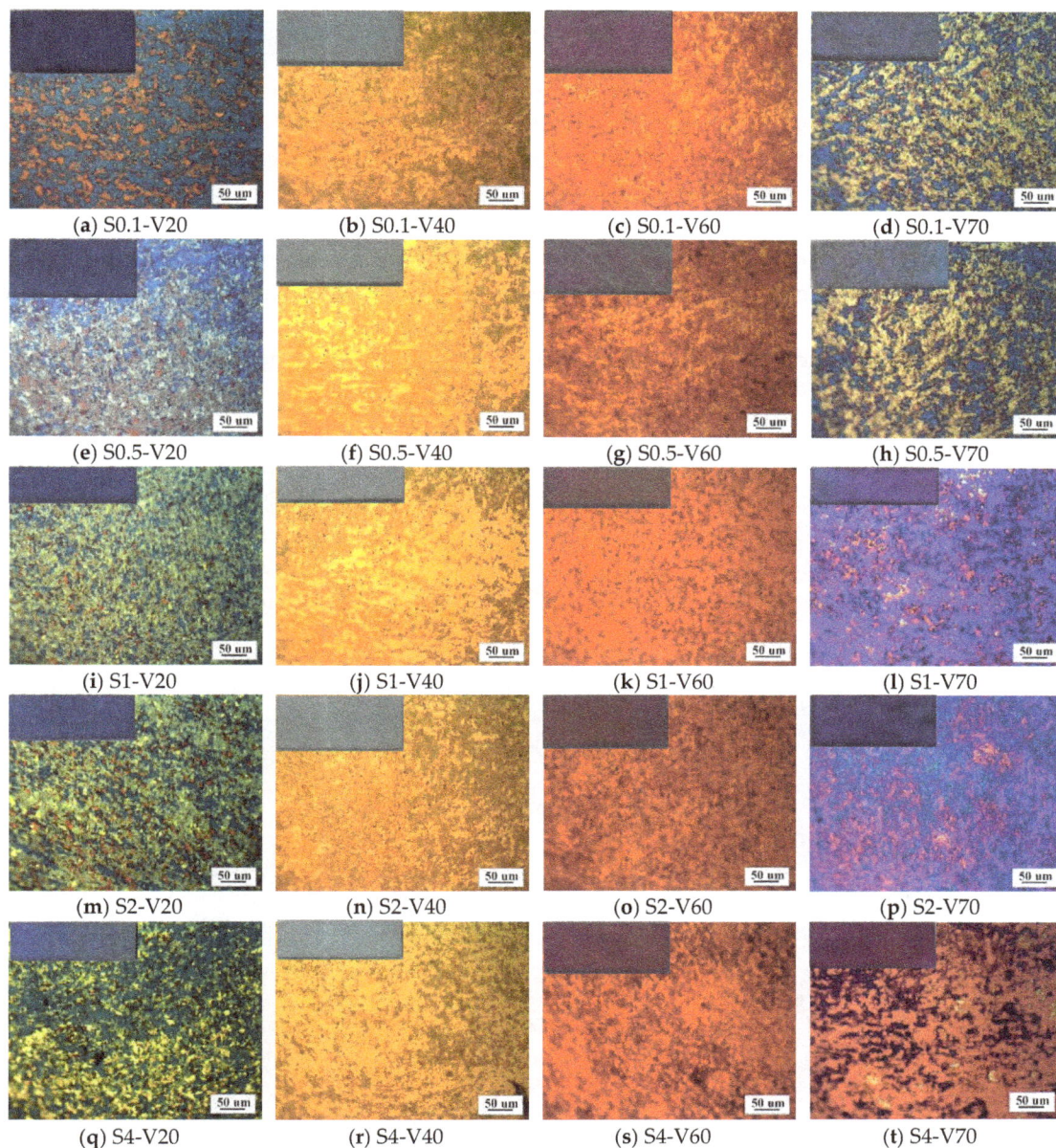

Figure 2. Macrographs (insets) and micrographs of anodic oxidized samples.

Macrographs of samples obtained at 60 V (Figure 2c,g,k,o,s) show two colors: Pink and yellow. In micrographs, the colors are red and orange. In both cases, variations in tonality are observed as the concentration increases up to 1 M (Figure 2k) when more uniformity is observed and pink or red predominated.

We mentioned previously that color depends on the thickness of the coatings, but color also changed with acid concentration, probably due to different oxide growth rate given different oxide stoichiometry [15]. Another explanation for this phenomenon could be that different fractions of crystalline structures can be formed which cause changes in the density and refractive index of the oxide films [15].

In the optical micrographs of each sample, two different color tonalities are observed, homogeneously distributed on the surface. These portions with different tonalities are of a size similar to the grain size, corresponding to the microstructure of the Ti-6Al-4V substrate, so this pattern of coloration can be attributed to a different growth rate of the oxide on the different crystal orientations of the phases of the Ti-6Al-4V substrate grains, which gives rise to oxides with slightly different thicknesses [13,23].

As observed in Figure 2, when the concentration of the H_2SO_4 increases (mainly from 1 M), the color of the coatings changed and became more intense, diminishing the difference between colors in the previously described pattern. This could be because the increase in concentration and conductivity of the electrolyte diminishes the differences in the growth rate of the oxides on the different phases or crystal orientations of the phases of the substrate.

Despite the small differences observed in the colors obtained using different concentrations of the electrolyte, the oxide colors can be used to make a quick qualitative identification of the thickness, from an established scale for each electrolyte. With 1 M H_2SO_4 in pre-spark conditions, a relationship between color, voltage, and thickness (2.4 nm/V) was previously established [13,24].

3.1.3. Morphology of the Coatings

Figure 3 shows SEM images of anodized samples with different voltages and concentration of electrolyte. Anodized samples at 20 V (not shown) and 40 V (Figure 3a,d,g,j) showed different surface morphologies but none of them presented porous surfaces. In SEM images of samples anodized at 60 V (Figure 3b,e,h,k), different isolated pore structures are observed. In S0.1-V70 and S0.5-V70 samples, irregular surfaces are observed, and in S1-V70 and S2-V70 samples, pores are larger and agglomerated.

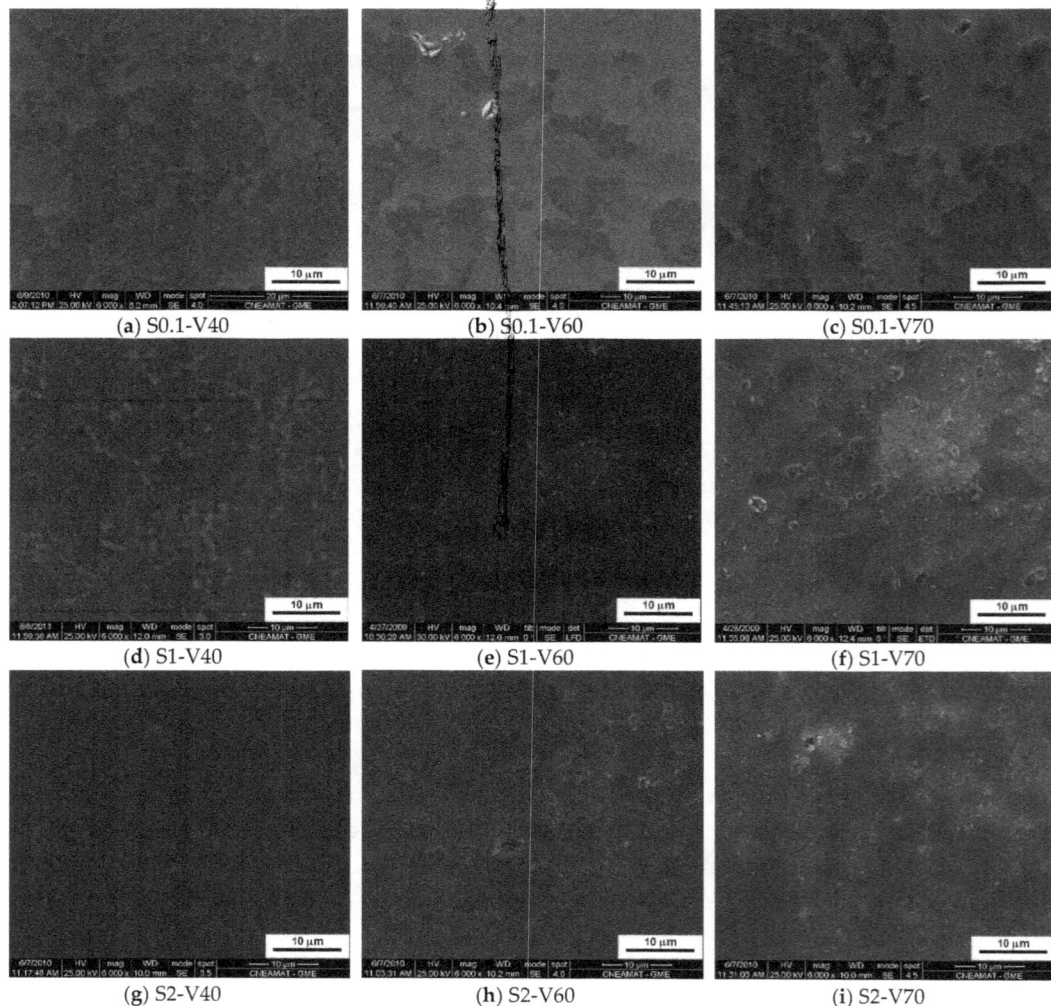

(a) S0.1-V40	(b) S0.1-V60	(c) S0.1-V70
(d) S1-V40	(e) S1-V60	(f) S1-V70
(g) S2-V40	(h) S2-V60	(i) S2-V70

Figure 3. *Cont.*

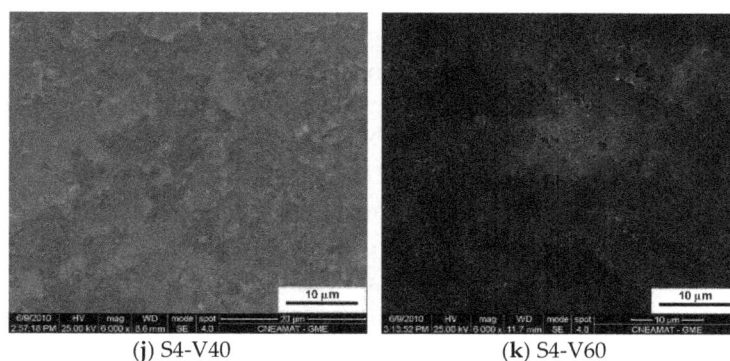

(j) S4-V40   (k) S4-V60

Figure 3. SEM micrographs of anodic oxidized samples.

As previously mentioned, many factors influence the morphology; Masahashi et al. [25] found that both roughness and surface area increase with the concentration of sulfuric acid. On the other hand, Kim and Ramaswamy [26] observed the appearance of microcracks in the oxides with the increase of the electrolyte concentration. However, in the present research, the most significant changes in morphology were observed with the increase of the voltage with each concentration of H_2SO_4 used, but not with the increase of the concentration independent of the voltage.

With respect to roughness, coatings have values of average roughness (R_a) in the range of 6 nm and 32 nm (Table 1), reproducing the surface roughness of the polished substrate (20 nm) and not changing appreciably with the concentration of H_2SO_4. That range of values complies with the desired values for the application of the manufacture of cardiovascular devices ($R_a \leq 50$ nm) [2,11]. The lower roughness values (R_a and R_z) were obtained for samples anodized at 20 V and 40 V. However, the increase of the roughness and porosity with the oxidation voltage was observed mainly in the values of R_z in the samples obtained at 60 V and 70 V with the solutions of H_2SO_4 of concentrations of 0.5 M, 1 M, and 2 M (Table 1). As the presence of porosity in samples made with voltages up to 60 V could promote clot formation, voltages lower than 60 V have to be used to obtain homogeneous and low roughness coating suitable for hemocompatibility.

Table 1. Roughness (R_a and R_z) and crystalline phases of the coatings.

Samples	R_a (nm)	R_z (nm)	Crystalline Phase
TiG5	20	100	α and β
S0.1-V20	7	54	amorphous
S0.1-V40	9	82	amorphous
S0.1-V60	12	98	amorphous
S0.1-V70	11	80	amorphous
S0.5-V20	10	90	amorphous
S0.5-V40	8	73	amorphous
S0.5-V60	18	121	amorphous
S0.5-V70	14	123	amorphous
S1-V20	19	31	amorphous
S1-V40	6	61	amorphous
S1-V60	13	132	amorphous
S1-V70	32	307	anatase (21 nm) *
S2-V20	15	95	amorphous
S2-V40	12	90	amorphous
S2-V60	16	151	amorphous
S2-V70	19	151	anatase (7 nm) *
S4-V20	13	93	amorphous
S4-V40	15	100	amorphous
S4-V60	14	111	anatase (15 nm) *

* Crystallite size of anatase phase is indicated in brackets.

3.1.4. Structure of the Coatings

The diffractograms of the samples obtained under glancing incidence of 1° are shown in Figure 4, grouped by voltages. In Figure 4a, it can be seen that the diffractograms of the samples obtained at 40 V do not show any anatase or rutile peak. Only the corresponding peaks of the alpha (α) and beta (β) phases of Ti-6Al-4V alloy were observed. The absences of the peaks of crystalline oxide phases may be due to the coatings being completely amorphous or to crystallite sizes being too small to be detected by XRD. The diffractograms of samples obtained at 60 V are shown in Figure 4b, where in the spectra of S4-V60 sample, anatase peak (101) at 2θ 25.29° is observed. This peak is clearer in the inset, where it is also observed that the anatase peak is not present in the S2-V60 sample. In Figure 4c, diffractograms of samples obtained at 70 V are shown, where the anatase peak appeared in samples obtained with 1 M and 2 M H_2SO_4 (S1-V70 and S2-V70).

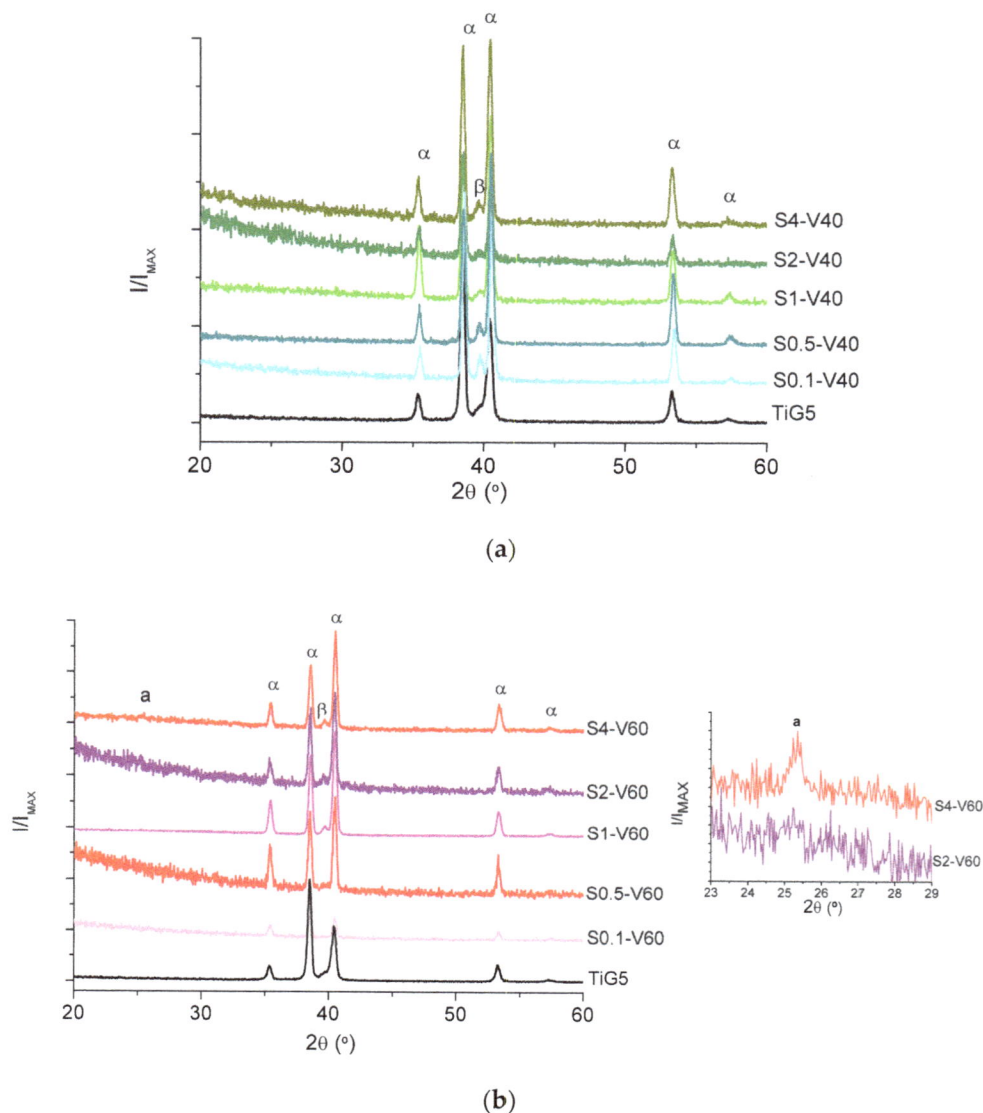

(a)

(b)

Figure 4. *Cont.*

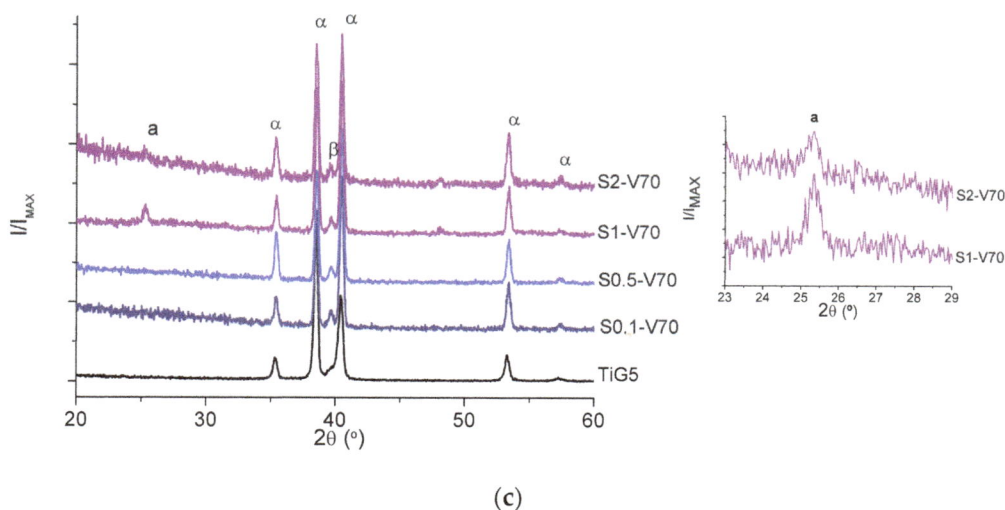

Figure 4. XRD patterns of the substrate and oxidized samples with different electrolytes at different voltages: (**a**) 40 V; (**b**) 60 V; (**c**) 70 V. a = anatase, α = α phase of TiG5, and β = β phase of TiG5.

Briefly, with 0.1 M and 0.5 M H_2SO_4, no crystalline coatings over the entire range of analyzed voltages (20–70 V) were obtained. With 1 M and 2 M H_2SO_4, crystalline coatings from 70 V were obtained; with 4 M, anatase appeared from 60 V. These results indicate that with an increase in the concentration of sulfuric acid, the voltage at which the transformation of the amorphous coatings to anatase occurs decreases, due to the increase of the conductivity with the concentration, which could favor the crystallization at lower voltage.

Thermal treatments will be necessary to crystallize amorphous coatings obtaining up to 60 V for 4 M H_2SO_4 and 70 V for lower H_2SO_4 concentrations [27].

Regarding the size of the anatase crystallites in the crystalline coatings (Table 1), it is known that anatase is a metastable phase whose thermodynamic stability is dependent on the size of the crystallite; the anatase is more stable than rutile when its crystallites are smaller than a critical size, which in this case seems to be larger than 21 nm [28,29].

4. Conclusions

The study of the anodic oxidation of Ti-6Al-4V alloy in sulfuric acid in different concentrations as electrolytes at constant voltages of 20 V to 70 V yielded the following conclusions:

Different interference colors of the coatings were obtained according to the applied voltage and the electrolyte used. Despite the differences observed in the colors obtained using different concentrations of the electrolyte, for all cases, the color of the oxides became more intense and tonality differences diminished with the increase of the concentration of H_2SO_4, especially from 1 M. Additionally, tonality differences diminished at 40 V and 60 V.

With respect to morphology, with all concentrations, porosity was observed in samples surfaces at 60 V, limiting the usable voltage. The lower roughness values (R_a and R_z) were obtained for samples anodizing at 20 V and 40 V.

Regarding the crystalline structure of the oxides, up to 60 V the coating was amorphous, and then starting at 70 V, it began to crystallize to the anatase phase in coatings obtained with 1 M and 2 M H_2SO_4. An increase in the concentration of H_2SO_4 decreased the voltage at which the transformation of amorphous to crystalline coatings occurred, and with 4 M H_2SO_4, the anatase phase appeared at 60 V.

To obtain non-porous, homogeneous, and low roughness coatings to avoid the promotion of blood clots, the voltage to work with is 40 V at a concentration of 1 M sulphuric acid. Those anodizing conditions produce coatings with the most appropriate characteristics for the manufacture of cardiovascular devices.

Acknowledgments: The authors wish to thank the financial support of Consejo Nacional de Investigaciones Científicas y Técnicas (CONICET) and Agencia Nacional de Promoción Científica y Tecnológica (ANPCyT) of Argentina; to Diego Lamas (CITEDEF) for DRX spectras. Ángeles Colaccio thanks to Comité Ejecutivo de Desarrollo e Innovación Tecnológica (CEDIT) of Misiones, Argentina for the scholarship.

Author Contributions: María Laura Vera and Ángeles Colaccio designed and performed the experiments; María Laura Vera, Mario Roberto Rosenberger and Carlos Enrique Schvezov analyzed the data; Mario Roberto Rosenberger, Carlos Enrique Schvezov and Alicia Esther Ares contributed reagents/materials/ analysis tools; María Laura Vera and Alicia Esther Ares wrote the paper.

Conflicts of Interest: The authors declare no conflict of interest.

References

1. Rosenberger, M.R.; Amerio, O.; Schvezov, C. Optimizing of the Design of a Prosthetic Heart Valve with Three Leaves. In Proceedings of the Forth International Congress of Cardiology on the Internet, Paraná, Argentina, 1 September 2005.

2. Amerio, O.N.; Rosenberger, M.R.; Favilla, P.C.; Alterach, M.A.; Schvezov, C.E. Prótesis valvular cardiaca trivalva asociada a última generación de materiales hemocompatibles. *Rev. Argent. Cir. Cardiovasc.* **2006**, *4*, 70–76. (In Spanish)

3. Huang, N.; Yang, P.; Leng, Y.X.; Chen, J.Y.; Sun, H.; Wang, J.; Wang, G.J.; Ding, P.D.; Xi, T.F.; Leng, Y. Hemocompatibility of titanium oxide films. *Biomaterials* **2003**, *24*, 2177–2187. [CrossRef]

4. Schvezov, C.E.; Alterach, M.A.; Vera, M.L.; Rosenberger, M.R.; Ares, A.E. Characteristics of haemocompatible TiO_2 nano-films produced by the sol-gel and anodic oxidation techniques. *JOM* **2010**, *62*, 84–87. [CrossRef]

5. Vera, M.L.; Schuster, J.; Rosenberger, M.R.; Bernard, H.; Schvezov, C.E.; Ares, A.E. Evaluation of the haemocompatibility of TiO_2 coatings obtained by anodic oxidation of Ti-6Al-4V. *Procedia Mater. Sci.* **2015**, *8*, 366–374. [CrossRef]

6. Leyens, C.; Manfres, P. *Titanium and Titanium Alloys: Fundamentals and Applications*; Wiley-VCH: Weinheim, Germany, 2003.

7. Lutjering, G.; Williams, J.C. *Titanium*; Springer: Berlin, Germany, 2007.

8. Khan, M.A.; Williams, R.L.; Williams, D.F. Conjoint corrosion and wear in titanium alloys. *Biomaterials* **1999**, *20*, 765–772. [CrossRef]

9. Diamanti, M.V.; Pedeferri, M.P. Effect of anodic oxidation parameters on the titanium oxides formation. *Corros. Sci.* **2007**, *49*, 939–948. [CrossRef]

10. Alajdem, A. Review—Anodic oxidation of titanium and its alloys. *J. Mater. Sci.* **1973**, *8*, 688–704.

11. Vera, M.L.; Ares, A.E.; Lamas, D.G.; Schvezov, C.E. Preparación y caracterización de recubrimientos de dióxido de titanio obtenidos por oxidación anodica de la aleación Ti-6Al-4V. Primeros resultados. *Anal. AFA* **2008**, *20*, 178–183. (In Spanish)

12. Maitz, M.F.; Pham, M.-T.; Wieser, E. Blood compatibility of titanium oxides with various crystal structure and element doping. *J. Biomater. Appl.* **2003**, *17*, 303–319. [CrossRef] [PubMed]

13. Vera, M.L. Obtención y Caracterización de Películas Hemocompatibles de TiO_2. Ph.D. Thesis, Universidad Nacional de General San Martín, Buenos Aires, Argentina, March 2013. (In Spanish)

14. Song, H.J.; Park, S.H.; Jeong, S.H.; Park, Y.J. Surface characteristics and bioactivity of oxide films formed by anodic spark oxidation on titanium in different electrolytes. *J. Mater. Process. Technol.* **2009**, *209*, 864–870. [CrossRef]

15. Sul, Y.-T.; Johansson, C.B.; Jeong, Y.; Albrektsson, T. The electrochemical oxide growth behaviour on titanium in acid and alkaline electrolytes. *Med. Eng. Phys.* **2001**, *23*, 329–346. [CrossRef]

16. Liu, X.; Chu, P.K.; Ding, C. Surface modification of titanium, titanium alloys, and related materials for biomedical applications. *Mater. Sci. Eng. R* **2004**, *47*, 49–121. [CrossRef]

17. Kuromoto, N.K.; Simao, R.A.; Soares, G.A. Titanium oxide films produced on commercially pure titanium by anodic oxidation with different voltages. *Mater. Charact.* **2007**, *58*, 114–121. [CrossRef]

18. Sharma, A.K. Anodizing titanium for space applications. *Thin Solid Films* **1992**, *208*, 48–54. [CrossRef]

19. Macdonald, D.D. The history of the point defect model for the passive state: A brief review of film growth aspects. *Electrochim. Acta* **2011**, *56*, 1761–1772. [CrossRef]

20. Capek, D.; Gigandet, M.-P.; Masmoudi, M.; Wery, M.; Banakh, O. Long-time anodisation of titanium in sulphuric acid. *Surf. Coat. Technol.* **2008**, *202*, 1379–1384. [CrossRef]

21. Darling, H.E. Conductivity of sulfuric acid solutions. *J. Chem. Eng.* **1964**, *9*, 421–426. [CrossRef]

22. Yang, B.; Uchida, M.; Kim, H.-M.; Zhang, X.; Kokubo, T. Preparation of bioactive titanium metal via anodic oxidation treatment. *Biomaterials* **2004**, *25*, 1003–1010. [CrossRef]

23. Vera, M.L.; Ares, A.E.; Lamas, D.G.; Rosenberger, M.R.; Schvezov, C.E. Influencia de la Textura y de la Microestructura de la Aleación Ti-6Al-4V en los Óxidos Obtenidos por Oxidación Anódica. In Proceedings of the Anales 9° Congreso Internacional de Metalurgia y Materiales SAM-CONAMET, Buenos Aires, Argentina, 19 October 2009; pp. 1951–1956. (In Spanish)

24. Vera, M.L.; Alterach, M.A.; Rosenberger, M.R.; Lamas, D.G.; Schvezov, C.E.; Ares, A.E. Characterization of TiO$_2$ nanofilms obtained by sol-gel and anodic oxidation. *Nanomaterials* **2014**, *4*, 1–11. [CrossRef]

25. Masahashi, N.; Mizukoshi, Y.; Semboshi, S.; Ohtsu, N. Enhanced photocatalytic activity of rutile TiO$_2$ prepared by anodic oxidation in a high concentration sulfuric acid electrolyte. *Appl. Catal. B* **2009**, *9*, 255–261. [CrossRef]

26. Kim, K.-H.; Ramaswamy, N. Electrochemical surface modification of titanium in dentistry. *Dent. Mater. J.* **2009**, *28*, 20–36. [CrossRef] [PubMed]

27. Vera, M.L.; Rosenberger, M.R.; Schvezov, C.E.; Ares, A.E. Fabrication of TiO$_2$ crystalline coatings by combining Ti-6Al-4V anodic oxidation and heat treatments. *Int. J. Biomater.* **2015**, *2015*, 395657. [CrossRef] [PubMed]

28. Chen, Y.; Kang, K.S.; Yoo, K.H.; Jyoti, N.; Kim, J. Cause of slow phase transformation of TiO$_2$ nanorods. *J. Phys. Chem. C* **2009**, *113*, 19753–19755. [CrossRef]

29. Li, W.; Ni, C.; Lin, H.; Huang, C.P.; Shah, S.I. Size dependence of thermal stability of TiO$_2$ nanoparticles. *J. Appl. Phys.* **2004**, *96*, 6663–6668. [CrossRef]

11

Electroplating of CdTe Thin Films from Cadmium Sulphate Precursor and Comparison of Layers Grown by 3-Electrode and 2-Electrode Systems

Imyhamy M. Dharmadasa [1], Mohammad L. Madugu [1,*], Olajide I. Olusola [1], Obi K. Echendu [2], Fijay Fauzi [3], Dahiru G. Diso [4], Ajith R. Weerasinghe [5], Thad Druffel [6], Ruvini Dharmadasa [6], Brandon Lavery [6], Jacek B. Jasinski [6], Tatiana A. Krentsel [6] and Gamini Sumanasekera [6]

[1] Materials & Engineering Research Institute, Faculty of Arts, Computing, Engineering and Sciences, Sheffield Hallam University, Sheffield S1 1WB, UK; Dharme@shu.ac.uk (I.M.D.); olajideibk@yahoo.com (O.I.O.)

[2] Physics Department, Federal University of Technology Owerri, Ihiagwa PMB 1526, Imo, Nigeria; oechendu@yahoo.com

[3] School of Electrical System Engineering, University of Malaysia Perlis, Pauh Putra Campus, 02600 Arau, Perlis, Malaysia; fijay@unimap.edu.my

[4] Kano University of Science and Technology, Wudil PMB 3244, Kano, Nigeria; dgdiso@yahoo.co.uk

[5] California State University, Fresno 2320 E. San Ramon Ave., Fresno, CA 93740, USA; ajithroshen@yahoo.com

[6] Conn Center for Renewable Energy Research, University of Louisville, Louisville, KY 40292, USA; thad.druffel@louisville.edu (T.D.); ruvinid8@gmail.com (R.D.); brandon.lavery@louisville.edu (B.L.); jbjasinski@gmail.com (J.B.J.); Tatiana.Krentsel@louisville.edu (T.A.K.); gamini.sumanasekera@louisville.edu (G.S.)

* Correspondence: maduguu@yahoo.com

Academic Editors: Massimo Innocenti and Alessandro Lavacchi

Abstract: Electrodeposition of CdTe thin films was carried out from the late 1970s using the cadmium sulphate precursor. The solar energy group at Sheffield Hallam University has carried out a comprehensive study of CdTe thin films electroplated using cadmium sulfate, cadmium nitrate and cadmium chloride precursors, in order to select the best electrolyte. Some of these results have been published elsewhere, and this manuscript presents the summary of the results obtained on CdTe layers grown from cadmium sulphate precursor. In addition, this research program has been exploring the ways of eliminating the reference electrode, since this is a possible source of detrimental impurities, such as K^+ and Ag^+ for CdS/CdTe solar cells. This paper compares the results obtained from CdTe layers grown by three-electrode (3E) and two-electrode (2E) systems for their material properties and performance in CdS/CdTe devices. Thin films were characterized using a wide range of analytical techniques for their structural, morphological, optical and electrical properties. These layers have also been used in device structures; glass/FTO/CdS/CdTe/Au and CdTe from both methods have produced solar cells to date with efficiencies in the region of 5%–13%. Comprehensive work carried out to date produced comparable and superior devices fabricated from materials grown using 2E system.

Keywords: electrodeposition; CdTe; $CdSO_4$ precursor; solar energy materials; CdS/CdTe solar cells

1. Introduction

Cadmium telluride (CdTe) thin films have received much attention due to their various applications in electronic devices, such as solar cells [1] and X- and γ-radiation detectors [2]. The increase in demand for clean and sustainable energy is a huge challenge for the photovoltaic

(PV) community to develop low-cost and high efficiency solar panels. The present sources of energy, which are mostly from fossil fuel, are harmful to the sustainability of our ecosystem. Alternative technologies, such as photovoltaics (PV), which convert sunlight into clean energy, have been the main research focus at present [3]. The II–VI semiconductor materials have been found suitable in complementing this effort. Among these semiconductors, CdTe stands out to be one of the most researched and promising semiconductor materials in the production of both laboratory-scale and large area optoelectronic devices, such as the solar panels. CdTe has a direct and near ideal bandgap of 1.45 eV for one bandgap and single p-n junction solar cells with a high absorption coefficient ($>10^4 \, \text{cm}^{-1}$) [4]. CdTe can be n- or p-type in electrical conduction [5,6] depending on the stoichiometry or intentionally-added dopants. These are among the properties that make it a suitable material for application in solar energy conversion. CdTe can be grown using low-cost techniques, and the material can absorb over 90% of photons with energy greater than $E_g = 1.45 \, \text{eV}$ using only about a 2.0 μm thick layer.

CdTe thin films have been grown using a large number of deposition techniques [7]. Some of the main techniques used are close-spaced sublimation (CSS) [2], sintering [8], electrodeposition (ED) [1], molecular beam epitaxy (MBE) [9], metalorganic chemical vapor deposition (MOCVD) [10], pulsed laser deposition (PLD) [11], etc. Electrodeposition is a simple and low-cost technique and offers the advantage of growing materials with both n-type and p-type electrical conductivity by simply changing the growth potential using a single electrolytic bath, and it produces electronic device-quality films for solar cell fabrication.

In electrodeposition, CdTe can be cathodically synthesized using either aqueous [1,12] or non-aqueous [13] electrolytes in acidic (pH = 1.00–3.00) or alkaline medium (pH = 8.40–10.70) [14,15]. The choice of acidic rather than alkaline medium for the electrodeposition of CdTe is due to the fact that Te is more soluble and stable in acidic medium. The history of CdTe growth based on electrodeposition using an aqueous electrolyte was first demonstrated by Mathers and Turner in 1928 [13]. A more elaborate work on the electrodeposition of CdTe was carried out by Panicker and Knaster in 1978 [12], and thereafter, many researchers [1,15–19] have electrodeposited CdTe from aqueous solutions. Usually, the electrodeposition of CdTe is carried out using $CdSO_4$ and TeO_2, which serve as the Cd and Te precursors, respectively. The research program continued by BP in the 1980s successfully demonstrated the scaling up of this technology by manufacturing nearly 1.0 m^2 in area solar panels with over 10% conversion efficiency [18]. The successful deposition of CdTe and comprehensive material characterization have also been carried out recently using $CdCl_2$ [14] and $Cd(NO_3)_2$ precursors [19] for selecting the best cadmium precursor for electroplating. The recent announcement of a 22.1% efficiency of CdTe solar cells by First Solar company [20] using CdTe grown by vapor transport deposition (VTD) is a giant stride in the PV field showing great potential in CdTe-based thin film solar cells. With all of these attractive properties, CdS/CdTe solar cell efficiency improvement was mostly hindered for decades by a number of challenges; limited know-how on material issues, processing steps and device physics has been the major bottle-necks. Achieving large grains, uniform, dense and pinhole-free thin films with minimum defects and optimum doping concentration have been identified as one of the challenges faced by the PV community.

During the past three decades, CdS/CdTe solar cell research was carried out assuming the CdTe layer in the device as a p-type material. Therefore, all experimental results were interpreted based on a simple p-n junction model. This has led to stagnation of the development of this device for a long period of time. The results presented in this paper aim to demonstrate that both p-type and n-type CdTe layers can be grown easily by changing its composition. Furthermore, this work aims to show that layers may remain n-type after post-growth heat treatments. Therefore, the new work aims to demonstrate that p-n junction devices and other complex n-n-n-Schottky barrier-type devices are possible from this material, and the latter seems more efficient in solar energy conversion.

The paper also compares the results obtained from CdTe layers grown by three-electrode (3E) and two-electrode (2E) configurations, in order to gauge the best approach for the minimization of impurities. Ag^+ ions and other group-I ions, such as Na^+ and K^+, are highly poisonous to CdTe solar cells [21], and the removal of the reference electrode from the electrolyte could be highly beneficial for achieving high performance devices. Reference electrodes usually have a saturated KCl solution outer jacket in commercially available products with a possibility of leaking K^+ into the electrolyte. The preliminary studies on the comparison of 3E and 2E systems were published recently by Echendu et al. [22]. This paper presents the results of a comprehensive study in order to confirm the conclusions arrived at the initial stages. Removal of the reference electrode from the electrodeposition system introduces several advantages in order to reduce the cost of CdS/CdTe solar cells further. Elimination of a possible impurity source and, hence, achieving high conversion efficiencies, system simplification, cost reduction and the ability to grow improved materials at temperatures higher than the reference electrode limit (70 °C) are some of the advantages.

In this paper, we present the summary of the results for CdTe thin films grown from aqueous and acidic electrolyte containing $CdSO_4$ and TeO_2 using potentiostatic cathodic electrodeposition. After deposition, the films were characterized using a wide range of analytical techniques for structural, morphological, optical and electrical properties. This paper presents the comparison of CdTe layers grown by 3E and 2E systems, and their effects on fully-fabricated devices. The paper also provides new insight into the physics of new devices based on CdTe thin films.

2. Experimental Details

2.1. Chemicals and Materials Used

In this work, cadmium sulfate ($CdSO_4$) powder with purity $\geq 99\%$ was used as the source of Cd while the source of tellurium (Te) was tellurium oxide (TeO_2) powder of high purity (99.999%). The substrates used were fluorine-doped tin oxide (FTO)-coated glasses with sheet resistance of 13 Ω/square. Dilute sulfuric acid (H_2SO_4) and ammonium hydroxide (NH_4OH) were used for pH adjustment in solutions. All chemicals and substrates were purchased from Sigma-Aldrich U.K. The solvent used for the electrolyte preparation was de-ionized water. For the 3E system, a saturated calomel electrode (SCE) was used as the reference with a platinum (Pt) anode. The saturated KCl solution in the outer jacket was replaced by a $Cd(SO_4)$ solution in order to avoid K leakage into the electrolyte. A carbon (C) rod was used as the anode for the 2E system. A clean glass/FTO substrate was used as the cathode for both systems. The source of power was a computerized Gill AC potentiostat, and the heating of the bath was provided by a hot-plate with a magnetic stirrer.

The containers used were Teflon beakers of 1000 mL for the electrolyte and 2000 mL Pyrex beakers for the outer jacket. Polytetrafluoroethylene (PTFE) tape was used to hold the glass/FTO substrates to the carbon connecting rod during deposition. The cleaning of the substrates was carried out using organic solvents (methanol and acetone) and de-ionized water.

2.2. Preparation of Electrolytic Cells

The electrolytic baths were prepared from aqueous solutions of 1 M $CdSO_4$ and TeO_2 solution in 800 mL of deionized water contained in a 1000 mL Teflon beaker. The solution containing only low purity $CdSO_4$ was electropurified for 50 h before adding the high purity TeO_2 dilute solution. The main reason for using low-purity $CdSO_4$ is the requirement of a large amount of $CdSO_4$ for this work and the very high cost of the high-purity $CdSO_4$. The TeO_2 was separately prepared by dissolving TeO_2 powder using dilute sulfuric acid for the addition to the electrolyte. The pH of the bath was adjusted to 2.00 ± 0.02 using H_2SO_4 acid or NH_4OH prior to CdTe deposition.

The choice of materials to be employed for this experiment is important since the medium of growth is acidic (pH = 2.00). The acidic electrolyte has been shown to absorb impurities from glass containers and the carbon anode in direct contact. This contamination can be minimized by reducing the contacts between the electrolyte and glass surfaces. [13]. Therefore, the Teflon beakers were used as the electrolyte-containing vessels for both 3E and 2E systems. This is to minimize impurities, which affect the final device parameters when incorporated into the CdTe layer during growth.

2.3. Analytical Techniques Used

A wide range of analytical techniques was used to investigate the electroplated CdTe layers. To study the structural properties, a Philips PW 3710 X' pert pro diffractometer (Philips Analytical, Almelo, The Netherlands) using the Cu-K$_\alpha$ excitation wavelength (λ = 1.542 Å) was employed, and the scan ranges of 2θ (20°–60°). The morphology of the film surfaces and the grain sizes were observed using the FEG NOVA NANO scanning electron microscope (SEM). The study of the molecular vibration to obtain unique fingerprints of the materials was carried out using a Renishaw Raman microscope (Renishaw Plc, Gloucestershire, UK) with a CCD detector and a 514 nm argon ion laser. The electrical conduction type of the films was detected using a photoelectrochemical (PEC) cell measurement system using an aqueous electrolyte of 0.1 M $Na_2S_2O_3$. Optical energy bandgaps were measured using a Cary 50 scan UV-Vis spectrophotometer (Agilent Technologies, Santa Clara, CA, USA). The DC electrical conductivity measurements were carried out using a fully-automated I–V system, including a Keithley 619 electrometer and multimeter (Keithley, Cleveland, OH, USA). Photoluminescence (PL) measurements to observe the defect levels in the bandgap were carried out using a Renishaw inVia Raman microscope (Renishaw, Hoffman Estates, IL, USA) with a CCD detector and a 632-nm He-Ne laser excitation source. Intense pulsed light (IPL) heat treatment was carried out in air using the Sintering 2000 (Xenon Corporation, Wilmington, MA, USA).

3. Experimental Results

3.1. Cyclic Voltammetry

Figure 1 shows typical voltammograms or current-voltage curves of the electrolytes with 3E and 2E electrode systems. Due to the voltage measurement with respect to the reference electrode in the 3E and the carbon anode in the 2E system, the magnitudes of the cathodic voltages are different in the graphs. Otherwise, the main features are similar in both cases.

The mechanism for the deposition of CdTe on the cathode from acidic aqueous electrolyte as proposed by Panicker et al. [12] is given below using Equations (1) and (2). The first process, which is a diffusion control, is the reduction of $HTeO_2^+$ to Te, which reacts with Cd^{2+} to form CdTe on the cathode. The formation of CdTe thin films on the cathode is highly influenced by the diffusion process. The overall equations for the formation of CdTe on the cathode are given below:

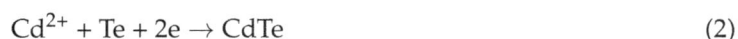

$$HTeO_2^+ + 3H^+ + 4e^- \rightarrow Te + 2H_2O \tag{1}$$

$$Cd^{2+} + Te + 2e \rightarrow CdTe \tag{2}$$

Since the redox potential of Te is +0.593 V with respect to the standard H_2 electrode, this element is easier to deposit on the cathode first. In the 3E system, Te deposition starts around 200 mV, and then, at higher cathodic voltages of around 550 mV, Cd starts to deposit (the redox potential of Cd is −0.403 V with respect to the standard H_2 electrode). As a result, when the cathodic potential is gradually increased, the sequence of deposition is: Te layer first, Te-rich CdTe second, stoichiometric CdTe third and, finally, Cd-rich CdTe layers. This transition takes place gradually, and at a certain voltage, the stoichiometric CdTe layer is formed. This voltage is labelled as the perfect potential of stoichiometry (PPS) in this paper. This is the voltage we expect to find for growing device-quality CdTe layers. In the reverse scan, when the cathodic voltage is gradually reduced, the material starts to dissolve in the solution. The dissolution order will be elemental Cd first, Cd from CdTe second and, finally, Te from the cathode surface. This dissolution process creates a current flow in the negative direction. The large peak in the negative current represents the dissolution of both Cd and Te from the cathode surface.

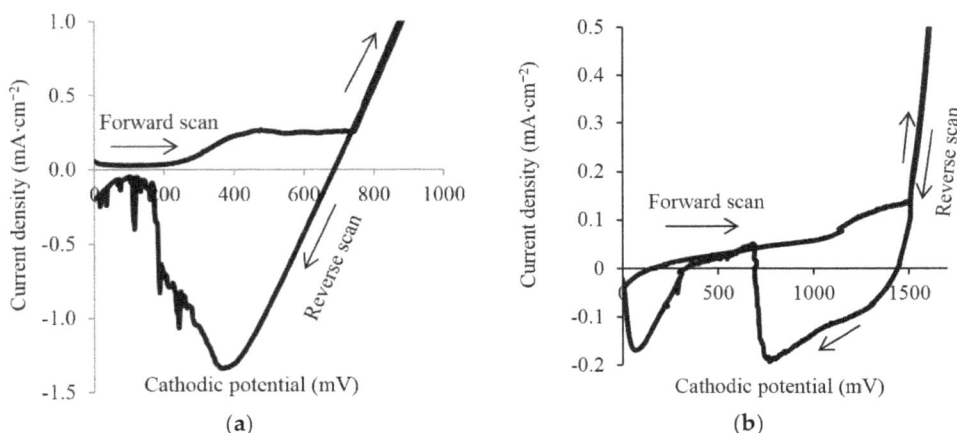

Figure 1. Cyclic voltammograms of aqueous solutions consisting of 1 M CdSO$_4$ and a low level of TeO$_2$ with glass/FTO cathode in (**a**) three-electrode (3E) and (**b**) two-electrode (2E) systems. The reference electrode used for the 3E system is a standard calomel electrode. The growth temperature was 70 °C and 85 °C for the 3E and 2E systems, respectively. The pH of both electrolytes was set to 2.00 ± 0.02 at the beginning of the experiment at room temperature.

The main features of the voltammogram for the 2E system are the same, but the absolute values of the cathodic voltages are different. It should be noted that these deposition voltages are very different from the redox potentials given above. The reason is that the redox potentials are given w.r.t. the standard H$_2$ electrode, and the measured values in these experiments are w.r.t. the carbon anode, for 2E, and the standard calomel electrode for the 3E system. Te deposition starts around 250 mV, and Cd deposition takes place around 1000 mV. In the reverse direction, Cd dissolution and Te dissolution are shown in two separate current peaks. This peak separation can be due to different experimental conditions, such as scan rate, different temperature and the stirring rates.

3.2. X-ray Diffraction

Careful observation of the voltammogram helps with estimating suitable cathodic voltages for growing stoichiometric CdTe. By growing layers at fixed voltages in this estimated region and observations using XRD, PEC and optical absorption methods, this estimated voltage range can be reduced to a very narrow range. In this study, these voltages were changed by 2 and 1 mV steps for the 3E and 2E systems, respectively, in order to pin-point the perfect potential of stoichiometry (PPS).

Figure 2 shows the XRD spectra measured for samples grown in the vicinity of PPS from both the 3E and 2E systems. Figure 2a shows the variation of XRD observed for CdTe grown by the 3E system. Every effort was taken to grow approximately equal thicknesses, and the variation of the intensity of the most intense (111) peak was closely monitored. It was found that the highest intensity was observed when the layers were grown at 834 mV. When heat treated in the presence of CdCl$_2$ (Figure 2c), the intensity variation remained the same, confirming that the highest crystallinity shifted to 830 mV w.r.t. the calomel reference and Pt anode. Therefore, the PPS value for the 3E system is confirmed as 830 mV with respect to SCE under the experimental condition used in this work. This voltage was taken as the PPS, since CdCl$_2$ treatment is used prior to the fabrication of CdS/CdTe solar cells.

Figure 2b,d shows similar results for CdTe grown by the 2E system. XRD features remain the same, and PPS for this experimental system can be determined as 1576 mV with respect to the carbon anode. All other XRD features are very similar, but the growth rate is usually faster in the 2E system, when compared to the 3E system. This can be taken as an advantage in the manufacturing process, reducing growth time. In the case of heat-treated samples in the presence of CdCl$_2$, the (220) and (311) peaks start to appear. This shows that the grains are losing their preferential orientation along (111).

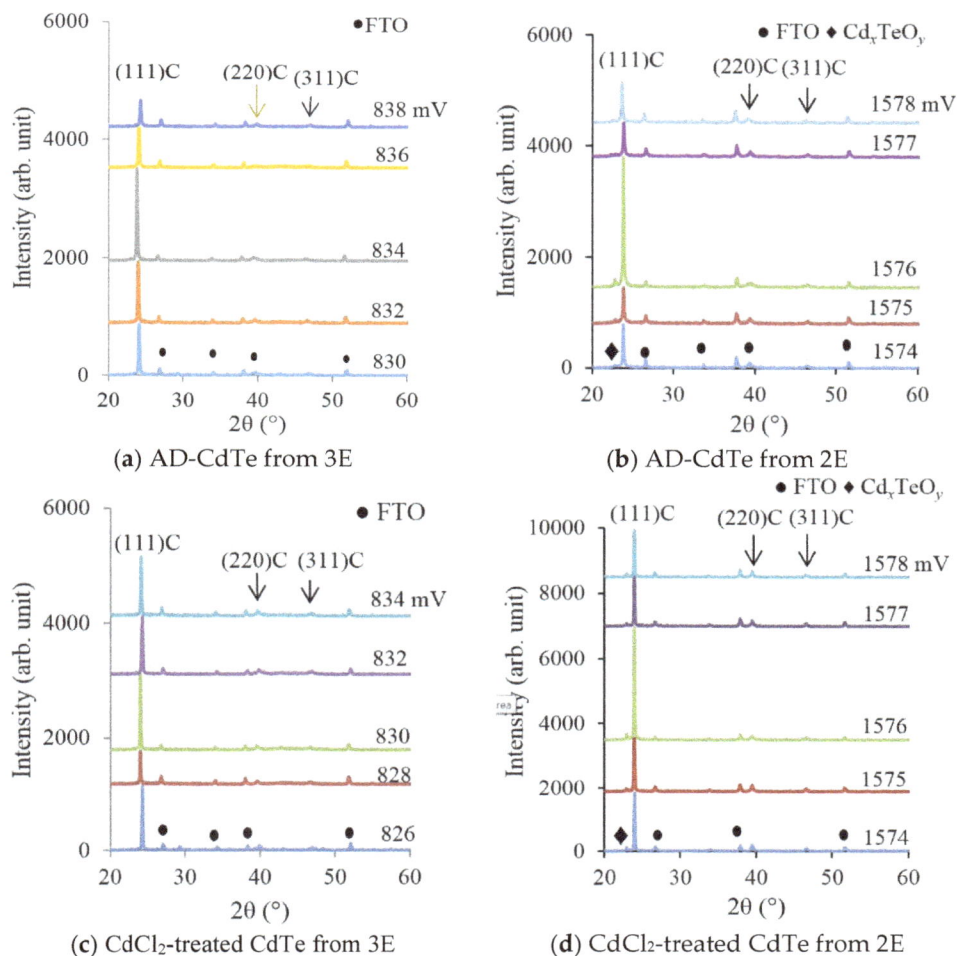

Figure 2. XRD patterns for as-deposited (AD) CdTe films deposited at different growth voltages using the (**a**) 3E and (**b**) 2E systems. (**c,d**) XRD patterns for CdCl$_2$-treated layers. All heat treatments were carried out at 400 °C for 15 min in air in the presence of CdCl$_2$.

In both cases, the (111) peak of as-deposited CdTe layers was used to estimate the crystallite sizes. The use of Scherrer's equation yields (20–65 nm) for the crystallite size for both material layers.

3.3. Scanning Electron Microscopy

Scanning electron microscopy (SEM) studies were carried out in order to investigate the surface morphology of electroplated CdTe layers using both the 3E and 2E systems. Typical results are shown in Figure 3 for both as-deposited and CdCl$_2$-treated layers. The as-deposited layers of 3E (Figure 3a) and 2E (Figure 3b) are covered with large clusters or agglomerations consisting of nano-crystallites. The clusters have varying sizes up to the sub-micron level for the largest ones. The small grains are crystalline CdTe, and their size varies from (20–65 nm) as determined by XRD measurements and Scherrer's equation. There is no noticeable difference in the morphology and crystallite sizes for 3E- and 2E-grown CdTe layers. Only the cluster size is larger in the layers grown by the 2E system.

Heat treatment at 450 °C for 20 min in air in the presence of CdCl$_2$ (Figure 3c,d) shows a dramatic change. Nano-crystallites within clusters have merged together to form large crystals, and the clusters become large grains of CdTe. This is thermodynamically expected since the surface to volume ratio is large for nano-crystallites. Grain sizes vary in the few microns size, and these layers show similarity in structure to CdTe grown by high temperature techniques, such close-spaced sublimation (CSS).

Prolong heating at temperatures 450 °C makes these grains larger due to Oswald ripening, but detrimental for device performance. This is due to the columnar growth producing larger grains exposing wider gaps between grains. In addition, the material losses due to sublimation have also been observed. Both of these processes increase pinhole or gap formation and therefore are detrimental for final device performance.

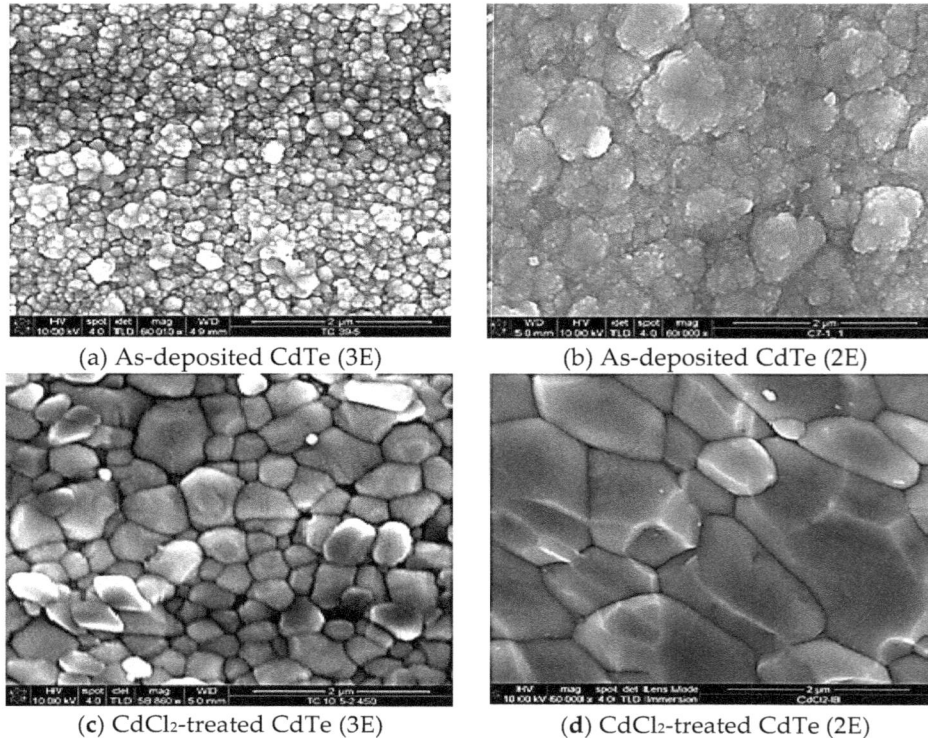

(a) As-deposited CdTe (3E)   (b) As-deposited CdTe (2E)

(c) CdCl₂-treated CdTe (3E)   (d) CdCl₂-treated CdTe (2E)

Figure 3. Typical SEM images for as-deposited and CdCl$_2$-treated CdTe layers grown by the 3E system (**a**,**c**) and the 2E system (**b**,**d**). CdCl$_2$ treatment was carried out in a conventional furnace at 450 °C for 20 min in air, in the presence of CdCl$_2$.

The heat treatment is crucial in device performance, and the PV solar cell developers are moving towards roll-to-roll production methods. Therefore, intense pulse light (IPL) annealing was also used to study the morphology changes during heat treatment. IPL is a rapid thermal annealing method applicable to materials grown on flexible substrates, and the CdTe layer can be heat treated from the top surface using pulses of white light. Heat energy released to the material layer can be controlled by the energy of a pulse and the number of pulses used to heat the layer. Full details of this technique are given in [23], and here, the main morphology changes of the surfaces are presented and discussed.

Figure 4a shows another SEM image of a CdTe layer grown by the 2E system. These surfaces are very similar to the ones shown in Figure 3a,b. Small crystallites are clustered together to form large agglomerations. Figure 4b,c,d shows the surface morphology after IPL treatments with energies of 1730 J·cm^{-2} (100 pulses $\times$ 17.3 J·cm^{-2}), 2160 J·cm^{-2} (100 pulses $\times$ 21.6 J·cm^{-2}) and 2588 J·cm^{-2} (100 pulses $\times$ 25.9 J·cm^{-2}), respectively.

From these SEM images, it is clear that the agglomerations become large crystalline grains, after mainly the grain boundary areas melting and subsequently freezing during cooling. Figure 4d clearly shows the large CdTe grains formed across the thin layer. Cross-section SEM images show that those grains have a columnar nature, spreading from the bottom to the top of the thin films. It also shows the melted grain boundaries frozen after heat treatment. As shown by the phase diagram of CdTe [7], the melting point of grain boundary materials with impurities, such as O, Cl and excess Cd from CdCl$_2$, are much lower (350–450 °C) than the melting point of pure CdTe grains (1093 °C).

Therefore during heat treatment, grain boundaries become a liquid, allowing CdTe grains to flow, coalesce and grow into large grains.

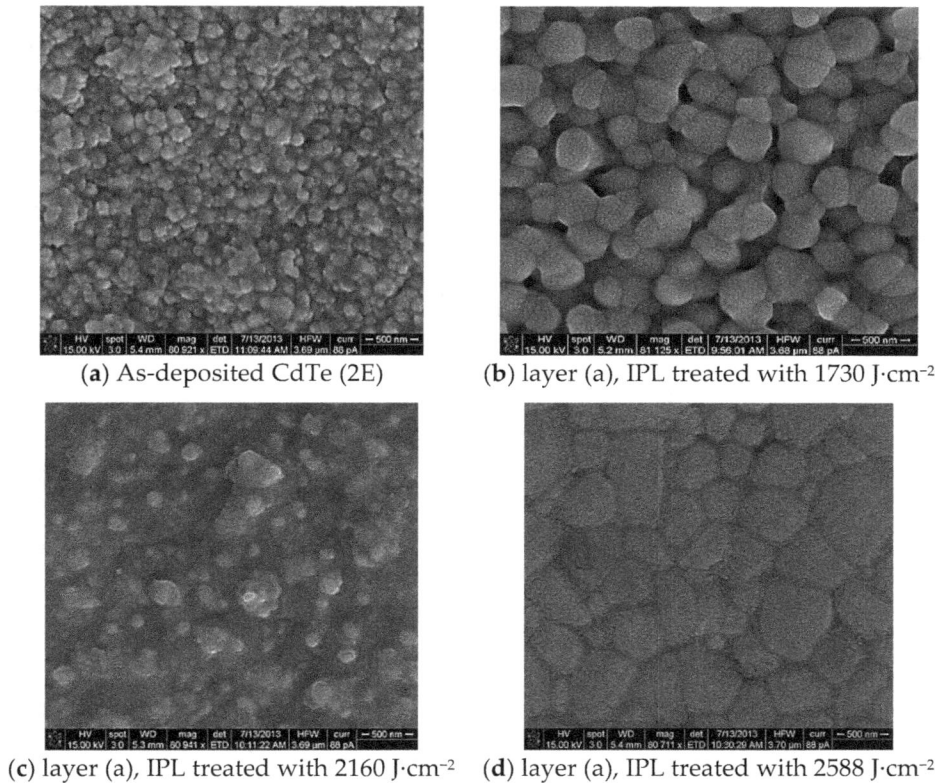

(a) As-deposited CdTe (2E) (b) layer (a), IPL treated with 1730 J·cm^{-2}

(c) layer (a), IPL treated with 2160 J·cm^{-2} (d) layer (a), IPL treated with 2588 J·cm^{-2}

Figure 4. A typical SEM image of the as-deposited CdTe layer from the 2E system and changes of morphology as the energy input is increased using intense pulse light (IPL) treatment.

3.4. Photoelectrochemical Cell Measurements

In order to use a material in any electronic device and interpret the experimental results, its electrical conductivity type must be accurately known. The most accurate method to find this is the conventional Hall effect measurements, but these studies cannot be performed on glass/FTO/CdTe layers due to the underlying conducting layer of FTO. Electrons find the lowest resistive path, and therefore, measurements are not at all reliable. For this reason, the photoelectrochemical (PEC) cell measurements have been used to find the electrical conduction type of these layers.

For these measurements, glass/FTO/CdTe was immersed in any suitable electrolyte (aqueous solution of $Na_2S_2O_3$ for example) in order to form a solid/liquid junction at the CdTe/electrolyte interface. This is equivalent to a weak Schottky diode, and its open circuit voltage is measured by measuring its voltage with respect to a counter electrode (graphite rod) immersed in the same electrolyte. The difference between the voltages measured under dark and illuminated conditions provides the PEC signal. After calibrating the system with a known material, the sign of the open circuit voltage or the PEC signal determines the electrical conductivity type of the CdTe layer. The magnitude of the PEC signal indicates the strength of the depletion region formed at the solid/liquid junction and, hence, an indirect and qualitative idea of the doping level of CdTe layer. Both insulating and metallic layers provide a zero PEC signal due to the non-formation of a healthy depletion region. In our calibrated system, the positive PEC signal indicates a p-type semiconductor, and a negative signal shows an n-type semiconductor.

Figure 5a,b shows the PEC signals observed for as-deposited and CdCl$_2$-treated CdTe layers grown from both the 3E and 2E systems, as a function of growth voltage. Both CdTe materials grown using the 3E and 2E systems show a similar behavior. At low cathodic voltages, Te-rich, p-type CdTe layers are grown, while at larger cathodic voltages, Cd-rich, n-type CdTe layers are grown. In between these two regions, there exists an inversion voltage (V_i) or perfect potential of stoichiometry (PPS), which produces stoichiometric CdTe with the highest crystallinity due to the existence of only one phase. Crystallinity of the layers reduces when grown away from V_i or PPS due to the presence of two phases within the layer. The absolute value of V_i varies depending on factors, such Te concentration, stirring rate and pH value.

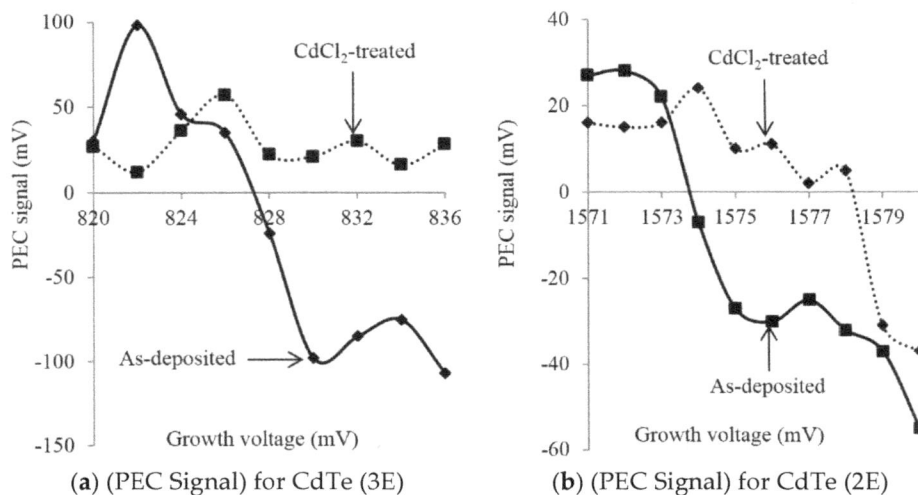

(a) (PEC Signal) for CdTe (3E)   (b) (PEC Signal) for CdTe (2E)

Figure 5. Photoelectrochemical (PEC) cell measurements for as-grown and CdCl$_2$-treated CdTe thin films deposited using (a) 3E and (b) 2E systems at different growth voltages. Samples were heat-treated at 400 °C for 15 min in air in the presence of CdCl$_2$.

Upon CdCl$_2$ treatment, both materials maintain a similar behavior. The p-type nature reduces and moves towards the n-type property, and the n-type nature reduces and moves towards the p-type property. It should be noted that this is due to the movement of the Fermi level within the bandgap, and complete type conversion may not occur. This shows the direction of movement of the Fermi level (FL) in the bandgap depending on the initial condition of the layer, doping effects and annealing out defects during the heat treatment. These measurements indicate only the electrical conductivity type of the top layers of CdTe materials, since the depletion region formed at the solid/liquid junction is limited only to the surface region.

3.5. Raman Studies

Raman scattering studies were carried out as a non-destructive and quick quality control method. This could be useful in a production line to check the quality of materials grown before processing devices. In our previous Raman studies [14,24,25], the peak arising from both CdTe and elemental Te was observed. The X- and γ-ray detectors research community has carried out comprehensive studies on Te-precipitation during CdTe growth [26,27]. In agreement with the reports, we also observe Raman peaks arising from elemental Te in electrodeposited CdTe layers.

Figure 6 shows typical Raman spectra of CdTe layers grown near the PPS, using both 3E ($V_g = 828$ mV) and 2E ($V_g = 1360$ mV) systems. In addition to the peak arising from CdTe at 161 cm^{-1}, peaks at 121 and 141 cm^{-1} can be identified as Te peaks. These elemental Te could appear as precipitates within the layer or Te thin film on the surface. As observed from PL (Section 3.7) and the overall devices' work (Section 3.9), excess Te in the CdTe layers deteriorates its electronic properties.

CdCl$_2$ treatment, however, tends to reduce the elemental Te in the layers by reacting with Cd from CdCl$_2$ and forming CdTe, in addition to other benefits, such as re-crystallization, reduction of grain boundaries, defects and doping. As shown in Figure 6, both CdTe layers from the 3E and 2E systems show a similar reduction of elemental Te from the electroplated CdTe layers.

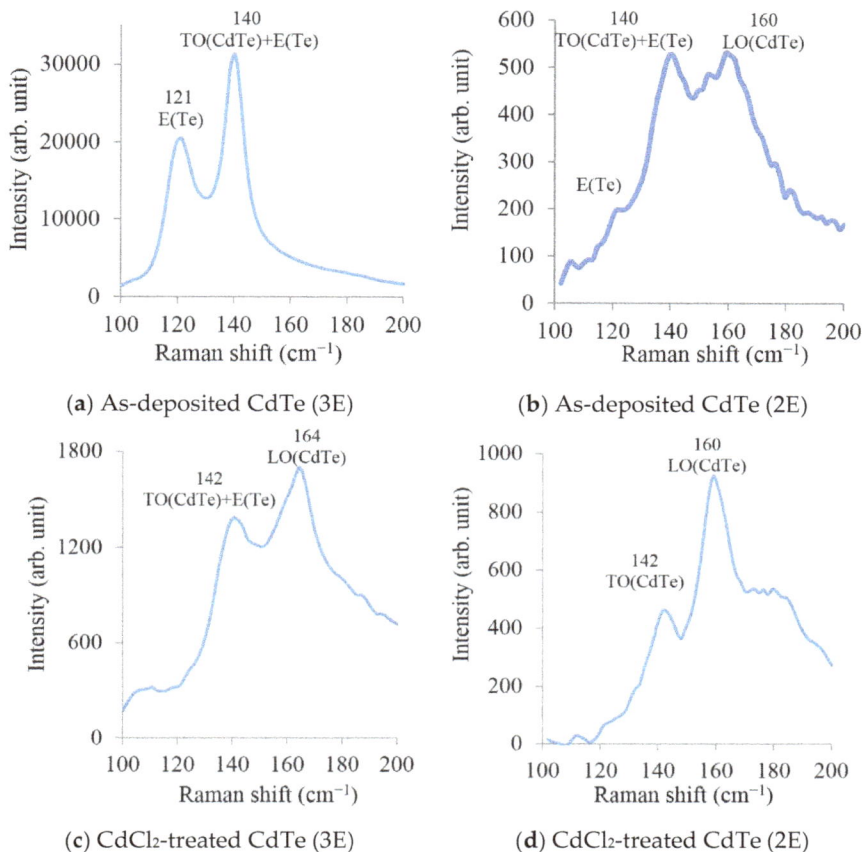

(a) As-deposited CdTe (3E)

(b) As-deposited CdTe (2E)

(c) CdCl₂-treated CdTe (3E)

(d) CdCl₂-treated CdTe (2E)

Figure 6. Plot of the Raman spectra of CdTe thin films for as-deposited and CdCl$_2$-treated layers. The samples were grown using (**a,c**) 3E and (**b,d**) 2E, respectively. The surfaces were thoroughly washed with deionized water after CdCl$_2$ treatment.

3.6. Optical Absorption

Optical absorption studies were carried out in order to study the absorption edge and estimate the optical energy gap of the material. This analysis usually plots ($\alpha h\upsilon^2$) versus photon energy ($h\upsilon$) or Tauc plots to determine energy gap values. In this research program, we have found that plots of the square of optical absorption (A^2) versus photon energy ($h\upsilon$) also produce very similar results, and therefore, Figure 7 shows the plots obtained for electroplated CdTe layers using the 3E and 2E systems.

It is a notable feature of the curves that when the material layers are grown at PPS, the optical absorption edge is sharp and produces accurate E_g values. Stoichiometric CdTe layers grown using both the 3E and 2E systems produce a bandgap of 1.48 eV very close to that of the bulk CdTe material. As the growth voltage deviates from the PPS, the slope of the absorption edge weakens and produces E_g values away from the bulk energy gap. The presence of several phases in the layer tends to produce smaller crystallites and, hence, produce slightly larger E_g values. Te precipitates, Te-rich CdTe and stoichiometric CdTe, when grown below PPS, are possible, and the layer consists of smaller grains and more pinholes. More pinholes means the passage of all wavelengths and, therefore, equivalent to lager bandgaps. The addition of more metallic cadmium when grown above PPS tends to reduce the bandgap due to the metallic property of Cd.

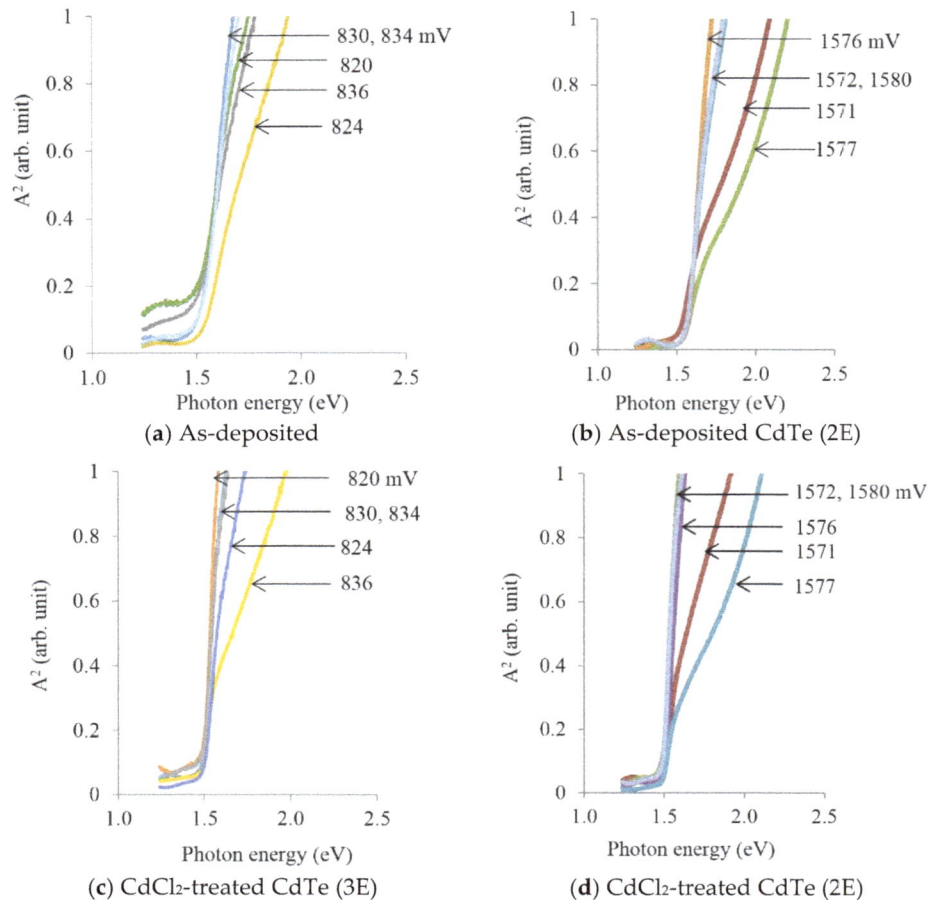

Figure 7. Optical absorption edges for as-deposited and CdCl$_2$-treated CdTe layers grown using the 3E (**a,c**) and 2E (**b,d**) systems at different growth potentials. The films were heat-treated at 400 °C for 15 min in air.

3.7. Photoluminescence Studies

Photoluminescence studies were carried out on both as-deposited and CdCl$_2$-treated CdTe layers obtained from the 3E and 2E electrodeposition systems. The aim was to observe the defect structure in the materials' bandgap. The PL system used is capable of detecting materials' defects between 0.55 eV below the conduction band and the top of the valence band. Full details of the experimental setup and other PL work are published elsewhere [28].

Figure 8 shows the experimentally-observed PL spectra for these CdTe layers for direct comparison. PL spectra from ED-CdTe layers grown from both the 3E and 2E systems show similar defect structures. Defect fingerprints consist of four main defect distributions (T_1–T_4) and the near band emission peak, labelled as E_g. The PL peaks are labeled as T_1–T_4, only to aid the discussion, but those broad peaks could include several peaks arising from closely-situated defects. The summary of the emission levels is shown in Table 1 with approximate energy distributions.

There are some important features to note in these PL spectra. The mid-gap peak labelled as T_2 is the most detrimental "killer center" in CdTe and related to Te richness in CdTe [29,30]. Therefore, these defects' distribution must be related to tellurium antisites (Te$_{Cd}$), tellurium interstitials (Te$_i$) and cadmium vacancies (V_{Cd}). This defects' distribution is very broad and spread over 0.30 eV, contributing to detrimental and effective recombination of photo-generated charge carriers. In the case of CdTe grown using the 2E system, these defects' reduction is more apparent after CdCl$_2$ treatment. Sharpening of the E_g peak is also clear for these layers. Therefore, based on the PL spectra, both materials are comparable; if not, the materials arising from the 2E systems are slightly superior. In order to produce high efficiency devices, these mid-gap "killer centers" (T_2) should be completely removed or at least minimized to reduce the recombination of photo-generated charge carriers.

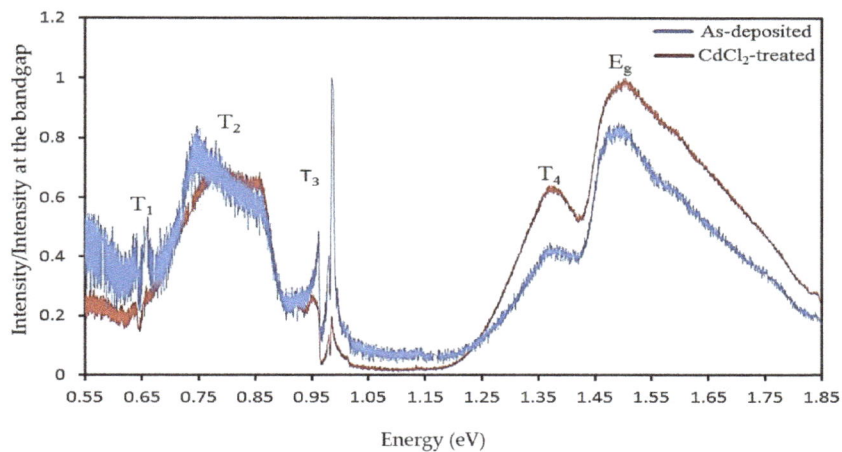

(a) CdTe-PL for 3E

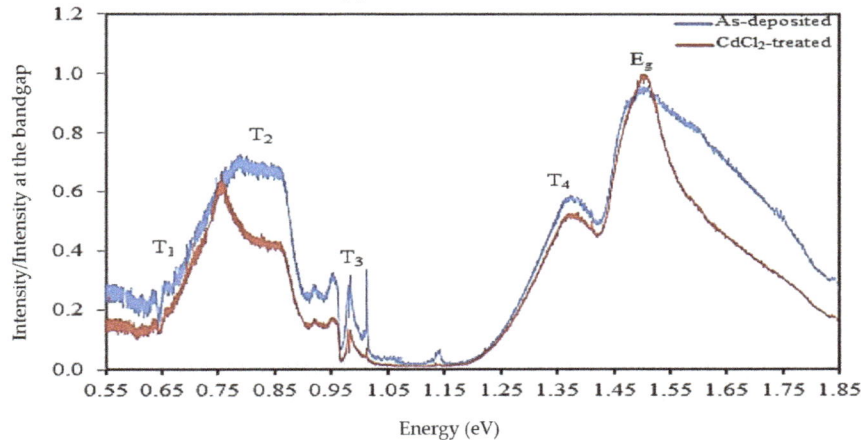

(b) CdTe-PL for 2E

Figure 8. Photoluminescence spectra obtained for as-deposited and CdCl$_2$-treated CdTe layers grown from (a) 3E and (b) 2E systems.

Table 1. Summary of electron traps observed at 80 K for CdTe layers electroplated using the 3E and 2E systems.

CdTe Layers	Energy Position Below the Conduction Band Minimum (eV)				
	$T_1 \pm 0.02$	$T_2 \pm 0.15$	$T_3 \pm 0.03$	$T_4 \pm 0.08$	E_g Peak
As-deposited (3E)	0.66	0.75	0.97	1.37	1.50
CdCl$_2$-treated (3E)	–	0.77	0.97	1.37	1.50
As-deposited (2E)	0.66	0.79	0.96	1.37	1.51
CdCl$_2$-treated (2E)	–	0.75	0.96	1.37	1.50

In [29,30], it has been clearly demonstrated that the CdTe layers can be produced with Te-rich or Cd-rich conditions. This work also showed that the largest Schottky barriers are produced on Cd-rich surfaces rather than Te-rich surfaces. Cd-rich CdTe removes mid-gap killer centers and pins the Fermi level close to the valence band maximum, producing excellent diodes. Therefore, high efficiency solar cells can be produced using a Cd-rich CdTe layer, as shown in Figure 9b.

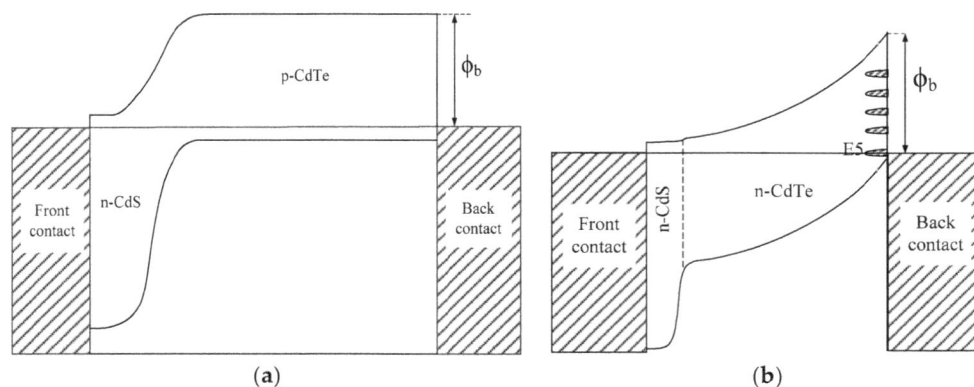

Figure 9. Possible device configurations when CdTe layers are used as (**a**) p-type and (**b**) n-type in electrical conduction.

3.8. Ultraviolet Photoelectron Spectroscopy

In the development of thin film solar cells based on CdTe, the usual superstrate device structure used is glass/FTO/CdS/CdTe/back electrical contact. Since CdTe exists both in n-type and p-type electrical conductivity, two PV devices are possible [31], as shown in Figure 9.

The above structure forms a simple p-n junction when the CdTe layer is p-type and forms an n-n heterojunction connected in parallel with a large Schottky barrier (SB) at the back metal contact (n-n + SB) when the CdTe layer is n-type. In both cases, the Fermi level (FL) should align very close to the valence band maximum, in order to produce high efficiency solar cells. However, strong FL pinning occurs at CdTe/metal interfaces as reported in [3]. Therefore, the knowledge on the positions of the FL at the CdTe back surface is useful in this device development program. In order to gather any useful information, ultraviolet photoelectron spectroscopy (UPS) studies were carried out. The detailed results on UPS and the observed results are published elsewhere [32].

After a comprehensive UPS study, it was found that the results need to be carefully analyzed depending on the surface preparations. UPS is a surface-sensitive technique (probing depth is one or two monolayers), and therefore, the results can depend on various factors, such as the length of exposure to the atmosphere. However, UPS results also provide valuable information on the interactions at the CdTe/Au electrical contacts. It measures the work function of the Au layer deposited on CdTe and that of the CdTe surface. Typical values measured are shown in Table 2.

Table 2. Summary of the work function measured for Au and CdTe surfaces using ultraviolet photoelectron spectroscopy (UPS).

Growth System	Material Used	Work Function of Au (eV)	Work Function of CdTe (eV)
CdTe from 3E	CdTe (CdCl$_2$) CdTe (CdSO$_4$)	4.31 4.42	3.89 4.00
CdTe from 2E	CdTe (CdSO4) CdTe (Cd(NO3)2)	4.39 4.60	4.35 3.61

The reported work functions for Au and Te are 5.10 [33] and 4.73 eV [34], respectively. It is clear that the average work function measured for Au contacts deposited on the CdTe layer is 4.40 eV and less than 5.10 eV (see the last column of Table 2). This confirms our previous conclusions based on soft-XPS results carried out at the synchrotron radiation laboratory in Daresbury. As Au is deposited on CdTe, a strong interaction takes place by Au alloying with Cd and releasing Te to the surface of the Au layer [3,35]. As a result, the Au layer is not pure Au and consists of floating Te or Au$_x$Te$_y$ alloy. Exposure to the atmosphere could oxidize some of the Te into Te$_x$O$_y$, as well. In fact, this can be visually observed in glass/FTO/CdS/CdTe/Au devices. As soon as the devices are made, the Au contacts show a shiny Au color. However, with time, the appearance of the Au contacts shows a dull and Cu color due to these microscopic interactions at the interface. This effect is shown on both Au/bulk CdTe and Au/thin films of CdTe thin films grown using either the 3E or 2E systems.

3.9. CdS/CdTe Solar Cells

In this research program, both 3E and 2E systems were used in parallel to electroplate CdTe from $CdSO_4$ precursor and to fabricate glass/FTO/CdS/CdTe/Au solar cells. Both materials are capable of producing PV active solar cells with varying conversion efficiencies in a wide range of 5%–13%. The efficiency differs from batch to batch, and the cell parameters show wide variation.

This wide variation of efficiency is not unique to devices fabricated with electroplated CdTe layers. This seems to be a common feature for devices made by different growth techniques. Therefore, this should be an inherent property of poly-crystallite CdTe and needs deep understanding of this material issue. Since the $CdCl_2$ treatment is a key processing step, full understanding is also essential to understand these efficiency variations. In addition, the existence of pinholes, non-uniformity during growth, different defect concentrations at different points and varying doping concentrations could contribute to this large variation.

In this comprehensive research program carried out over the past two decades, the growth was performed by more than six researchers using the 3E system. However, the device efficiencies observed were variable, and the highest efficiency observed was 6.9%. Although the efforts devoted to CdTe growths using the 2E system are less than half that of the 3E system, the probability of achieving better performance is higher with 2E-grown materials. Table 3 shows the parameters observed for the best solar cells to date using CdTe grown from the 2E and 3E systems.

Table 3. Summary of the device parameters obtained from the I–V characteristics for devices fabricated with CdTe grown by the 2E and 3E systems.

Device Parameter	RF	n	I_o (nA)	Φ_b (eV)	R_s (Ω)	R_{sh} (MΩ)
I–V Parameters Measured under Dark Condition						
2E	104.3	1.88	0.79	>0.81	1351	$\to\infty$ (81)
3E	104.7	2.40	1.00	>0.80	277	$\to\infty$ (95)
Device Parameter	V_{oc} (V)	J_{sc} (mA·cm^{-2})	FF	η (%)	R_s (Ω)	R_{sh} (Ω)
I–V Parameters measured under AM 1.5 illumination						
2E	0.670	41.5	0.46	12.8	134	3819
3E	0.670	22.0	0.47	6.9	185	5270

The highest efficiency value observed to date for 2E is 12.8%, as shown in Figure 10 and Table 3. The I–V characteristics of dark and illuminated (AM 1.5) conditions for a device with CdTe grown by the 2E system are shown in Figure 10. This structure also has a thin layer (100 nm) of n-ZnS as a buffer layer, forming a three-layer graded bandgap device. All possible device properties have been extracted from these I–V characteristics and summarized in Table 3. These excellent properties, including very high short circuit current densities, show the high potential of graded bandgap device structures [36,37]. These high efficiencies have been observed for CdTe materials grown using 2E systems. Although it is pre-mature to draw a firm conclusion, materials grown from 2E systems are comparable, if not superior to those grown from 3E systems.

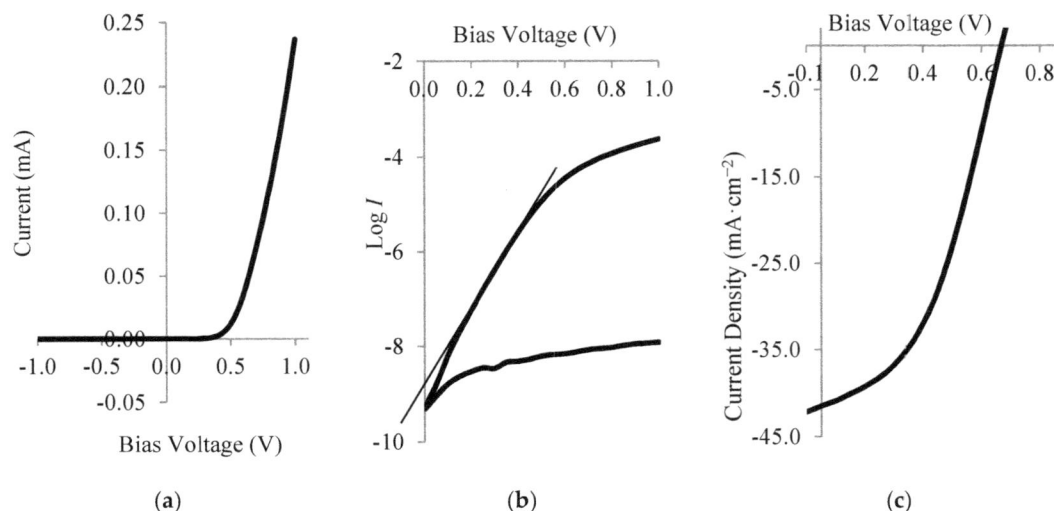

Figure 10. Typical *I–V* characteristics observed for the glass/FTO/n-ZnS/n-CdS/n-CdTe/Au device with CdTe layers grown using the 2E system. (**a**) Linear-linear *I–V* & (**b**) Log-Linear *I–V* under dark condition, and (**c**) Linear-linear *I–V* under AM 1.5 illumination condition.

4. Conclusions

The results presented in this paper lead to the drawing of several conclusions.

- Both 3E and 2E systems produce CdTe layers with similar structural properties. CdTe material is poly-crystalline, with a cubic crystal structure and grown with a (111) preferential orientation. The only difference observed is the higher growth rate when the 2E system is used, and this is an added advantage in a production line.

- The morphologies of the CdTe layers are very similar. In as-deposited layers, the FTO substrate is covered by large CdTe agglomerations, consisting of small crystallites. The sizes of the crystallites vary from 20 to 65 nm, as estimated by XRD measurements. Upon $CdCl_2$ treatment, these crystallites merge into large crystals ranging in the few microns size. Then, the layers are comparable with materials grown by high temperature techniques, such as closed-space sublimation. IPL treatment provides a convenient rapid thermal annealing method suitable for flexible substrates and roll-to-roll production methods. SEM images also indicate the melting of grain boundary regions during heat-treatment in the presence of $CdCl_2$.

- CdTe layers grown from both methods seem to have elemental Te as precipitates or a surface layer. $CdCl_2$ treatment removes this excessive Te and makes the material more stoichiometric. Comprehensive device work also shows that Te richness is detrimental for devices, and stoichiometric or Cd-rich CdTe is more suitable for enhanced device performance.

- Both the 3E and 2E systems allow the growth of p- and n-type CdTe layers. Te richness produces p-type material, and Cd richness produces n-type CdTe. Both methods produce high crystallinity at the perfect point of stoichiometry (PPS), and the energy bandgap measured for stoichiometric material is close to the bandgap of bulk CdTe (1.45 eV).

- PL spectra of CdTe layers grown from the 3E and 2E systems have similar fingerprints. $CdCl_2$ treatment shows the reduction of some defects, and the materials arising from the 2E system show improved quality in terms of defect concentrations.

- UPS results summarized and soft-XPS results reported before show a strong intermixing at the CdTe/Au interface. This may lead to the degradation of devices, and therefore, a reaction barrier should be introduced in order to improve the stability and lifetime of the devices.

- From a large amount of device fabrication experience, both CdTe layers grown by the 3E and 2E systems show wide variation of the efficiency values between 5% and 13%. Although premature to draw firm conclusions, the highest efficiency values observed to date in our research laboratories have been fabricated using CdTe layers grown by the 2E system. Therefore, the elimination of a possible impurity source (reference electrode) introduces several advantages, such as improving the growth rate, system simplification, cost reduction and ability to grow improved materials at elevated temperatures and, hence, the fabrication of comparable or better performing devices.

Acknowledgments: Authors would like to thank A.P. Samantilleke, Nandu Chaure, G. Muftah, Jayne Wellings, N. A. Abdul-Manaf, A.A. Ojo and H.I. Salim for their contributions to this work.

Author Contributions: SHU-Solar Energy Group members (Imyhamy M. Dharmadasa, Mohammad L. Madugu and Olajide I. Olusola): conceived of the work and did the design and the writing of the paper, the electrodeposition of CdS and CdTe, materials' characterization, device fabrication and assessment, interpretation of the results, understanding the science behind the materials and devices, the drafting of the manuscript, the drawing of diagrams and the completion of the paper. Obi K. Echendu, Fijay Fauzi, Dahiru G. Diso and Ajith R. Weerasinghe: The development of the materials and optimization. Conn Center for Renewable Energy Research, University of Louisville, Louisville, Kentucky 40292, USA Group members (Thad Druffel, Ruvini Dharmadasa, Brandon Lavery, Jacek B. Jasinski, Tatiana A. Krentsel and Gamini Sumanasekera): Carried out the photoluminescence (PL) experiment and analysis.

Conflicts of Interest: The authors declare no conflicts of interest.

References

1. Basol, B.M. High-efficiency electroplated heterojunction solar cell. *J. Appl. Phys.* **1984**, *55*, 601–603. [CrossRef]
2. Gnatenko, Y.P.; Bukivskij, P.M.; Opanasyuk, S.; Kurbatov, D.I.; Kolesnyk, M.M.; Kosyak, V.V.; Khlyap, H. Low-temperature photoluminescence of II–VI films obtained by close-spaced vacuum sublimation. *J. Lumin.* **2012**, *132*, 2885–2888. [CrossRef]
3. Dharmadasa, I.M. Recent developments and progress on electrical contacts to CdTe, CdS and ZnSe with special reference to barrier contacts to CdTe. *Prog. Cryst. Growth Charact. Mater.* **1998**, *36*, 249–290. [CrossRef]
4. Bube, R.H. *Photovoltaic Materials*; Newman, R.C., Ed.; Imperial College Press: London, UK, 1998.
5. Takahashi, M.; Uosaki, K.; Kita, H.; Suzuki, Y. Effects of heat treatment on the composition and semiconductivity of electrochemically deposited CdTe films. *J. Appl. Phys.* **1985**, *58*, 4292. [CrossRef]
6. Chaure, N.B.; Samantilleke, A.P.; Dharmadasa, I.M. The effects of inclusion of iodine in CdTe thin films on material properties and solar cell performance. *Sol. Energy Mater. Sol. Cells* **2003**, *77*, 303–317. [CrossRef]
7. Dharmadasa, I.M.; Bingham, P.A.; Echendu, O.K.; Salim, H.I.; Druffel, T.; Dharmadasa, R.; Sumanasekera, G.U.; Dharmasena, R.R.; Dergacheva, M.B.; Mit, K.A.; et al. Fabrication of CdS/CdTe-Based thin film solar cells using an electrochemical technique. *Coatings* **2014**, *4*, 380–415. [CrossRef]
8. Panthani, M.G.; Kurley, J.M.; Crisp, R.W.; Dietz, T.C.; Ezzyat, T.; Luther, J.M.; Talapin, D.V. High efficiency solution processed sintered CdTe nanocrystal solar cells: The role of interfaces. *Nano Lett.* **2014**, *14*, 670–675. [CrossRef] [PubMed]
9. Sugiyama, K. Properties of CdTe films grown on InSb by molecular beam epitaxy. *Thin Solid Films* **1984**, *115*, 97–107. [CrossRef]
10. Mora-Seró, I.; Tena-Zaera, R.; González, J.; Muñoz-Sanjosé, V. MOCVD growth of CdTe on glass: Analysis of in situ post-growth annealing. *J. Cryst. Growth* **2004**, *262*, 19–27. [CrossRef]
11. Ghosh, B.; Hussain, S.; Ghosh, D.; Bhar, R.; Pal, A.K. Studies on CdTe films deposited by pulsed laser deposition technique. *Phys. B Condens. Matter* **2012**, *407*, 4214–4220. [CrossRef]
12. Panicker, M.P.R. Cathodic deposition of CdTe from aqueous electrolytes. *J. Electrochem. Soc.* **1978**, *125*, 566. [CrossRef]
13. Pandey, R.K.; Sahu, S.B.; Chandra, S. *Handbook of Semiconductor Electrodepotion*; Marcel Dekker, Inc.: New York, NY, USA, 1996.
14. Abdul-Manaf, N.; Salim, H.; Madugu, M.; Olusola, O.; Dharmadasa, I. Electro-plating and characterisation of CdTe thin films using $CdCl_2$ as the cadmium source. *Energies* **2015**, *8*, 10883–10903. [CrossRef]
15. Arai, K.; Kawaguchi, J.; Murase, K.; Hirato, T.; Awakura, Y. Effect of chloride ions on the electrodeposition behavior of CdTe from ammoniacal basic electrolytes. *J. Surf. Finish. Soc. Jpn.* **2006**, *57*, 70–76. [CrossRef]

16. Bhattacharya, R.N.; Rajeshwar, K. Heterojunction CdS/CdTe solar cells based on electrodeposited p-CdTe thin films: Fabrication and characterization. *J. Appl. Phys.* **1985**, *58*, 3590–3593. [CrossRef]

17. Lincot, D. Electrodeposition of semiconductors. *Thin Solid Films* **2005**, *487*, 40–48. [CrossRef]

18. Cunningham, D.; Rubcich, M.; Skinner, D. Cadmium telluride PV module manufacturing at BP Solar. *Prog. Photovolt. Res. Appl.* **2002**, *10*, 159–168. [CrossRef]

19. Salim, H.I.; Patel, V.; Abbas, A.; Walls, J.M.; Dharmadasa, I.M. Electrodeposition of CdTe thin films using nitrate precursor for applications in solar cells. *J. Mater. Sci. Mater. Electron.* **2015**, *26*, 3119–3128. [CrossRef]

20. First Solar's Cells Break Efficiency Record. Available online: https://www.technologyreview.com/s/600922/first-solars-cells-break-efficiency-record (accessed on 19 January 2016).

21. Dennison, S. Dopant and impurity effects in electrodeposited CdS/CdTe thin films for photovoltaic applications. *J. Mater. Chem.* **1994**, *4*, 41–46. [CrossRef]

22. Echendu, O.K.; Okeoma, K.; Oriaku, C.I.; Dharmadasa, I.M. Electrochemical deposition of CdTe semiconductor thin films for solar cell application using two-electrode and three-electrode configurations: A comparative study. *Adv. Mater. Sci. Eng.* **2016**, *2016*, 3581725. [CrossRef]

23. Dharmadasa, R.; Lavery, B.W.; Dharmadasa, I.M.; Druffel, T. Processing of CdTe thin films by intense pulsed light in the presence of $CdCl_2$. *J. Coat. Technol. Res.* **2015**, *12*, 835–842. [CrossRef]

24. Diso, D.G. Research and Development of CdTe Based Thin Film PV Solar Cells. Ph.D. Thesis, Sheffield Hallam University, Sheffield, UK, 2011.

25. Dharmadasa, I.M.; Ehendu, O.K.; Fauzi, F.; Abdul-Manaf, N.A.; Olusola, O.I.; Salim, H.I.; Madugu, M.L.; Ojo, A.A. Improvement of composition of CdTe thin films during heat treatment in the presence of $CdCl_2$. *J. Mater. Sci. Mater. Electron.* **2016**. [CrossRef]

26. Cheung, J.T. Role of atomic tellurium in the growth kinetics of CdTe (111) homoepitaxy. *Appl. Phys. Lett.* **1987**, *51*, 1940–1942. [CrossRef]

27. Fernandez, P. Defect structure and luminescence properties of CdTe based compounds. *J. Optoelectron. Adv. Mater.* **2003**, *5*, 369–388.

28. Dharmadasa, I.M.; Echendu, O.K.; Fauzi, F.; Abdul-Manaf, N.A.; Salim, H.I.; Druffel, T.; Dharmadasa, R.; Lavery, B. Effects of $CdCl_2$ treatment on deep levels in CdTe and their implications on thin film solar cells: A comprehensive photoluminescence study. *J. Mater. Sci. Mater. Electron.* **2015**, *26*, 4571–4583. [CrossRef]

29. Dharmadasa, I.M.; Thornton, J.M.; Williams, R.H. Effects of surface treatments on Schottky barrier formation at metal/n-type CdTe contacts. *Appl. Phys. Lett.* **1989**, *54*, 137–139. [CrossRef]

30. Sobiesierski, Z.; Dharmadasa, I.M.; Williams, R.H. Correlation of photoluminescence measurements with the composition and electronic properties of chemically etched CdTe surfaces. *Appl. Phys. Lett.* **1988**, *53*, 2623–2625 [CrossRef]

31. Dharmadasa, I.M. *Advances in Thin-Films Solar Cells*; Pan Stanford Publishing Pte Ltd.: Singapore, 2012.

32. Dharmadasa, I.M.; Echendu, O.K.; Fauzi, F.; Salim, H.I.; Abdul-Manaf, N.A.; Jasinski, J.B.; Sherehiy, A.; Sumanasekera, G. Study of fermi level position before and after $CdCl_2$ treatment of CdTe thin films using ultraviolet photoelectron spectroscopy. *J. Mater. Sci. Mater. Electron.* **2016**. [CrossRef]

33. *Handbook of Chemistry and Physics*; CRC Press: Cleveland, OH, USA, 1977.

34. Okuyama, K.; Tsuhako, J.; Kumagai, Y. Behavior of metal contacts to evaporated tellurium films. *Thin Solid Films* **1975**, *30*, 119–126. [CrossRef]

35. Forsyth, N.M.; Dharmadasa, I.M.; Sobiesierski, Z. An investigation of metal contacts to II–VI compounds: CdTe and CdS. *Vacuum* **1988**, *38*, 369–371. [CrossRef]

36. Dharmadasa, I.M.; Ojo, A.; Salim, H.; Dharmadasa, R. Next generation solar cells based on graded bandgap device structures utilising rod-type nano-materials. *Energies* **2015**, *8*, 5440–5458. [CrossRef]

37. Echendu, O.; Dharmadasa, I.M. Graded-bandgap solar cells using all-electrodeposited ZnS, CdS and CdTe thin Ffilms. *Energies* **2015**, *8*, 4416–4435. [CrossRef]

12

Strain Effects by Surface Oxidation of Cu$_3$N Thin Films Deposited by DC Magnetron Sputtering

Abhijit Majumdar [1,2], **Steffen Drache** [1], **Harm Wulff** [1], **Arun Kumar Mukhopadhyay** [2], **Satyaranjan Bhattacharyya** [3], **Christiane A. Helm** [1] **and Rainer Hippler** [1,*]

[1] Institute of Physics, University of Greifswald, Felix Hausdorff Str. 6, Greifswald 17489, Germany; majuabhijit@googlemail.com (A.M.); Drachesteffen@gmail.com (S.D.); wulff@uni-greifswald.de (H.W.); helm@uni-greifswald.de (C.A.H.)

[2] Indian Institute of Engineering Science and Technology, Shibpur, Howrah-3, West Bengal 711103, India; akmphy_dac@rediffmail.com

[3] Surface Physics & Materials Science Division, Saha Institute of Nuclear Physics, 1/AF Bidhan Nagar, Kolkata 700 064, India; satya.bhattacharyya@saha.ac.in

* Correspondence: hippler@uni-greifswald.de

Academic Editors: Klaus Pagh Almtoft and Alessandro Lavacchi

Abstract: We report the self-buckling (or peeling off) of cubic Cu$_3$N films deposited by DC magnetron sputtering of a Cu target in a nitrogen environment at a gas pressure of 1 Pa. The deposited layer partially peels off as it is exposed to ambient air at atmospheric pressure, but still adheres to the substrate. The chemical composition of the thin film as investigated by means of X-ray photoelectron spectroscopy (XPS) shows a considerable surface oxidation after exposure to ambient air. Grazing incidence X-ray diffraction (GIXRD) confirms the formation of a crystalline Cu$_3$N phase of the quenched film. Notable are the peak shifts in the deposited film to smaller angles in comparison to stress-free reference material. The X-ray pattern of Cu$_3$N exhibits clear differences in the integral width of the line profiles. Changes in the film microstructure are revealed by X-ray diffraction, making use of X-ray line broadening (Williamson–Hall and Stokes–Fourier/Warren–Averbach method); it indicates that the crystallites are anisotropic in shape and show remarkable stress and micro-strain.

Keywords: thin film deposition; copper nitride; magnetron sputtering; crystal structure

1. Introduction

The interest in copper nitride (Cu$_3$N) thin films has increased in recent years due to their potential applications for recording media [1–3] and as precursor material for microscopic copper lines by mask-less laser writing [4]. Various methods have been employed to obtain copper nitride films, such as RF-sputtering [5–9], RF-plasma chemical reactor [10], reactive pulsed laser deposition [11], and activated reactive evaporation [12]. Despite the promising properties of Cu$_3$N, large discrepancies reported in the literature about its measured physical properties have hampered the implementation of reliable technological devices.

The control of the structure and properties of copper nitride films is an interesting topic. In the present work, buckling of the deposited Cu$_3$N film after exposure to ambient air is observed. The deposited thin copper nitride films partially peel off from the silicon substrate within several minutes of air exposure. The crystalline structure and the surface oxidation of the deposited films are investigated and analysed.

2. Materials and Methods

Copper nitride films were prepared by DC magnetron sputtering [13–16]. The 2″ sputter target was made of oxygen-free copper with a purity of 99.9%. Films were deposited on commercially available p-Si (100) substrates. Alternatively, glass substrates were employed. All substrates were ultrasonically cleaned in acetone. The vacuum chamber was evacuated by a turbo molecular pump to a base pressure of less than 10^{-5} Pa. The working gas was 99.999% pure N_2. The chamber pressure during magnetron operation was maintained at 1 Pa. Typical discharge power during sputter deposition was 100–130 W (discharge voltage 325 V, plasma current 0.4 A). Deposition took place at room temperature and with electrically floating substrates. Films were deposited for 30 min (glass substrate) or 60 min (Si substrate). After deposition, the substrates were taken out of the chamber. The films were further examined using a scanning electron microscope (SEM). The SEM (FEI Company, Hillsboro, OR, USA, Quanta 200F) was operated with an energy of 10 keV at tilt angles of 0.6° and 30° and with typical magnifications of 100.

Film thickness was investigated with the help of spectroscopic ellipsometry employing a phase-modulated ellipsometer (HORIBA Jobin-Yvon Inc., Edison, NJ, USA, UVISEL). The investigated wavelength region was 380–830 nm with energy steps of less than 0.5 nm. The experiments were carried out under an incidence angle of 70° corresponding to the Brewster angle of the Si (100) substrate. A film thickness of 3 μm was derived for the Cu_3N/Si film (deposition time 60 min).

Grazing incidence X-ray diffraction (GIXRD, asymmetric Bragg case) measurements were done to determine the phase composition of deposited films; the employed methods are described in Reference [17]. All measurements were performed with a θ–2θ Diffractometer (D5000, Bruker AXS GmbH, Karlsruhe, Germany) using Cu Kα radiation (40 kV, 40 mA). The scanned 2θ range was 20°–50° at a constant incidence angle $\omega = 1.0°$. The crystal structure data of Cu_3N were used to identify the crystallographic phases [18].

X-ray photoelectron spectroscopy (XPS) measurements of the Cu_3N films were performed on a multi-technique 100 mm hemispherical electron analyser (Fisons Instruments VG Microtech, Uckfield, UK, CLAM2), using Mg Kα radiation (photon energy 1253.6 eV) as the excitation source and the binding energy (BE) of Au (Au $4f_{7/2}$: 84.00 eV) as the reference. XPS measurements were performed at two different detector angles of 90° and 20° with respect to the substrate [17].

3. Results and Discussion

Photographs of the deposited Cu_3N/Si film after exposure to ambient air are displayed in Figure 1. Photographs were taken at intervals of about 10–15 min. Gradual changes of the surface morphology with time are noted. Similar changes are observed for films deposited on glass substrates (not shown here). Figure 1a was taken right after the film was exposed to ambient air at room temperature (≈295 K). Self-buckling of the deposited films was observed after about 12 min (Figure 1b). The buckled area increased gradually with time (Figure 1b–e). Almost the entire film surface was buckled after 90 min, as is obvious from Figure 1f. The buckled area as a function of time is displayed in Figure 2.

Figure 3 shows the SEM picture of the buckled Cu_3N surface area. Random ripple structures are observed at tilt angles of 0.6° and 30°. This shows that the film surface is partially taking off towards upwards directions from the substrate, and empty spaces or voids in between the film and the underlying substrate are created. The observed ripple structure is caused by this empty space. Nevertheless, the film still firmly adheres to the substrate. Apparently, the film properties change when exposed to air at atmospheric pressure, which leads to buckling and a partial peeling off the substrate.

Figure 4 shows the X-ray diffraction pattern of the polycrystalline Cu_3N film deposited on Si and glass substrates. XRD measurements were carried out shortly after deposition in the case of the Cu_3N/Si film and after 5 years in ambient air in the case of the Cu_3N/glass film. No significant differences between the two films were observed (see Table 1), demonstrating the good stability of the deposited Cu_3N films. The dashed line shows the simulated X-ray patterns of the Cu_3N/Si film, which were calculated on the basis of single crystal structure data [18,19]. A lattice parameter $a_0 = 0.3819$ nm was used as a reference value for a stoichiometric and stress-free sample. Free parameters of the simulation (fit) are background, lattice parameter, profile function, and line profile width. Copper nitride has a cubic anti-ReO$_3$ type structure (SG: Pm-3m, No. 221) which is reproduced in the insert of Figure 4. No other crystalline phases were detected. In the ideal crystal structure, the Cu atoms completely occupy the 3c Wyckoff-position with $0 \frac{1}{2} \frac{1}{2}$; $\frac{1}{2} 0 \frac{1}{2}$; $\frac{1}{2} \frac{1}{2} 0$. The N atom is in 1b position, with $\frac{1}{2} \frac{1}{2} \frac{1}{2}$. This implies that the (100) and (200) lattice planes differ concerning their occupation with Cu and N atoms. The intensity ratios correspond to bulk values with a statistical distribution of crystallites (Table 1). The X-ray pattern (Figure 4) does not show any preferred orientation. The lattice parameter of the deposited Cu$_3$N films was measured as $a_0 = 0.3837$ nm (Table 1)—significantly larger than that measured by Zachwieja et al. [19] and Navio et al. [20]. Moreno-Armenta et al. [21] reported non-stoichiometric copper nitride (Cu$_3$N) with a lattice parameter $a_0 = 0.384$ nm, which is in reasonable agreement with the present results. Pierson proved the existence of non-stoichiometric copper nitride with lattice constants ranging from as low as 0.375–0.384 nm which are attributed to sub-stoichiometric and over-stoichiometric Cu$_3$N, respectively [22]. Stoichiometric Cu$_3$N means that all regular lattice positions are completely occupied—i.e., the site occupation factor (SOF) is 1 for both lattice positions. Over-stoichiometry means that additional nitrogen atoms are inserted into the Cu$_3$N lattice, probably as interstitials. Both findings indicate that defects play an important role in the development of increasing lattice parameters during the film deposition process, and may lead to the formation of intrinsic stress.

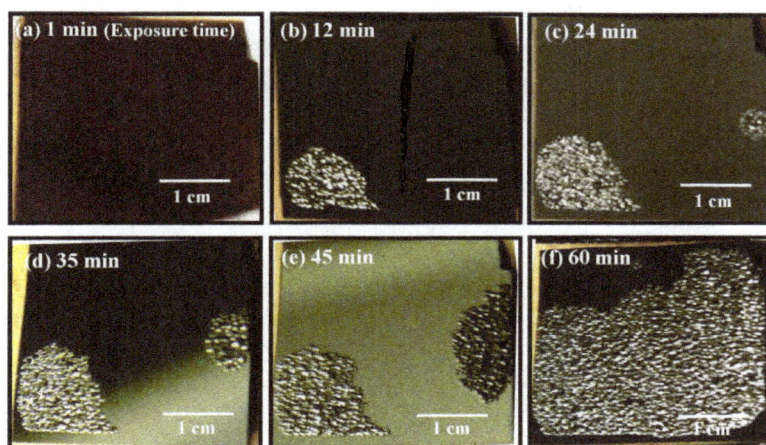

Figure 1. Photographs of deposited Cu$_x$N film on Si substrates taken (**a**) 1 min; (**b**) 12 min; (**c**) 24 min; (**d**) 35 min; (**e**) 45 min; and (**f**) 60 min after the film was taken out of the deposition chamber and exposed to ambient air.

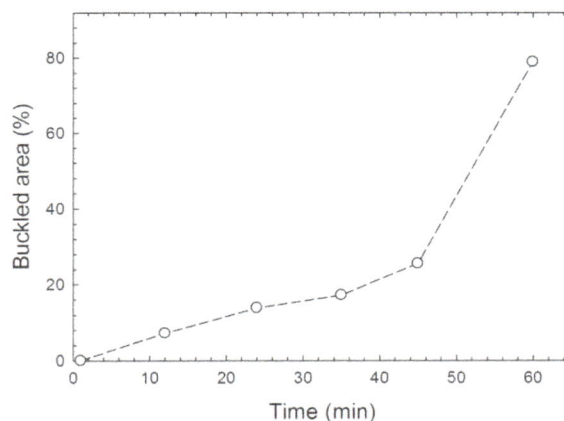

Figure 2. Buckled film area (in percent of the total area) versus exposure time.

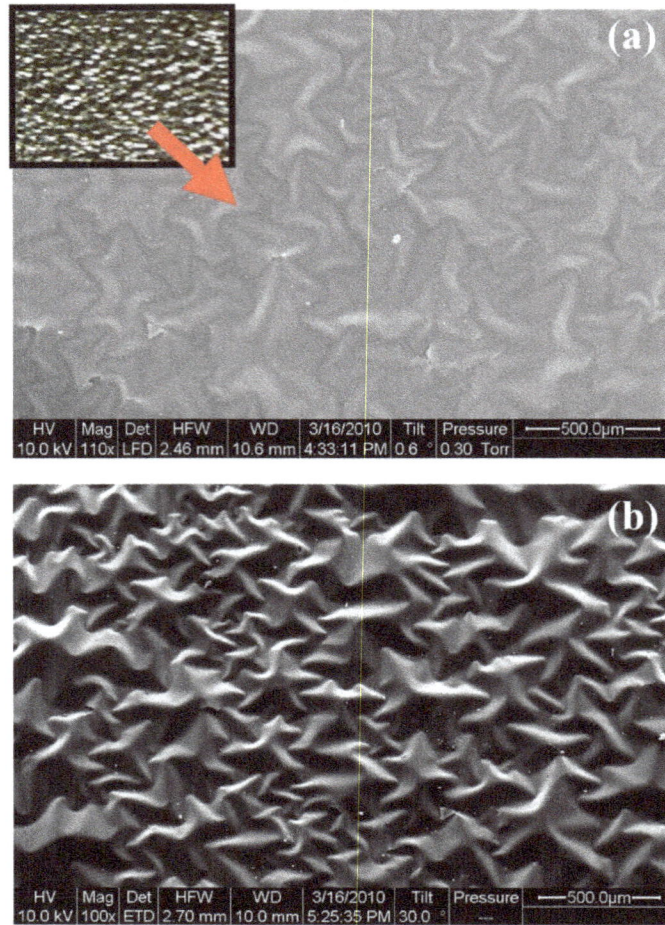

Figure 3. SEM images of deposited Cu$_x$N film taken at tilt angles of (**a**) 0.6° and (**b**) 30°.

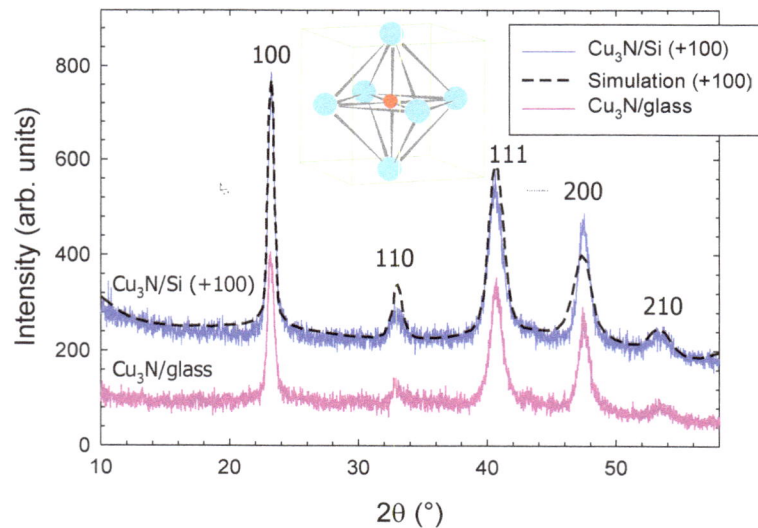

Figure 4. Measured and simulated (solid line) X-ray diffraction pattern of polycrystalline Cu$_3$N films deposited on a Si and on a glass substrate. Simulation results (dashed line) are also shown for the Cu$_3$N/Si data. To ease comparison, the Cu$_3$N/Si data spectra have been shifted upwards by +100.

Table 1. XRD results of deposited Cu_3N films on (**a**) Si and (**b**) glass substrates.

Sample	HKL	2θ (°)	D (nm)	Relative Intensity	FWHM (°)
	100	23.166	0.3836(4)	63.0	0.621
	110	32.993	0.2712(8)	21.9	0.938
(**a**) Cu_3N/Si	111	40.702	0.2215(0)	100.0	1.083
	200	47.353	0.1918(2)	60.3	1.161
	210	53.355	0.1715(7)	22.2	1.200
	100	23.159	0.3837(6)	61.7	0.657
	110	32.983	0.2713(6)	21.7	0.928
(**b**) Cu_3N/glass	111	40.690	0.2215(6)	100.0	1.103
	200	47.338	0.1918(8)	61.0	1.243
	210	53.338	0.1716(2)	22.7	1.364

The problem of the cubic Cu_3N films is the poor adhesion strength to the substrate, which causes buckling and a partial peeling off the layers. Mechanical properties strongly depend on the micro-structure, which can be extracted from the XRD pattern. The behaviour can result from intrinsic stress and micro-strain in the films. Thin films are almost invariably in a state of stress. The present investigation therefore focuses on intrinsic stress and strain properties of the deposited film (uniform and non-uniform strain). X-ray data are influenced by lattice defects. Imperfections of the first type, such as point defects, displacement disorders or substitution disorders, shift the position of the diffraction line. The imperfections may result in intrinsic stress and strain, which also influences the diffraction line shape (width). Domain sizes and dislocations also influence the diffraction line shape. Notably, all the peaks shift to smaller angles in the deposited film in comparison to stress-free material [19]. Local changes of the atomic ordering of the cubic Cu_3N lattice during deposition and crystallization result in a change of its molar volume. In this context, the term "molar volume" is meant to indicate the volume occupied by 1 mol of Cu_3N plus associated defects. Point defects, displacement disorder, and substitution disorders, but also anisotropic grain growth affect the molar volume of the material. Changes of this imperfection concentration should be accompanied by a change in uniform film stress. In the case of tensile stress, the atoms are farther apart compared to the annealed (stress-free) case, while it is the opposite in the case of compressive stress [23]. It is evident that a molar volume increase of the films bonded to a rigid substrate will result in an overall increase in film stress. The intrinsic stress can be estimated from Bragg-angle shifts. Assuming an isotropic defect concentration, the intrinsic stress $\Delta\sigma$ due to an increase $\Delta V/V$ of the molar volume (excess volume) can be estimated as [24]:

$$\Delta\sigma = \frac{1}{3}\frac{E}{(1-\nu)}\frac{\Delta V}{V} \tag{1}$$

where E is Young's modulus and ν is Poisson's ratio of the film. Unfortunately, we could not find experimental data of E and ν for Cu_3N. An estimate based on the measured $\Delta V/V$ using theoretical data for $E = 156$ GPa and $\nu = 0.28$ [25] yields a large intrinsic stress $\Delta\sigma = 0.96$ GPa of our Cu_3N films. Tensile stress will relieve itself by micro-cracking of the film [23]. On the other hand, in the case of compressive stress, the atoms are closer to each other compared to the relaxed (stress-free) case. Compressive stress is relieved by buckling [23]. Hence, the present results could imply that the observed lattice expansion is the result of over-stoichiometry [22]. The observed buckling is then caused by compressive stress in the film, which is relieved by surface oxidation after its exposure to ambient air.

The increase of lattice parameters obtained from smooth and rough film parts is not explainable by the stress and elastic anisotropy alone, since it persists after the film's partial delamination from the substrate. Besides the uniform strain, we also observed a non-uniform strain. A comparison of the integral width β of the Cu_3N sample and a reference standard material (LaB_6) that contains no defects, strain, or particle size broadening (NIST, Gaithersburg, MD, USA, SRM 660b) measured under the same conditions shows an appreciable broadening effect in the Cu_3N profiles (Figure 5a). Here we define the dimensionless integral width β as the ratio of the integral intensity of a line profile and the peak intensity at the peak's maximum. The corrected integral width $\beta_{cor} = \beta_{Cu3N} - \beta_{LaB6}$ is obtained by subtracting the width of the reference material (LaB_6) from the measured Cu_3N width.

The Williamson–Hall (WH) analysis [26] immediately gives qualitative information about the size and shape of the crystallites and the presence of non-uniform lattice strain (Figure 5b). The WH-plot has the following main features: (i) there is a wide scatter of points and (ii) a line connecting the data points (dashed line) displays a positive slope. The scattered data points indicate that the crystallites are anisotropic in shape. The positive slope is an indication of domains with many crystallographic defects which, for example, may arise from deviations in stoichiometry. Making use of [26],

$$\beta_{corr} \cos \beta = \frac{\lambda}{T} + 4\varepsilon \sin \beta \qquad (2)$$

we can estimate the *volume-weighted* micro-strain $\varepsilon = |d - d_0| / d_0$, where $\lambda = 0.1542$ nm is the X-ray wavelength, $T \approx 50$ nm (see below) is the assumed mean particle size, and d_0 is the interplanar spacing derived from the peak position. The extracted micro-strain is reasonably large (about 1%–2%) and strongly depends on the lattice orientation (Figure 5b).

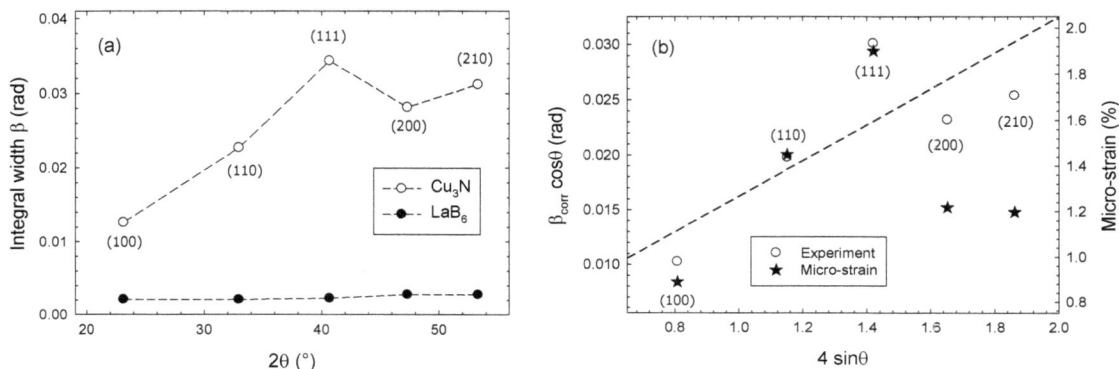

Figure 5. (a) Integral width β of Cu_3N and reference material LaB_6 vs. diffraction angle. LaB_6 (SRM 660b) is used for the determination of the instrument profile function; (b) Williamson–Hall analysis and the extracted micro-strain of the Cu_3N thin film. Dashed line is added to guide the eye only.

Anisotropic line broadening of (100) and (200) line profiles may be caused by non-stoichiometry of the Cu_3N films (see above). Therefore, it is necessary to investigate both these profiles in more detail. The microstructure and lattice defects are directly extracted from diffraction pattern in terms of a Fourier transform of the scattered intensity without using any peak shape function [27]. Based on the Warren–Averbach (WA) theory [28,29], Klimanek developed a single line profile method [30]. We use this method because the diffraction vector is different for the (100) and (200) reflection in GIXRD.

The Fourier coefficients $F(L)$ of the physical line profile contain information on particle size T and mean strain S, which are constant within a crystallite or sub-grain, and dislocation density D:

$$F(L) = F_T(L) \times F_S(L) \times F_D(L) \qquad (3)$$

where L (in units of nm) is defined as

$$L = \frac{n\lambda}{2(\sin \theta_2 - \sin \theta_1)}$$

and where θ_1 and θ_2 define the angular range for the experimentally observable line profile, λ is the X-ray wavelength of Cu Kα radiation, and where the integers n are the harmonics of the Fourier coefficients. The logarithm of $F(L)$ can be expressed as [30]

$$-\frac{\ln F(L)}{nL} = \frac{1}{T} + K\langle e^2(L)\rangle \tag{4}$$

where K is a constant which depends on the interplanar spacing. The linear part of a graph of $\ln F(L)/nL$ allows for a determination of the mean particle size T from the intercept and the *area-weighted* mean squared micro-strain $\langle e^2(L)\rangle$ due to internal stress from the slope. The dislocation density can be estimated from the linear branch of a Krivoglaz–Wilkens plot [31,32],

$$-\frac{\ln F(L)}{L^2} = \frac{1}{T}L + (K\langle e^2(L)\rangle + B\ln L_0) - B\ln L \tag{5}$$

where the factor B is proportional to the mean total dislocation density and L_0 is a length which is proportional to the core radius r_0 of the strain field of a dislocation.

Figure 6 shows the WA-plot of the (100) and (200) reflections. We must keep in mind that both these planes differ concerning their occupation with nitrogen atoms. The extracted effective particle sizes are $T \approx 50$ nm, but there is a significant difference in the area-weighted micro-strain. This implies that deviations in stoichiometry are responsible for the observed line broadening. There is no indication that dislocation densities determined from Krivoglaz–Wilkens plot are different in the (100) and (200) reflections. We calculate dislocation densities of $1.6 \times 10^{12}/\text{cm}^2$. In the order of magnitude, they correspond to plastic deformation of the films.

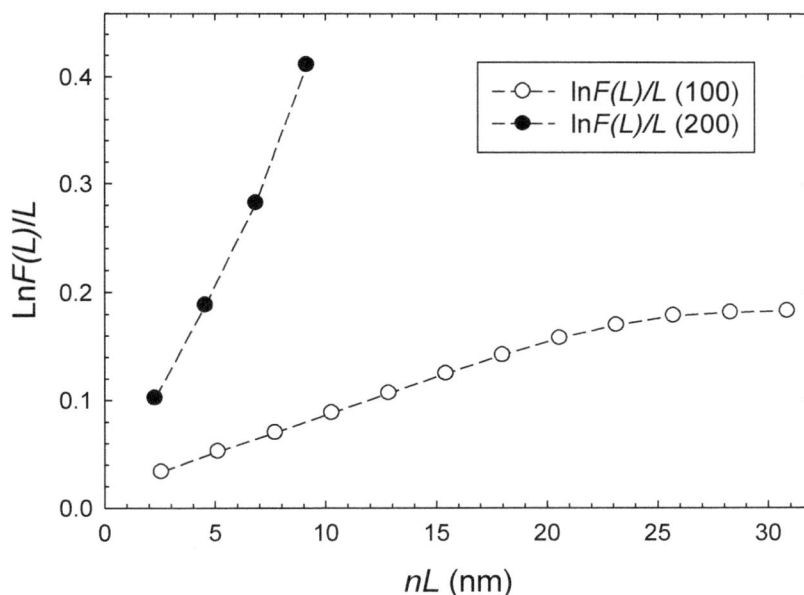

Figure 6. Warren–Averbach plot of the (100) and (200) reflections.

Non-uniform and uniform strain are large to the extent of deviations from stoichiometry. A possible explanation is related to film inhomogeneity in dependence on stoichiometry and the orientation of the crystallites. Possible reasons for this inhomogeneity are non-randomly-oriented two-dimensional lattice defects. The occurrence of these defects in dependence on the grain orientation with respect to the specimen surface can cause the inhomogeneity of the films. The parts of a film after peeling-off have a plate-like shape.

The recorded X-ray photoelectron spectra of the copper nitride film show photoelectron emission from the Cu-2p, N-1s, O-1s, and C-1s states (Figure 7). Relative surface compositions of the deposited

Cu/N films are given in Table 2. The large amount of carbon and oxygen and the relatively small amounts of Cu and N needs some explanation. XPS is a surface analytical technique, and is thus sensitive to surface contaminations—in particular, adventitious carbon. An analysis of the carbon peak (Figure 7d) shows that it is composed of three contributions, which are attributed to C–C, C–O, and C=O bonds [33,34]. Quantitatively, about 30% of the carbon has bonds with oxygen. The O-1s peak can be decomposed into two peaks which can be associated with oxygen bound to either copper or carbon [35–37]. The nitrogen peak (Figure 7c) resembles a single peak with a peak width of ≈ 2 eV.

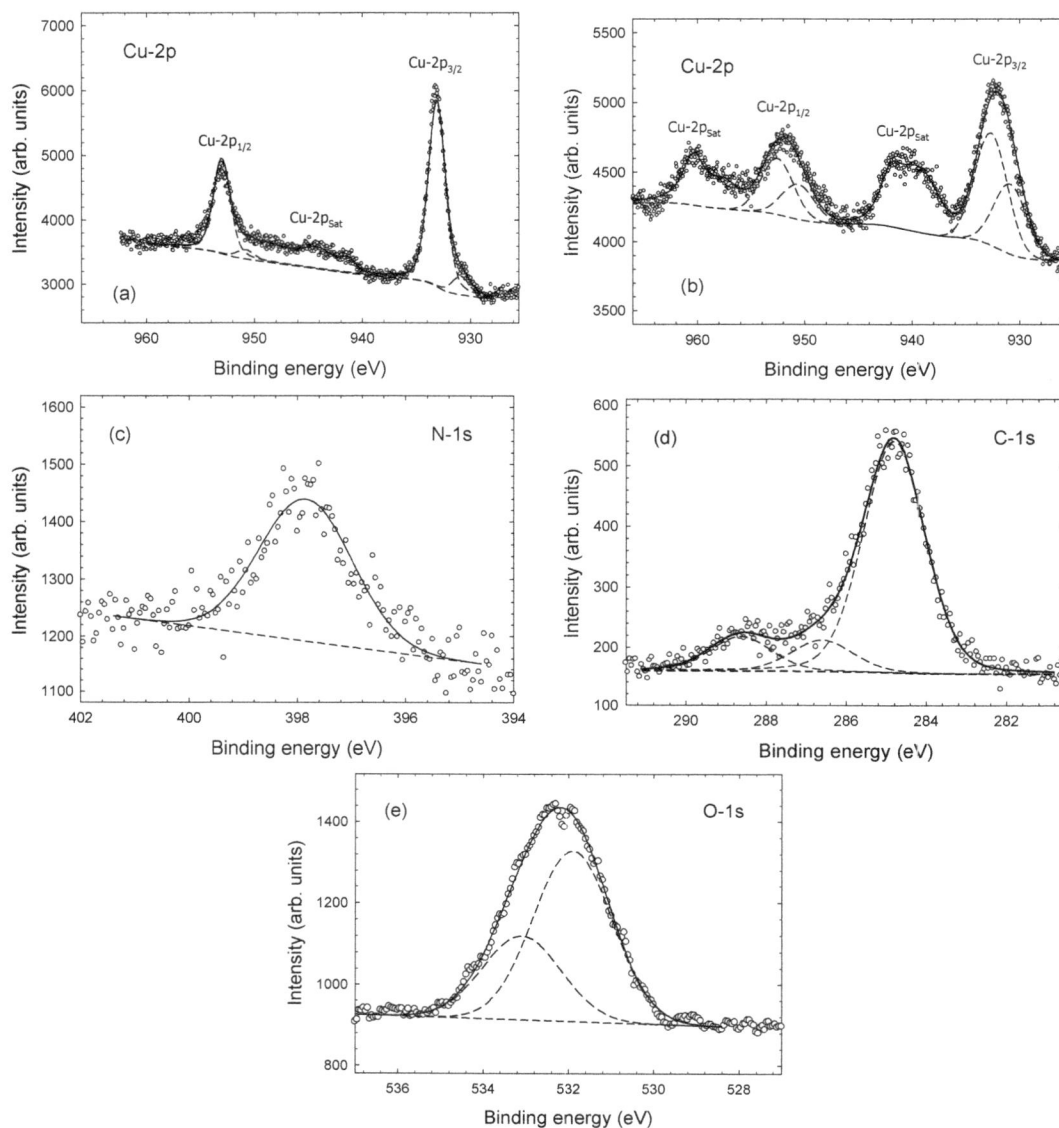

Figure 7. X-ray photoelectron spectroscopy (XPS) spectra of deposited Cu_xN films (detection angle $90°$). (**a**) Cu-2p (Si substrate); (**b**) Cu-2p (glass substrate); (**c**) N-1s (Si substrate); (**d**) C-1s (Si substrate); (**e**) O-1s (Si substrate).

Table 2. Relative surface composition of deposited Cu_xN films as derived from XPS analysis.

Sample	Angle	Cu (%)	N (%)	Cu/N	C (%)	O (%)
Cu_xN/Si	$90°$	30.4	13.8	2.2	20.4	35.5
Cu_xN/glass	$20°$	10.2	4.6	2.2	41.0	44.2
Cu_xN/glass	$90°$	14.0	5.8	2.4	32.6	47.6

The relatively small amount of Cu and N detected by XPS is hence explained by surface contamination with carbon and oxygen of the copper nitride films. The argument is further supported by the observed differences between the copper nitride films deposited on Si or glass samples. The larger contamination of the Cu_xN-on-glass with carbon and oxygen is explained by the much longer exposure to ambient air. Further proof comes from XPS measurements of the Cu_xN/glass sample being performed at two different detector angles of 90° and 20° with respect to the substrate. Measurements at smaller detection angles are more sensitive to surface contributions. The enhanced carbon contribution of the 20° measurement indicates that much or all of the carbon is sitting on the surface. The extracted Cu/N ratio is in the range of 2.2–2.4, and hence significantly smaller than the expected ratio of 3 for a Cu_3N film. Similar ratios of 2.2–2.7 were measured by Navio et al. [20].

Figure 7a displays the Cu-$2p$ spectrum for the as-deposited Cu_xN film on a Si substrate; it was measured after the sample had been taken out of the deposition chamber and exposed to ambient air. Figure 7b displays the same Cu-$2p$ region for a Cu_xN film on a glass substrate after five years in ambient air. Both spectra show the Cu-$2p_{1/2}$ and Cu-$2p_{3/2}$ peaks. In addition, there are Cu shake-up satellite lines which are attributed to copper oxide Cu(II)O [35]. Some differences are immediately noted. Most significantly, the width of the Cu-$2p$ peak is 2 eV and 4.4 eV (Figure 7a,b, respectively), significantly different for the two samples. In addition, the intensity of the satellite peaks is larger in Figure 7b compared to Figure 7a. The observations are attributed to the larger oxygen content at the film surface of the Cu_xN/glass sample, presumably due to the much longer time in air. The strong appearance of the satellite lines provides evidence that a significant fraction of the surface copper is oxidised.

While the XRD results (Figure 4) leave little doubt that much of the deposited film is composed of crystalline Cu_3N, the XPS results provide clear evidence that the film has become oxidised on its surface during its exposure to ambient air. We believe that this oxidation is the main reason for the observed stress in the films and for the peeling-off. This idea is supported by the observed time dependence of the peeling-off process (Figure 1), which takes several minutes to complete. Deviations from a stoichiometric Cu_3N composition seem to play a significant role in the observed lattice parameter expansion. The nature of this departure from stoichiometry is not fully clear, however. In particular, the question remains whether oxygen is chemisorbed only on the film surface in a thin superficial layer, or to which extent the whole lattice is affected.

4. Conclusions

Cu_xN films have been deposited by DC magnetron sputtering. After exposure to ambient air, the films partially peel-off, but remain firmly attached to the substrate. XRD measurements show that the bulk material is crystalline Cu_3N. However, the X-ray pattern shows remarkable deviations from the ideal poly-crystalline structure. Information about uniform and non-uniform strain is obtained from line shifts and line broadening. Films are characterized by a large intrinsic stress of about 1 GPa and an anisotropic micro-strain of about 1%–2%, presumably caused by stoichiometric deviations in the films. XPS measurements reveal a considerable surface oxidation and the presence of Cu(II)O phases. We suspect that surface oxidation is responsible for the observed stress and micro-strain in the film. Possible reasons for the observed non-uniform and uniform strain are deviations from perfect stoichiometry introduced by surface oxidation. More experiments will be required to prove a causal relationship, however.

Acknowledgments: The work was partly supported by the Deutsche Forschungsgemeinschaft (DFG) through Sonderforschungsbereich SFB/TRR24.

Author Contributions: Abhijit Majumdar and Rainer Hippler conceived and designed the experiments; Abhijit Majumdar performed the experiments; Abhijit Majumdar, Steffen Drache, Harm Wulff, Arun Kumar Mukhopadhyay, Satyaranjan Bhattacharyya, Christiane A. Helm and Rainer Hippler analyzed the data; Abhijit Majumdar, Harm Wulff and Rainer Hippler wrote the paper.

Conflicts of Interest: The authors declare no conflict of interest.

References

1. Maruyama, T.; Morishita, T. Copper nitride and tin nitride thin films for write-once optical recording media. *Appl. Phys. Lett.* **1996**, *69*, 890–891. [CrossRef]

2. Ma, X.D.; Bazhanov, D.I.; Fruchart, O.; Yildiz, F.; Tokoyama, T.; Przybyiski, M.; Stepanyuk, V.S.; Hergert, W.; Kirschner, M. Strain relief guided growth of atomic nanowires in a Cu_3N-Cu(110) molecular network. *Phys. Rev. Lett.* **2009**, *102*, 205503. [CrossRef] [PubMed]

3. Borsa, D.M.; Grachev, S.; Presura, C.O.; Boerma, D.O. Growth and properties of Cu_3N films and Cu_3N/γ'-Fe_4N bilayers. *Appl. Phys. Lett.* **2002**, *80*, 1823–1825. [CrossRef]

4. Maya, L. Covalent nitrides for maskless laser writing of microscopic metal lines. *MRS Proc.* **1992**, *282*, 203–208. [CrossRef]

5. Maruyama, T.; Morishita, T. Copper nitride thin films prepared by radio-frequency reactive sputtering. *J. Appl. Phys.* **1995**, *78*, 4104–4107. [CrossRef]

6. Kamat, H.; Wang, X.; Parry, J.; Qin, Y.; Zeng, H. Synthesis and characterization of copper-iron-nitride thin films. *MRS Adv.* **2016**, *1*, 203–208. [CrossRef]

7. Ji, X.; Ju, H.; Zou, T.; Luo, J.; Hong, K.; Yang, H.; Wang, H. Effects of sputtering pressure on Cu_3N thin films by reactive radio frequency magnetron sputtering. *Adv. Mater. Res.* **2015**, *1105*, 74–77. [CrossRef]

8. Leng, J.; Chen, L.; Zhu, X.; Sun, Z. Structure and photoelectric properties of Cu_3N thin films by reactive magnetron sputtering. *Mater. Sci. Forum* **2015**, *814*, 620–624. [CrossRef]

9. Chen, L.; Leng, J.; Yang, Z.; Meng, Z.; Sun, B. Influence of sputtering power on the structure, optical and electric properties of Cu_3N films. *Mater. Sci. Forum* **2015**, *814*, 596–600. [CrossRef]

10. Soukup, L.; Sicha, M.; Fendrych, F.; Jastrabik, L.; Hubicka, Z.; Studnicka, V.; Wagner, T.; Novak, M. Copper nitride thin films prepared by the RF plasma chemical reactor with low pressure supersonic single and multi-plasma jet system. *Surf. Coat. Technol.* **1999**, *116*, 321–326. [CrossRef]

11. Gallardo-Vega, C.; de la Cruz, W. Study of the structure and electrical properties of the copper nitride thin films deposited by pulsed laser deposition. *Appl. Surf. Sci.* **2006**, *252*, 8001–8004. [CrossRef]

12. Sahoo, G.; Meher, S.R.; Jain, M.K. Room temperature growth of high crystalline quality Cu_3N thin films by modified activated reactive evaporation. *Mater. Sci. Eng. B* **2015**, *191*, 7–14. [CrossRef]

13. Wrehde, S.; Quaas, M.; Bogdanowicz, R.; Steffen, H.; Wulff, H.; Hippler, R. Reactive deposition of TiN_x layers in a DC-magnetron discharge. *Surf. Interface Anal.* **2008**, *40*, 790–793. [CrossRef]

14. Hippler, R.; Steffen, H.; Quaas, M.; Röwf, T.; Tun, T.M.; Wulff, H. Plasma-Assisted Deposition and Crystal Growth of Thin Indium-Tin-Oxide (ITO) Films. In *Advances in Solid State Physics 44*; Kramer, B., Ed.; Springer: Heidelberg, Germany, 2004; p. 299.

15. Bräuer, G.; Szyszka, B.; Vergöhl, M.; Bandorf, R. Magnetron sputtering—Milestones of 30 years. *Vacuum* **2010**, *84*, 1354–1359. [CrossRef]

16. Pflug, A.; Siemers, M.; Melzig, T.; Schäfer, L.; Bräuer, G. Simulation of linear magnetron discharges in 2D and 3D. *Surf. Coat. Technol.* **2014**, *260*, 411–416. [CrossRef]

17. Wulff, H.; Steffen, H. Characterization of Thin Films. In *Low Temperature Plasmas*; Hippler, R., Kersten, H., Schmidt, M., Schoenbach, K.-H., Eds.; Wiley-VCH: Berlin, Germany, 2008; pp. 329–362.

18. *Inorganic Crystal Structure Database (ICSD)*; Version 1.9.8; FIZ: Karlsruhe, Germany, 2016.

19. Zachwieja, U.; Jacobs, H. Ammonothermalsynthese von Kupfernitrid, Cu_3N. *J. Less Common Met.* **1990**, *161*, 175–184. [CrossRef]

20. Navío, C.; Capitán, M.J.; Álvarez, J.; Yndurain, F.; Miranda, R. Intrinsic surface band bending in Cu_3N(100) ultrathin films. *Phys. Rev. B* **2007**, *76*, 085105. [CrossRef]

21. Moreno-Armenta, M.G.; Soto, G.; Takenchi, N. Ab initio calculations of non-stoichiometric copper nitride, pure and with palladium. *J. Alloys Comp.* **2011**, *509*, 1471–1476. [CrossRef]

22. Pierson, J.F. Structure and properties of copper nitride films formed by reactive magnetron sputtering. *Vacuum* **2002**, *66*, 59–64. [CrossRef]

23. Janssen, G.C.A.M. Stress and strain in polycrystalline thin films. *Thin Solid Films* **2007**, *515*, 6654–6664. [CrossRef]

24. Weihnacht, V.; Brückner, W. Abnormal grain growth in {111} textured Cu thin films. *Thin Solid Films* **2002**, *418*, 136–144. [CrossRef]

25. Rahmati, A.; Ghoohestani, M.; Badehian, H.; Baizaee, M. Ab initio study of the structural, elastic, electronic and optical properties of Cu_3N. *Mater. Res.* **2014**, *17*, 303–310. [CrossRef]

26. Williamson, G.K.; Hall, W.H. X-ray line broadening from filed aluminium and wolfram. *Acta Metall.* **1953**, *1*, 22–31. [CrossRef]

27. Stokes, R. A Numerical Fourier-analysis method for the correction of widths and shapes of lines on X-ray powder photographs. *Proc. Phys. Soc.* **1948**, *61*, 382–391. [CrossRef]

28. Warren, B.E.; Averbach, B.L. The effect of cold-work distortion on X-ray patterns. *J. Appl. Phys.* **1950**, *21*, 595–599. [CrossRef]

29. Warren, B.E. *X-ray Diffraction*; Addison-Wesley: Boston, MA, USA, 1969.

30. Klimanek, P. X-ray diffraction analysis of substructures in plastically deformed BCC materials. *J. Phys. IV* **1993**, *3*, 2149–2154. [CrossRef]

31. Krivoglaz, M.A. *X-ray and Neutron Diffraction in Nonideal Crystals*; Springer: Berlin, Germany, 1966.

32. Wilkens, M. The determination of density and distribution of dislocations in deformed single crystals from broadened X-ray diffraction profiles. *Phys. Stat. Sol. a* **1970**, *2*, 359–370. [CrossRef]

33. Thejaswini, H.C.; Bogdanowicz, R.; Danilov, V.; Schäfer, J.; Meichsner, J.; Hippler, R. Deposition and characterization of organic polymer thin films using a dielectric barrier discharge with different C_2H_m/N_2 (m = 2, 4, 6) gas mixtures. *Eur. Phys. J. D* **2015**, *69*, 1–6.

34. Majumdar, A.; Das, G.; Basvani, K.R.; Heinicke, J.; Hippler, R. Role of nitrogen in the formation of HC-N films by CH_4/N_2 barrier discharge plasma: Aliphatic tendency. *J. Phys. Chem. B* **2009**, *113*, 15734–15741. [CrossRef] [PubMed]

35. Biesinger, M.C.; Lau, L.W.M.; Gerson, A.R.; Smart, R.S.C. Resolving surface chemical states in XPS analysis of first row transition metals, oxides and hydroxides: Sc, Ti, V, Cu and Zn. *Appl. Surf. Sci.* **2010**, *257*, 887–898. [CrossRef]

36. Wurth, W.; Schneider, C.; Treichler, R.; Umbach, E.; Menzel, D. Evolution of adsorbate core-hole states after bound and continuum primary excitation: Relaxation versus decay. *Phys. Rev. B* **1987**, *35*, 7741–7744. [CrossRef]

37. Wagner, C.D.; Zatko, D.A.; Raymond, R.H. Use of the Oxygen KLL Auger lines in identification of surface chemical states by electron spectroscopy for chemical analysis. *Anal. Chem.* **1980**, *52*, 1445–1451. [CrossRef]

13

Development of a Fabrication Process Using Suspension Plasma Spray for Titanium Oxide Photovoltaic Device

Hsian Sagr Hadi A * and Yasutaka Ando

Ashikaga Institute of Technology, 268-1 Omae, Ashikaga, Tochigi, 326-8558, Japan; yando@ashitech.ac.jp
* Correspondence: sagr@hotmail.co.jp

Academic Editor: Bill Clyne

Abstract: In order to reduce the high costs of conventional materials, and to reduce the power necessary for the deposition of titanium dioxide, titanium tetrabutoxide has been developed in the form of a suspension of TiO_2 using water instead of expensive ethanol. To avoid sedimentation of hydroxide particles in the suspension, mechanical milling of the suspension was conducted in order to create diffusion in colloidal suspension before using it as feedstock. Consequently, through the creation of a colloidal suspension, coating deposition was able to be conducted without sedimentation of the hydroxide particles in the suspension during the deposition process. Though an amorphous as-deposited coating was able to be deposited, through post heat treatment at 630 °C for 60 min, the chemical structure became anatase rich. In addition, it was confirmed that the post heat treated anatase rich coating had enough photo-catalytic activity to decolor methylene-blue droplets. From these results, this technique was found to have high potential in the low cost photo-catalytic titanium coating production process.

Keywords: thermal plasma; suspension plasma spray; titanium oxide; photo-catalysis

1. Introduction

Thin titanium oxide films for photo-catalytic application (anatase TiO_2, etc.) can be deposited by various deposition techniques such as sputtering [1], metalorganic chemical vapour deposition (MOCVD) [2], spray pyrolysis [3], the sol-gel process [4], thermal spray [5–8], or by various other methods. However, there are several engineering problems associated with these deposition processes, such as the necessity of vacuum equipment, low deposition rates, deposition time, the requirement of special and costly equipment, as well as feedstock powder [9]. Recently, some atmospheric thermal plasma processes for titania deposition have been successfully developed [10]. It is still difficult to develop the atmospheric thermal plasma process, especially in cases where low power DC arc discharge equipment is being used. In our previous study, in order to reduce equipment costs, a thermal spray process for photo-catalytic titanium oxide coating was fabricated using 1 kW class Atmospheric Solution Precursor Plasma Spray (ASPPS) equipment. Titanium oxide film deposition was then carried out [9,11]. Although photo-catalytic anatase rich titanium oxide films were successfully deposited, the feedstock cost could not be decreased because of the necessity of expensive anhydrous ethanol which is required to create the low viscosity titanium iso butoxide solution (TTIB, $Ti(OC_4H_9)_4$) without hydration of TTIB due to residual water in the ethanol. Although the anhydrous ethanol-titanium-oxide-suspension method is a possible solution to the above problems, it is difficult to use this method in practice. This method creates large secondary particles as a result of hydrolysis due to the presence of TTIB as well as the fact that titanium dioxide is insoluble in water. It has large particles, which solidify when in contact with water (hydrolysis). These properties prevent the suspension from being stable and mixing with nanoparticles in water. It is also difficult to inject or spray such a mixture. Titanium tetra-butoxide was developed in the form of a titanium hydroxide suspension by using nothing but stable water to prepare the titanium hydroxide suspension and then by milling and filtering the suspension. The suspension plasma spray (SPS) that developed new feedstock material seems to be highly promising as a low cost starting material and in the rapid formation of functional thin films.

2. Experimental Procedure

2.1. Method

The tools used for mixing titanium oxide with water are shown in Figure 1. They consist of a micro filter, a mixer, and a special container (mortar and pestle) for mixing as well as for crushing and grinding. An amount of 2 mL of TTIB is diluted with stable water, and 20 mL of water is gradually added to the TTIB. It is then mixed from one to 4 min before performing suspension filtration, in which we force large particles to pass through the holes of the filter and take the shape and size of the filter holes in the form of submicron particles. In order to reduce the particles to a smaller size than the holes of the filter, we use milled particles. We continue mixing the suspension after this grinding/milling process, until the suspension is ready to use and is then injected into the plasma jet. Titanium oxide deposited by the suspension plasma spray is shown in Figure 2. Heat treatment is then performed on all samples using an electric furnace. The heat treatment is performed by gradually increasing the heat of the samples from 30 to 630 °C for approximately 60 min.

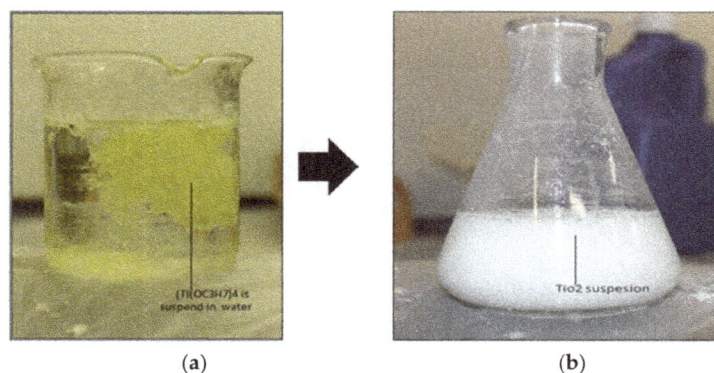

(a)   (b)

Figure 1. Creation of the titanium hydroxide suspension. (**a**) Before hydrolysis; (**b**) After hydrolysis.

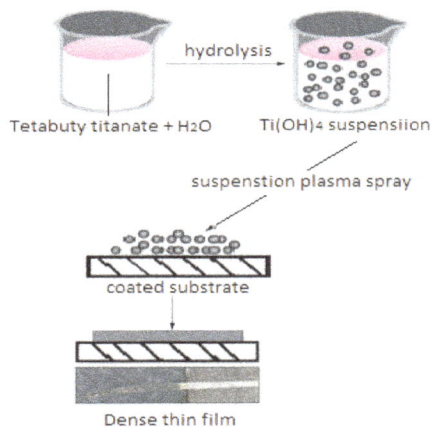

Figure 2. Illustration of the estimated film deposition mechanism in this suspension plasma spray (SPS) process.

The equipment used consists of a plasma torch, a DC power feeding system, an air pressure spray (starter material feed), and a working Ar gas feeding system. The plasma torch is cooled by a water flow around the nozzle. The anode, which has a suspension feeding port at its head, has a constrictor that is 6 mm in diameter. The nozzle shown in Figures 3 and 4 explain how this equipment was made for the experiment in this study. In order to promote the vaporization of the starting material and to raise the temperature of titanium particles, Ar was used as the working gas. Mass flow rate of the gas was fixed at 5–10 SLM, and the discharge current was fixed at 50–60 A. The titanium hydroxide ($Ti(OH)_4$) suspension submicron particles were suspended in water.

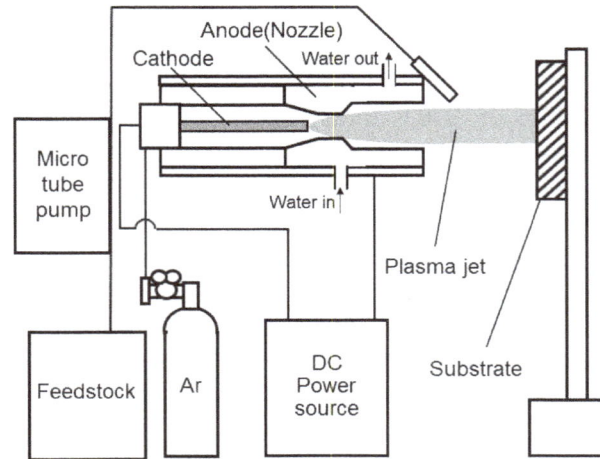

Figure 3. Schematic diagram of the SPS equipment.

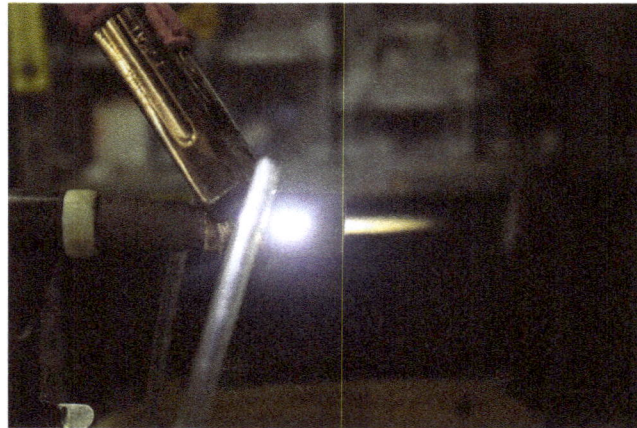

Figure 4. Plasma jet generated by our thermal spray equipment.

2.2. Plasma Spray Parameters

Titanium oxide deposited by suspension plasma spraying is shown in Table 1. The anode, which has a suspension feeding port at its head, has a constrictor that is 6 mm in diameter. The nozzle shown in Figure 4 demonstrates how this equipment was configured for the experiment in this study. The anode is 450 mm in length and 20 mm in width.

Table 1. Suspension plasma spray parameters used to deposit the TiO_2 coatings in this study.

Process Parameter	Value
Sandblasting realized	Always
Working gas composition	Ar
Working gas flow rate, L/min	5–15
Spray distance, mm	100–150
Discharge current, A	60–50
Constrictor diameter of plasma torch nozzle, mm	6
Feedstock	$Ti(OH)_4$ suspension *
Feedstock feed rate, mL/min	11
Feedstock injection port, diameter, mm	0.4
Deposition distance, mm	60–80

* Created by hydrolysis of titanium tetraisobutoxide (TTIB, $Ti(OC_4H_9)_4$) (volume ratio of TTIB/H_2O = 2/20).

2.3. Suspension Preparation

Ethanol has solvent properties and is a versatile solvent, miscible with water and titanium tetra butoxide among many other materials. However, it is expensive. One way to solve the problem of cost is to use a starting material that does not require ethanol and which uses only distilled water. In this process, the reaction between the alkoxide precursor and the desired amount of water occurred in an anhydrous alcohol medium. The hydrolysis and condensation reactions can be summarized as follows:

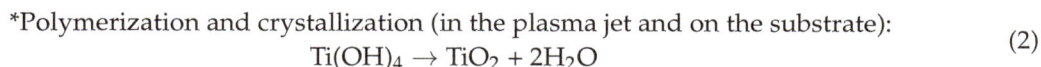

$$*\text{Hydrolysis: } Ti(OC_4H_9)_4 + 4H_2O \rightarrow Ti(OH)_4 + 4C_4H_9OH \tag{1}$$

$$*\text{Polymerization and crystallization (in the plasma jet and on the substrate):}$$
$$Ti(OH)_4 \rightarrow TiO_2 + 2H_2O \tag{2}$$

It is difficult to inject or spray the suspension before the filtration process because of the size of the particles. Therefore, it is necessary to use equipment to facilitate mixing titanium oxide with water. The equipment consists of a filter funnel, a mixer, and a special container (mortar and pestle), which is used for crushing, grinding, and mixing, as shown in Figure 5. Then begins the process of mixing for one to four minutes before the commencement of suspension filtration in which we force large particles to pass through the holes of the filter and take the shape and size of the holes to form of submicron particles. In order to obtain a smaller particle size than the holes of the filter, we use the milled particles, and mix the suspension after this process until the suspension is ready for use. At this point, it is injected into the plasma jet.

Figure 5. The shape of submicron particles.

2.4. Injection Method

A radial injection of the suspension feedstock was made by using two different modes: (i) an internal injection mode (Figure 6a) using an airbrush atomization feeding system, which was 0.4 mm in diameter in Table 1, and an internal port 6 mm in diameter; (ii) an external injection mode (Figure 6b) using the same system, where the suspension is first atomized and then injected into the plasma jet.

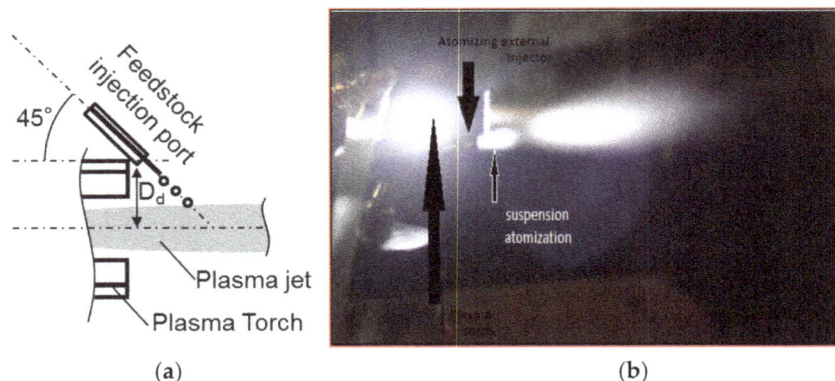

(a)   (b)

Figure 6. Configuration of the suspension plasma spray torch. (**a**) Illustration of the feedstock injection port; (**b**) Appearance of the plasma jet.

3. Results and Discussion

3.1. SPS Titanium Oxide Film

The coated samples before heat treatment seemed to be in an amorphous phase (Figure 7). When the deposition was performed with a low substrate temperature, the starting material did not have enough mobility to form a crystalline structure. Following that, heat treatment was conducted on all of the titanium oxide films that were deposited on the substrate. The films were heated gradually from 30 to 630 °C with a temperature increase of 10 °C·min^{-1} for approximately 60 min. In respect to the XRD patterns of the heat treated films, anatase crystalline peaks appeared for all samples and became sharper by increasing distances from $d = 60$ mm to 80 mm. Figure 8 shows the XRD pattern of the samples deposited at 80 mm, 60 mm deposition distance. At 80 mm in deposition distance, the starting material heating time was longer than that of the 60 mm sample. Therefore, the quantity of deposited particles at 80 mm was thought to be higher than the quantity of the 60 mm sample. From these results, it was proven that crystallization of the titania films occurred during heat treatment.

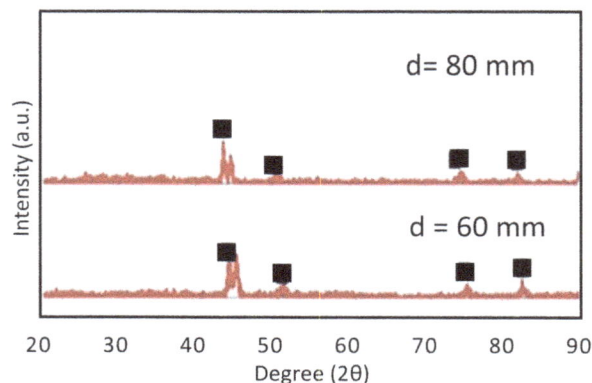

Figure 7. XRD patterns of SPS films before heat treatment. (d = deposition distance, ■: Fe (substrate)).

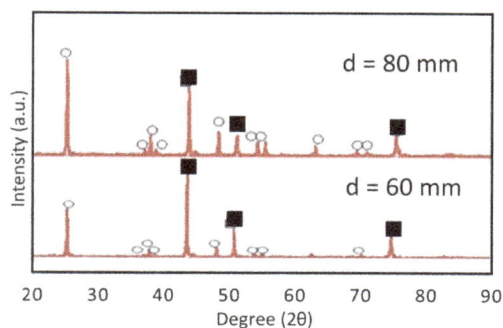

Figure 8. XRD patterns of SPS films after heat treatment (d = deposition distance, ○: Anatase, ■: Fe (substrate)).

The top-surface and cross-section micrographs of suspension sprayed coating using differing distances are shown in Figures 9 and 10. The microstructural investigations revealed that the suspension plasma coating presented a morphology, which was caused by the way that the suspension was injected into the plasma jet. The coating, in the case of 60 mm in spray distance (d = 60 mm), is homogeneous and has a grainy structure. The top-surface microscope analyses revealed the presence of agglomerates of grains that are porous, loosely bound, and have a thick film. On the other hand, in the case of d = 80 mm, the film had a grainy and porous structure. The difference in the coatings' microstructure and surface shape can be related to the type of suspension and evolution of the suspension droplets in the plasma jet. According to the particle sizes shown, the film deposition rate dramatically increased with increasing deposition distance (Figure 8).

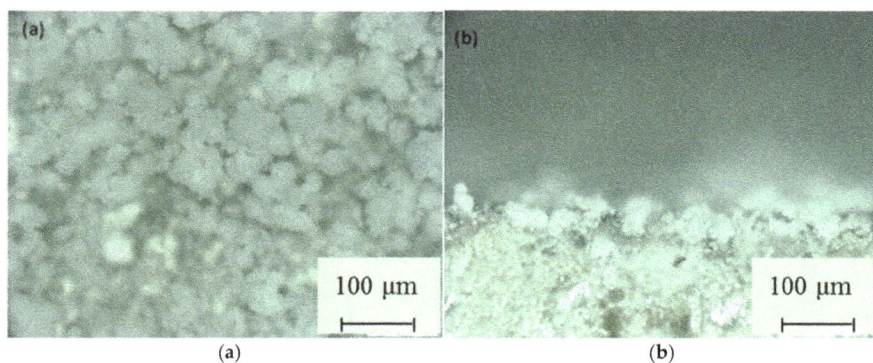

Figure 9. Micrographs of suspension sprayed coating deposited at d = 60 mm. (**a**) Top-surface; (**b**) Cross-section.

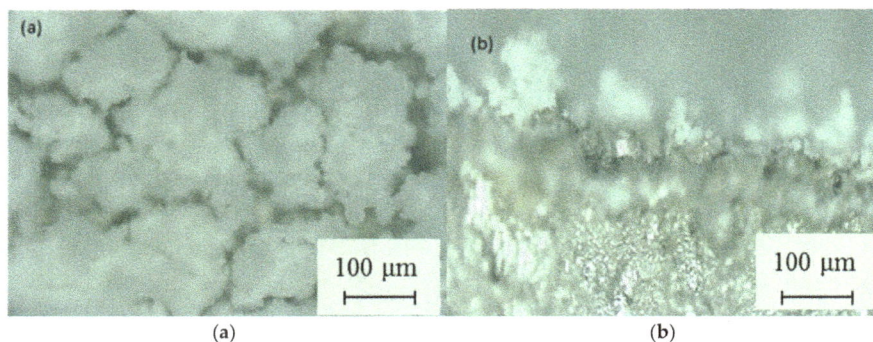

Figure 10. Micrographs of suspension sprayed coating deposited at d = 80 mm. (**a**) Top-surface; (**b**) Cross-section.

3.2. Photocatalytic Activity of TiO$_2$ Coating

In order to confirm the photo-catalytic property of the coatings, photo-catalytic activity of the sprayed TiO$_2$ suspension coating was evaluated by using UV irradiation equipment and measurements of the degradation of an aqueous solution of methylene-blue were conducted (decoloration test) (Figure 11) [9]. Figure 12 shows the results of the methylene-blue decoloration test in the case of the coating deposition at a distance of d = 80 mm. In both cases of the coatings at the distances of d = 60 mm and 80 mm, decoloration could be confirmed after 1 week of UV irradiation. From these results, it was confirmed that this technique has enough potential to deposit photo-catalytic titanium oxide films.

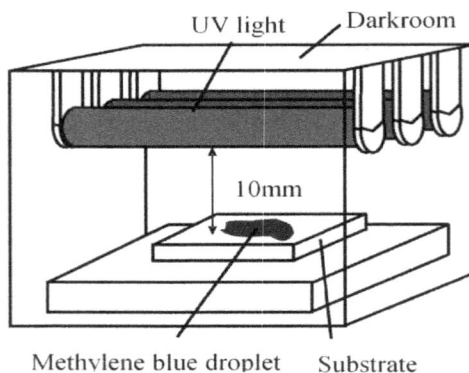

Figure 11. Schematic diagram of the equipment used in methylene-blue decoloration test.

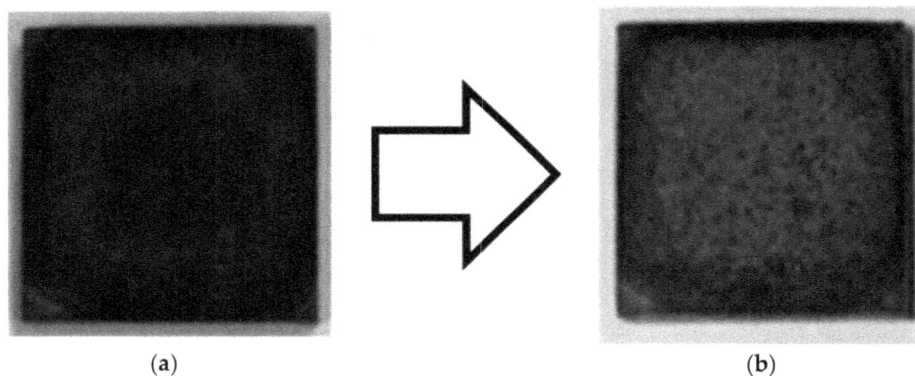

Figure 12. Results of methylene blue decoloration test when the coating was deposited at d = 80 mm (d = deposition distance). (**a**) Before UV irradiation; (**b**) After 7-day UV irradiation.

4. Conclusions

In order to develop low cost materials that can be used for the titanium dioxide film deposition process, deposition of high-rate film using low cost materials was carried out. Thick photo-catalyst film was obtained using an atmospheric suspension plasma spray with spray injectors as a fabrication process. Consequently, anatase film was obtained, and it was confirmed that the anatase films had photo-catalytic properties by using a methylene-blue droplet test and its decoloration test. From these results, this low cost starting material with atmospheric SPS has the potential for a high rate and low cost, functional, oxide film deposition.

Author Contributions: H. Sagr Hadi A created the SPS equipment for this study, conducted TiO$_2$ film deposition, investigated macrostructures of deposited films by optical microscope and XRD. Y. Ando designed and created the advanced prototype of the SPS equipment using a commercial arc welding equipment.

Conflicts of Interest: The authors declare no conflict of interest.

References

1. Sima, C.; Waldhauser, W.; Lacknar, J.; Kahn, J.; Kahn, M.; Nicolae, I.; Viespe, C.; Grigoriu, C.; Manea, A. Properties of TiO_2 thin films deposited by RF magnetron sputtering. *J. Optoelectron. Adv. Mater.* **2007**, *9*, 1446–1449.

2. Cimpean, A.; Popescu, S.; Ciofrangeanu, C.M.; Gleizes, A.N. Effects of LPMOCVD prepared TiO_2 thin films on the in vitro behavior of gingival fibroblasts. *Mater. Chem. Phys.* **2011**, *125*, 485–492. [CrossRef]

3. Oja, I.; Mere, A.; Krunks, M.; Solterbeck, C.-H.; Es-Souni, M. Properties of TiO_2 Films Prepared by the Spray Pyrolysis Method. *Solid State Phenom.* **2004**, *99*, 259–264. [CrossRef]

4. Kajitvichyanukula, P.; Ananpattarachaia, J.; Pongpomb, S. Sol–gel preparation and properties study of TiO_2 thin film for photocatalytic reduction of chromium(VI) in photocatalysis process. *Sci. Technol. Adv. Mater.* **2005**, *6*, 352–358. [CrossRef]

5. Zhao, X.; Liu, X.; Ding, C.; Chu, P.K. In vitro bioactivity of plasma-sprayed TiO_2 coating after sodium hydroxide treatment. *Surf. Coat. Technol.* **2006**, *200*, 5487–5492. [CrossRef]

6. Li, C.-J.; Yang, G.-J.; Wang, Y.-Y.; Li, C.-X.; Ye, F.-X.; Ohmori, A. Phase Formation during Deposition of TiO_2 Coatings through High Velocity Oxy-Fuel Spraying. *Mater. Trans.* **2006**, *47*, 1690–1696. [CrossRef]

7. Yang, G.-J.; Li, C.-J.; Han, F.; Li, W.-Y.; Ohmori, A. Low Temperature Deposition and Characterization of TiO_2 Photocatalytic Film through Cold Spray. *Appl. Surf. Sci.* **2008**, *254*, 3979–3982. [CrossRef]

8. Yamada, M.; Isago, H.; Nakano, H.; Fukumoto, M. Cold Spraying of TiO_2 Photocatalyst Coating with Nitrogen Process Gas. *J. Therm. Spray Technol.* **2010**, *19*, 1218–1223. [CrossRef]

9. Ando, Y.; Tobe, S.; Tahara, H. Photo-catalytic TiO2 film deposition by atmospheric TPCVD. *Vacuum* **2006**, *80*, 1278–1283. [CrossRef]

10. Popescu, S.; Jerby, E.; Meir, Y.; Barkay, Z.; Ashkenazi, D.; Mitchell, J.B.A.; Le Garrec, J.L.; Narayanan, T. Plasma column and nano-powder generation from solid titanium by localized microwaves in air. *J. Appl. Phys.* **2015**, *118*, 023302. [CrossRef]

11. Ando, Y. Titanium Oxide Film Deposition on Acrylic Resin by Atmospheric TPCVD. *IEEE Trans. Plasma Sci.* **2009**, *37*, 2202–2206. [CrossRef]

14

Using an Atmospheric Pressure Chemical Vapor Deposition Process for the Development of V_2O_5 as an Electrochromic Material

Dimitra Vernardou

Center of Materials Technology and Photonics, School of Applied Technology, Technological Educational Institute of Crete, 710 04 Heraklion, Crete, Greece; dimitra@iesl.forth.gr

Academic Editors: Mingheng Li and Alessandro Lavacchi

Abstract: Vanadium pentoxide coatings were grown by atmospheric pressure chemical vapor deposition varying the gas precursor ratio (vanadium (IV) chloride:water) and the substrate temperature. All samples were characterized by X-ray diffraction, Raman spectroscopy, scanning electron microscopy, cyclic voltammetry, and transmittance measurements. The water flow rate was found to affect the crystallinity and the morphological characteristics of vanadium pentoxide. Dense stacks of long grains of crystalline oxide are formed at the highest amount of water utilized for a substrate temperature of 450 °C. Accordingly, it was indicated that for higher temperatures and a constant gas precursor ratio of 1:7, the surface morphology becomes flattened, and columnar grains of uniform size and shape are indicated, keeping the high crystalline quality of the material. Hence, it was possible to define a frame of operating parameters wherein single-phase vanadium pentoxide may be reliably expected, including a gas precursor ratio of 1:7 with a substrate temperature of >450 °C. The as-grown vanadium pentoxide at 550 °C for a gas precursor ratio of 1:7 presented the best electrochemical performance, including a diffusion coefficient of 9.19×10^{-11} cm$^2 \cdot$s^{-1}, a charge density of 3.1 mC·cm^{-2}, and a coloration efficiency of 336 cm$^2 \cdot$C^{-1}. One may then say that this route can be important for the growth of large-scale electrodes with good performance for electrochromic devices.

Keywords: atmospheric pressure chemical vapor deposition (APCVD); vanadium pentoxide; electroactive material

1. Introduction

A lot of attention is given to materials that can be used in "smart windows". Responsive materials can regulate a more comfortable living environment, while saving energy usually consumed for appropriate lighting or air conditioning [1]. A typical example is the electrochromic device, which shows reversible optical changes in response to an applied voltage. One of the most studied cathodes for electrochromic devices is vanadium pentoxide (V_2O_5) [2–5]. A foreign atom such as Li$^+$ can be intercalated or deintercalated from its lattice, switching reversibly from a bleached to a colored state through the following reaction [6]:

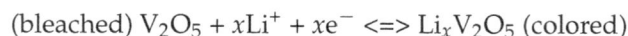

$$\text{(bleached) } V_2O_5 + x\text{Li}^+ + x\text{e}^- <=> \text{Li}_x V_2O_5 \text{ (colored)}$$

V_2O_5 is the only oxide that can show both anodic and cathodic coloration [7]. The reversible Li$^+$ intercalation/deintercalation processes of V_2O_5 lead to both reversible optical and multicolor changes for aesthetics in the voltage range ± 1 V [8–10].

Electrochromic V_2O_5 coatings have been prepared by electrodeposition [11,12], "Doctor Blade" [13], reactive sputtering [14], polyol process [15], sol–gel [16], and hydrothermal growth [17]. Strategies based on solution can hardly be up-scaled, due to the long reaction period, toxic reducing agents, and large amount of solvent required. Additionally, methods such as sputtering are not compatible with on-line manufacture. Hence, a low-cost, simple, and easily integrated into float-glass production lines technique is required.

As with many coatings, chemical vapor deposition (CVD) routes are more attractive for the production of vanadium oxides on glass than other conventional techniques, since the stoichiometry and the morphology of the coatings can be simply controlled by tuning the vapor flows in the coating zone [18,19]. In addition, the simplicity of CVD—particularly when performed at atmospheric pressure (APCVD)—makes such a process compatible with on-line glass manufacturing processes. Nevertheless, there are very few reports related to APCVD electrochromic V_2O_5 [18].

The aim of this work is to define a frame of APCVD operating parameters wherein single-phase V_2O_5 may be reliably expected to form. Additionally, a correlation of the morphological characteristics with the corresponding electrochemical/electrochromic performance of the coatings is highlighted, which is important for the understanding and the enhancement of Li^+ intercalation into the vanadium oxide matrix.

2. Materials and Methods

The APCVD reactor used in this work is an in-house design, and consists of a cold-wall reactor connected to an arrangement of stainless-steel heated pipes, valves, and bubblers, as shown in Figure 1 [18]. Bubbler 1 was used for the vanadium (IV) chloride, VCl_4 (Aldrich, Munich, Germany, 99%), and bubbler 2 for the deionised H_2O. The carrier gas was N_2, which was passed through the apparatus during all operations of the reactor. During the growth, two series of experiments were performed varying the precursor ratio of $VCl_4:H_2O$ (1:1, 1:2, 1:3, 1:5, and 1:7) at 450 °C and the substrate temperature (500 and 550 °C) for constant ratio of $VCl_4:H_2O$, 1:7. The deposition time was 2 min. The substrates used during the APCVD experiments were commercial fluorine-doped SnO_2 (FTO)-precoated glass substrates (Pilkington, Manchester, UK), all of dimensions 2 cm × 2 cm × 0.3 cm. Prior to coating, all substrates were cleaned with H_2O and detergent, rinsed thoroughly with H_2O and deionised H_2O, and allowed to dry.

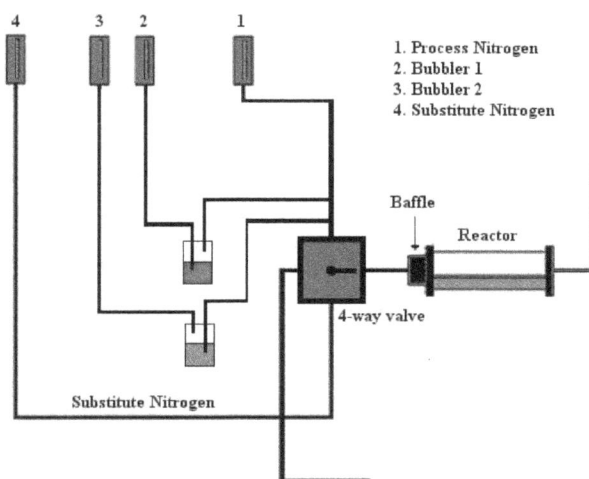

Figure 1. A schematic presentation of the atmospheric pressure chemical vapor deposition (APCVD) reactor.

X-ray diffraction (XRD) measurements were carried out in a Siemens D5000 Diffractometer (SCIMED, Manchester, UK) for 2-θ = 10.00°–60.00°, step-size = 0.02°, and step time = 30 s/°. Raman measurements were performed with a Nicolet Almega XR micro-Raman system (CRAIC, Hertfordshire, UK) operating at wavenumber range of 100–1100 cm^{-1} using a 473 nm laser. Scanning electron microscopy (SEM) was utilized for the morphology observation of the as-grown coatings through a Jeol JSM-6390LV electron microscope (JEOL, Freising, Germany). For the SEM characterization, all samples were over-coated with a thin film of gold to make them more conductive. UV-Vis transmittance spectra were obtained using a Perkin Elmer Lambda 950 spectrometer over the wavelength range of 300–1100 nm. Finally, cyclic voltammetry experiments were performed using an electrochemical cell with a tri-electrode configuration and a computer-controlled AUTOLAB potentiostat/galvanostat [19–23]. Vanadium oxide-coated glass substrates acted as the working electrode biased in the range between -1 and $+1$ V. Ag/AgCl and a Pt foil were employed as the reference and the counter electrode, respectively. Additionally, cyclic voltammograms were obtained at scan rates of 2, 5, 10, 20, 30, 40, 50, 75, 100, 150, and 200 mV·s^{-1}. To study Li$^+$ intercalation/deintercalation process with respect to time, chronoamperometry for a step of 200 s was carried out at -1 and $+1$ V. In all cases, the electrolyte was 1 M, LiClO$_4$/polypropylene carbonate.

Finally, the coating's thickness was estimated using a profilometer A-step TENCOR (KLA Tencor, Dublin, Ireland) by etching the vanadium oxide coatings off the FTO glass substrate in 1:3 H$_2$O$_2$ (30%):HCl. FTO remained intact, and the thickness was deduced from the measured step height.

3. Results and Discussion

The coatings produced in the APCVD reaction of VCl$_4$ and deionised H$_2$O were yellow, adhesive, and they passed the Scotch tape test. Additionally, they had similar properties (structural, morphological, optical, and electrochemical) after approximately six months, indicating their stability with time.

3.1. Structure

Figure 2 shows a representative XRD pattern of the as-grown APCVD coatings. It exhibits characteristic 2-θ values of 13.5°, 20.5°, 26.6°, 30.3°, 41.2°, and 54.8° with respective Miller indices (200), (001), (110), (301), (002), and (021), which are consistent with crystalline V$_2$O$_5$ [24].

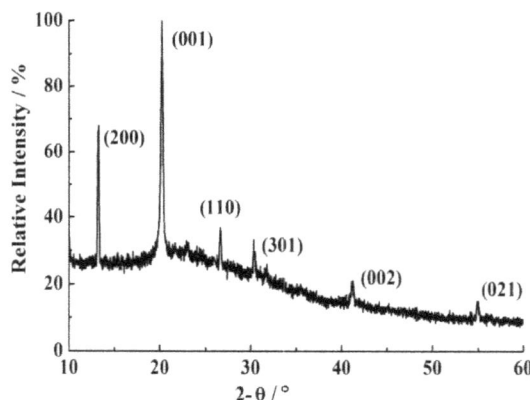

Figure 2. XRD pattern of as-grown vanadium oxide thin films by APCVD on fluorine-doped SnO$_2$ (FTO) glass substrates at 450 °C, using gas precursor ratio of 1:3.

Figure 3 displays the Raman spectra of V$_2$O$_5$ coatings deposited on FTO glass substrates for gas precursor ratio of VCl$_4$:H$_2$O 1:7 at 450, 500, and 550 °C. The peak positions agree with the literature spectra to within ± 2 cm^{-1}. The high-frequency Raman peak at 995 cm^{-1} corresponds to the terminal oxygen (V=O) stretching mode, which results from unshared oxygen [25]. Peaks at 530 and 705 cm^{-1} are attributed to stretching modes of the V–O–V bridging bonds, with the bending motions of these bonds assigned to 486 cm^{-1} [26]. Peaks located at 283 and 406 cm^{-1} are assigned to the bending vibrations of V=O bonds [27]. Two more low-frequency Raman peaks at 145 and 197 cm^{-1} can also be distinguished, which correspond to the lattice vibrations [28].

Such outcomes suggest that by manipulating the substrate temperature, one can synthesize V_2O_5 with some oxygen deficiency. This happens because the high temperature (>450 °C) enhances the surface mobility and the crystallization due to highly reactive atomic and anionic oxygen, which can act as crystallization agents towards the composition of V_2O_5 [29]. Additionally, XRD analysis gave no indication of oxygen deficiency, but Raman study revealed the presence of additional peaks along with those of V_2O_5 at the lowest substrate temperature. This is due to the sensitivity of Raman spectroscopy to short-range vibrational modes of bond configurations, while XRD responds to long-range order crystallinity of materials [30]. Hence, the observed behavior suggests that the samples are mainly V_2O_5 retaining, however, a short-range oxygen deficiency that degrades as the substrate temperature increases.

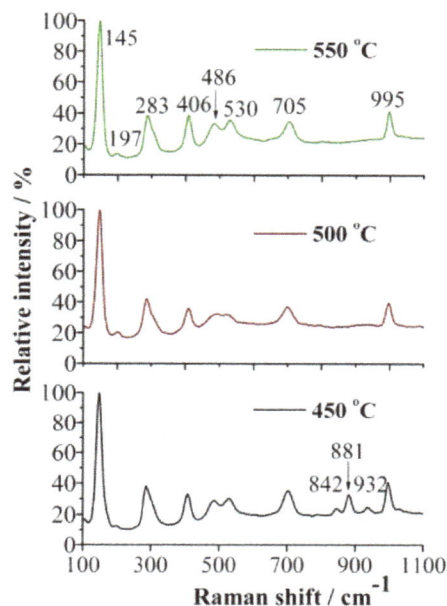

Figure 3. Raman spectra of as-grown vanadium oxide films by APCVD for a deposition time of 2 min and gas precursor ratio of VCl_4:H_2O of 1:7 at 450, 500, and 550 °C.

From the above, it is possible to define a frame of APCVD operating parameters wherein single-phase V_2O_5 may be reliably expected to form: these conditions involve a precursor ratio equal to 1:7, with a substrate temperature of >450 °C. This observation contradicts previous studies [31] due to the different designs of APCVD reactors utilized, the size of the substrates, the total gas flow rates (on the order of 1.5 L·min^{-1} in previous work [31], while in the present study it was 12 L·min^{-1}), and the precursor delivery systems employed. These parameters are expected to directly influence the nature and extent of the reaction between the precursors.

3.2. Morphology

SEM images of vanadium oxide coatings deposited at 450 °C for gas precursor ratios of VCl_4:H_2O being 1:1, 1:5, and 1:7 are shown in Figure 4. As the gas precursor ratio increases to 1:7, dense stacks of long grains are formed. This behavior may originate from the mobility of species on the developing coating surface promoted by the increased H_2O ratio.

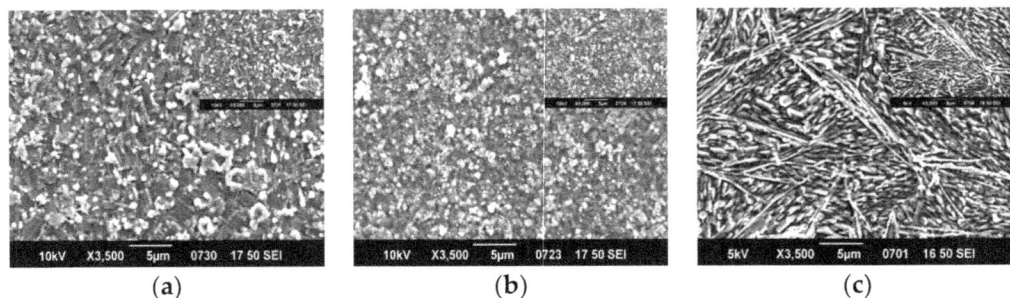

Figure 4. SEM images of the as-grown vanadium oxide coatings by APCVD at 450 °C for gas precursor ratios of VCl_4:H_2O of (**a**) 1:1, (**b**) 1:5, and(**c**) 1:7.

As the growth temperature increases from 450 to 550 °C (Figure 5), surface diffusion is activated, the surface morphology becomes flattened, and columnar grains of uniform size and shape are formed.

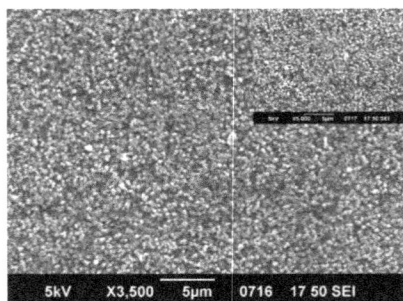

Figure 5. SEM images of the as-grown vanadium oxide coating by APCVD at 550 °C for gas precursor ratio VCl_4:H_2O of 1:7.

The growth rate of the as-grown coatings was estimated from the deposition time and the coating's thickness as derived from the profilometer. The highest value was 80 nm·min^{-1} for the deposited V_2O_5 at 550 °C. As the substrate temperature decreased to 450 °C, the growth rate was calculated to be 35 nm·min^{-1}. A proportional relationship was also seen between the amount of H_2O and the growth rate, having a maximum value of 72 nm·min^{-1} for the excess amount of H_2O introduced into the reactor (the growth rate was 15 nm·min^{-1} for the lowest amount of H_2O).

Considering these outcomes and the surface morphology of the samples, it may be suggested that either a film decomposition [32] or precursor pre-reaction with the oxygen source [33] takes place to some extent, such that the concentration of the precursor is depleted before it reaches the substrate. Since film decomposition does not occur, it is suggested that a depletion of the gas phase concentration of VCl_4 with the corresponding reduction in growth rate occurs.

3.3. Electrochemical Performance

3.3.1. Cyclic Voltammetry

The effect of substrate temperature for a constant gas precursor ratio VCl_4:H_2O of 1:7 on the electrochemical performance of the as-grown coatings was evaluated by cyclic voltammetry, as indicated in Figure 6. The potential range was −1 to +1 V at a scan rate of 10 mV·s^{-1}. All curves were normalized to the area of the working electrode, resulting in units of A·cm^{-2}. It can be observed that they all consist of four redox peaks centered at −0.19 V/+0.35 V and −0.71 V/−0.05 V (vs. Ag/AgCl), which can be assigned to the reversible Li$^+$ intercalation/deintercalation reaction. Initially, the coatings had a yellow color when they were cathodically polarized in LiClO$_4$, which became green and then blue with increasing cathodic potential. Then, the blue layers turned yellow again when anodically polarized. Additionally, it becomes evident from Figure 6 that the current density of the as-grown coating at 550 °C is the highest of all, presenting an enhanced electrochemical activity. One may then suggest that the amount of incorporated charge is enhanced by the increased crystalline quality and the visibly empty space between the columnar grains rather than those left by the stacks of long grains at lower temperature. Regarding the as-grown coatings at 450 °C for gas precursor ratios of VCl_4:H_2O being 1:1, 1:2, 1:3, and 1:5, the current density was four degrees of magnitude (maximum current density of 0.000001 mA·cm^{-2}) lower than the one for 1:7, due to the immediate detachment of the oxide by the electrolyte (the working electrode was optically the same with the substrate prior to deposition). Since the electrochemical cell is made up of glass, these changes could be observed during the electrochemical measurements.

The maximum current density obtained for the as-grown coating at 550 °C (Figure 6) was lower than the colloidal crystal-assisted electrodeposited amorphous three-dimensional ordered macroporous (0.5 mA·cm^{-2} [4]), sol–gel (0.6 mA·cm^{-2} [16]), aerosol-assisted CVD (1.5 mA·cm^{-2} [34]) V_2O_5, and was comparable with APCVD (0.01 mA·cm^{-2} [18]) and hydrothermal growth (0.05 mA·cm^{-2} [17]) of V_2O_5. This value was also lower than the one obtained from the vacuum deposited (0.6 mA·cm^{-2} [35]) tungsten trioxide (WO_3), and comparable with APCVD (0.02 mA·cm^{-2} [36]), evaporation-induced self-assembly (0.06 mA·cm^{-2} [37]), low-pressure CVD (0.08 mA·cm^{-2} [38]), and hydrothermal (0.04 mA·cm^{-2} [39]) growth WO_3. Overall, this APCVD route is advantageous, since it produces higher current density compared with growth techniques which require longer reaction periods and higher substrate temperatures. On the other hand, for the cases where the current density is higher than the one reported in this work, templates and more complicated equipment are required, restricting their compatibility for large-scale manufacturing.

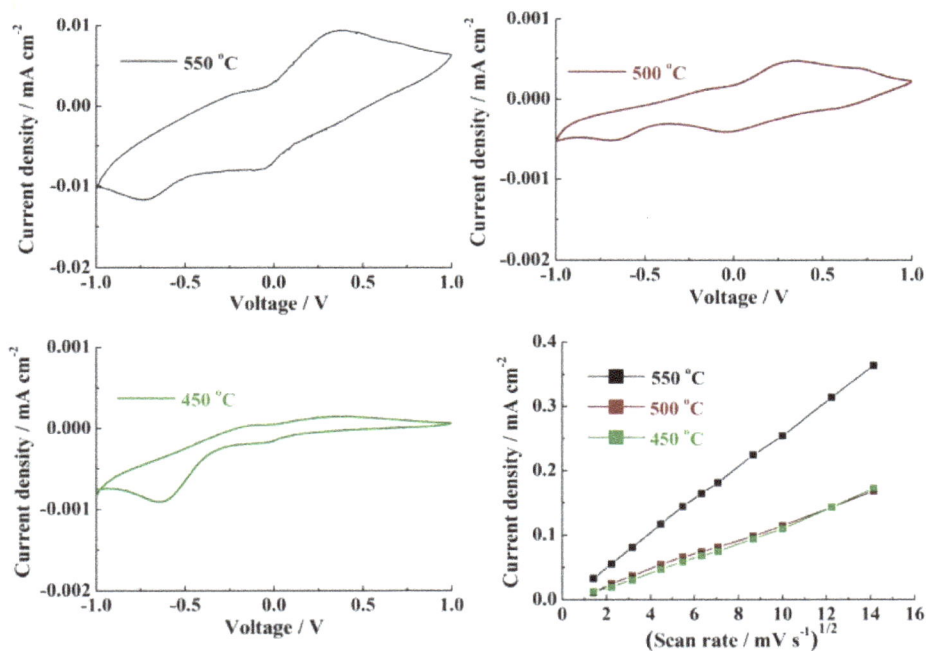

Figure 6. Cyclic voltammograms of APCVD V_2O_5 at 450, 500, and 550 °C recorded at 10 mV·s^{-1} and anodic peak current density as a function of the square root of the scan rates.

Figure 6 shows the dependence of the anodic peak current density on the square route of the scan rate, indicating a diffusion-controlled process. The diffusion coefficient of Li$^+$ can be obtained according to the following equations [40]:

$$I_p = D^{1/2}2.72 \times 10^5 n^{3/2} A C v^{1/2} \tag{1}$$

$$D^{1/2} = \frac{a}{2.72 \times 10^5 n^{3/2} AC} \qquad (2)$$

where n is the number of electrons, I_p is the peak current (A), D is the diffusion coefficient ($cm^2 \cdot s^{-1}$), A is the area (cm^2), C is the concentration of Li^+ ($mol \cdot cm^{-3}$), and a is the slope obtained in Figure 6. As is expected, the Li^+ diffusion coefficient of the as-grown V_2O_5 at 550 °C is the highest (9.19×10^{-11} $cm^2 \cdot s^{-1}$), which is in accordance with the above discussion. Additionally, it is well known that the Li^+ intercalation/deintercalation response is dependent on the diffusion coefficient of the species; consequently, a large diffusion coefficient will result in a fast response [40]. The diffusion coefficient of APCVD V_2O_5 estimated in this work is higher than V_2O_5 grown by atomic layer CVD [41], solution process [42], spin-coating [43], and pulsed spray pyrolysis [44], strengthening the growth of large area electrodes by APCVD with good electrochemical performance.

3.3.2. Chronoamperometry

To calculate the amount of Li^+ interchanged between the V_2O_5 and the electrolyte, chronoamperometric measurements were performed, switching the potential between −1 V and +1 V at an interval of 200 s for a total period of 2000 s, as indicated in Figure 7. It is observed that the intercalated is lower than the deintercalated charge, indicating that Li^+ remains in the material. The deintercalated charge density was estimated to be 3.1 mC·cm^{-2} for the as-grown V_2O_5 by APCVD at 550 °C (Figure 7), which is the highest of all due to the larger volume of active material available (i.e., highest thickness estimated).

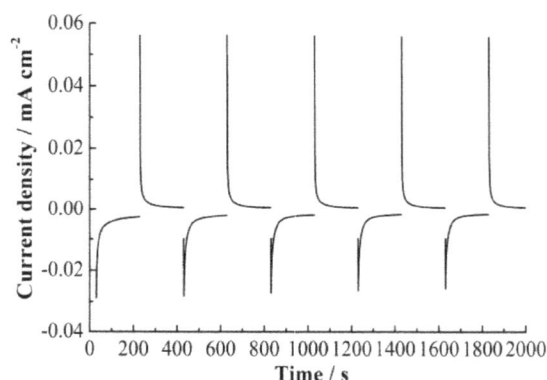

Figure 7. The chronoamperometric response at −1 and +1 V for an interval of 200 s and a total period of 2000 s of the as-grown APCVD V_2O_5 at 550 °C.

3.3.3. Ex-Situ Transmittance Measurements

The electrochromic response of the APCVD V_2O_5 at 550 °C was tested by performing transmittance measurements in the visible range and comparing the normally oxidized state at +1 V with the reduced state at −1 V, imposed by the Li^+ intercalated charge as indicated in Figure 8. It is observed that the transmittance is reduced upon cation insertion (colored state) compared to the bleached state. The contrast ratio between the two is significant, reaching a value of approximately 11 at 630 nm. Additionally, a parameter used to evaluate an electrochromic material is its coloration efficiency (CE), defined as the change in optical density (ΔOD) per unit inserted charge density ΔQ (see Equation 3 [45]),

$$CE = \frac{\Delta OD}{\Delta Q} \qquad (3)$$

The CE was found to be 336 cm$^2 \cdot C^{-1}$ at 630 nm, which is one of the highest values reported if one compares it with the electrodeposited V_2O_5 (28 cm$^2 \cdot C^{-1}$ [11]), the evaporation-induced self-assembly crystalline mesoporous WO_3 (40 $cm^2 \cdot C^{-1}$ [37]), the APCVD ($\approx$83 $cm^2 \cdot C^{-1}$ [38]), and the vacuum deposition method of WO_3 (28 $cm^2 \cdot C^{-1}$ [35]).

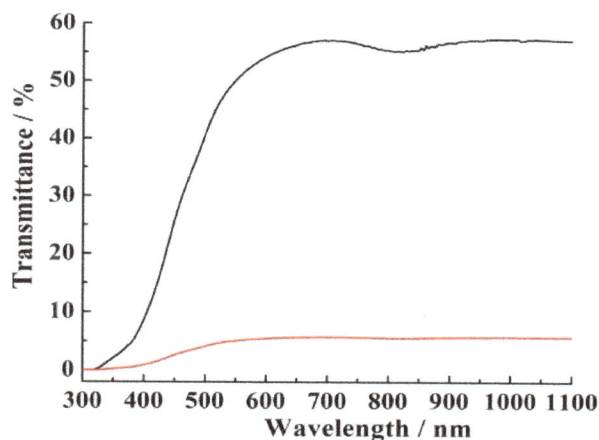

Figure 8. Transmittance spectra of APCVD V_2O_5 at 550 °C in the colored (red line) at −1 V and bleached (black line) states at +1 V.

4. Conclusions

V_2O_5 coatings were grown by APCVD, varying the gas precursor ratio of VCl_4:H_2O and the substrate temperature. From XRD and Raman spectroscopy, it was found that the samples are mainly V_2O_5, retaining, however, a short-range oxygen deficiency that degrades as the substrate temperature increases. As the amount of oxygen source introduced into the reactor increases to the maximum value, dense stacks of long grains are formed. On the contrary, the increase of substrate temperature results in surface diffusion activation, and columnar grains of uniform size and shape are formed. It was shown that the electrochemical properties of the as-grown coatings are correlated with the crystalline quality of V_2O_5, and consequently, the morphological characteristics of the samples. From the basic characterization obtained, it was possible to define a frame of APCVD operating parameters wherein single-phase V_2O_5 may be reliably expected to form: these conditions involve a precursor ratio equal to 1:7, with a substrate temperature of >450 °C. Crystalline V_2O_5 grown at 550 °C for a ratio VCl_4:H_2O of 1:7 consisting of empty space between the columnar grains was found to exhibit improved electrochemical/electrochromic performance. The diffusion coefficient was estimated to be 9.19×10^{-11} $cm^2 \cdot s^{-1}$, and the charge density was 3.1 $mC \cdot cm^{-2}$, with a coloration efficiency of 336 $cm^2 \cdot C^{-1}$.

Conflicts of Interest: The author declares no conflict of interest.

References

1.   Brigouleix, C.; Topart, P.; Bruneton, E.; Sabary, F.; Nouhaut, G.; Campet, G. Roll-to-Roll Pulsed dc Magnetron Sputtering Deposition of WO_3 for Electrochromic Windows. *Electrochim. Acta* **2001**, *46*, 1931–1936. [CrossRef]

2.   Zhai, T.; Liu, H.; Li, H.; Fang, X.; Liao, M.; Li, L.; Zhou, H.; Koide, Y.; Bando, Y.; Golberg, D. Centimeter-Long V_2O_5 Nanowires: From Synthesis to Field-Emission, Electrochemical, Electrical Transport, and Photoconductive Properties. *Adv. Mater.* **2010**, *22*, 2547–2552. [CrossRef] [PubMed]

3.   Tong, Z.; Lv, H.; Zhang, X.; Yang, H.; Tian, Y.; Li, N.; Zhao, J.; Li, Y. Novel Morphology Changes from 3D Ordered Macroporous Structure to V_2O_5 Nanofiber Grassland and its Application in Electrochromism. *Sci. Rep.* **2015**, *5*, 16864. [CrossRef] [PubMed]

4.   Vernardou, D. State-of-the-art of Chemically Grown Vanadium Pentoxide Nanostructures with Enhanced Electrochemical Properties. *Adv. Mater.* **2013**, *4*, 798–810.

5.   Drosos, C.; Vernardou, D. Perspectives of Energy Materials Grown by APCVD. *Sol. Energy Mater. Sol. Cells* **2015**, *140*, 1–8. [CrossRef]

6.   Talledo, A.; Granqvist, C.G. Electrochromic Vanadium-Pentoxide-Based Films: Structural, Electrochemical and Optical Properties. *J. Appl. Phys.* **1995**, *77*, 4655–4666. [CrossRef]

7.   Li, L.; Steiner, U.; Mahajan, S. Improved Electrochromic Performance in Inverse Opal Vanadium Oxide Films. *J. Mater. Chem.* **2010**, *20*, 7131–7134. [CrossRef]

8.   Scherer, M.R.J.; Li, L.; Cunha, P.M.S.; Scherman, O.A.; Steiner, U. Enhanced Electrochromism in Gyroid-Structured Vanadium Pentoxide. *Adv. Mater.* **2012**, *24*, 1217–1221. [CrossRef] [PubMed]

9.   Yang, Y.; Kim, D.; Schmuki, P. Electrochromic Properties of Anodically Grown Mixed V_2O_5-TiO_2 Nanotubes. *Electrochem. Commun.* **2011**, *13*, 1021–1025. [CrossRef]

10. Chernova, N.A.; Roppolo, M.; Dillon, A.C.; Whittingham, M.S. Layered Vanadium and Molybdenum Oxides: Batteries and Electrochromics. *J. Mater. Chem.* **2009**, *19*, 2526–2552. [CrossRef]

11. He, W.; Liu, Y.; Wan, Z.; Jia, C. Electrodeposition of V_2O_5 on TiO_2 Nanorod Arrays and Their Electrochromic Properties. *RSC Adv.* **2016**, *6*, 68997–69006. [CrossRef]

12. Tong, Z.; Zhang, X.; Lv, H.; Li, N.; Qu, H.; Zhao, J.; Li, Y.; Liu, X.-Y. From Amorphous Macroporous Film to 3D Crystalline Nanorod Architecture: A New Approach to Obtain High-Performance V_2O_5 Electrochromism. *Adv. Mater. Interfaces* **2015**, *2*, 1500230. [CrossRef]

13. Mjejri, I.; Manceriu, L.M.; Gaudon, M.; Rougier, A.; Sediri, F. Nano-Vanadium Pentoxide Films for Electrochromic Displays. *Solid State Ion.* **2016**, *292*, 8–14. [CrossRef]

14. Lin, Y.-S.; Tsai, C.-W. Reactive Sputtering Deposition of V_2O_{5-z} on Flexible PET/ITO Substrates for Electrochromic Devices. *Surf. Coat. Technol.* **2008**, *202*, 5641–5645. [CrossRef]

15. Salek, G.; Bellanger, B.; Mjejri, I.; Gaudon, M.; Rougier, A. Polyol Synthesis of Ti-V_2O_5 Nanoparticles and Their Use as Electrochromic Films. *Inorg. Chem.* **2016**, *55*, 9838–9847. [CrossRef] [PubMed]

16. Ma, X.; Lu, S.; Wan, F.; Hu, M.; Wang, Q.; Zhu, Q.; Zakharova, G.S. Synthesis and Electrochromic Characterization of Graphene/V_2O_5/MoO_3 Nanocomposite Films. *ECS J. Solid State Sci. Technol.* **2016**, *5*, P572–P577. [CrossRef]

17. Vernardou, D.; Loudoudakis, D.; Spanakis, E.; Katsarakis, N.; Koudoumas, E. Electrochemical Properties of Vanadium Oxide Coatings Grown by Hydrothermal Synthesis on FTO Substrates. *New J. Chem.* **2014**, *38*, 1959–1964. [CrossRef]

18. Vernardou, D.; Paterakis, P.; Drosos, H.; Spanakis, E.; Povey, I.M.; Pemble, M.E.; Koudoumas, E.; Katsarakis, N. A Study of the Electrochemical Performance of Vanadium Oxide Thin Films Grown by Atmospheric Pressure Chemical Vapour Deposition. *Sol. Energy Mater. Sol. Cells* **2011**, *95*, 2842–2847. [CrossRef]

19. Vernardou, D.; Pemble, M.E.; Sheel, D.W. In-Situ Fourier Transform Infrared Spectroscopy Gas Phase Studies of Vanadium (IV) Oxide Coating by Atmospheric Pressure Chemical Vapour Deposition Using Vanadyl (IV) Acetylacetonate. *Thin Solid Films* **2008**, *516*, 4502–4507. [CrossRef]

20. Vernardou, D.; Pemble, M.E.; Sheel, D.W. In-Situ FTIR Studies of the Growth of Vanadium Dioxide Coatings on Glass by Atmospheric Pressure Chemical Vapour Deposition for VCl_4 and H_2O System. *Thin Solid Films* **2007**, *515*, 8768–8770. [CrossRef]

21. Vernardou, D.; Louloudakis, D.; Spanakis, E.; Katsarakis, N.; Koudoumas, E. Functional Properties of APCVD VO_2 Layers. *Int. J. Thin Films Sci. Technol.* **2015**, *4*, 187–191.

22. Vernardou, D.; Apostolopoulou, M.; Katsarakis, N.; Koudoumas, E.; Drosos, C.; Parkin, I.P. Electrochemical Properties of APCVD α-Fe_2O_3 Nanoparticles at 300 °C. *Chem. Sel.* **2016**, *1*, 2228–2234.

23. Christou, K.; Louloudakis, D.; Vernardou, D.; Savvakis, C.; Katsarakis, N.; Koudoumas, E.; Kiriakidis, G. Effect of Solution Chemistry on the Characteristics of Hydrothermally Grown WO_3 for Electroactive Applications. *Thin Solid Films* **2015**, *594*, 333–337. [CrossRef]

24. Su, Q.; Huang, C.K.; Wang, Y.; Fan, Y.C.; Lu, B.A.; Lan, W.; Wang, Y.Y.; Liu, X.Q. Formation of Vanadium Oxides with Various Morphologies by Chemical Vapor Deposition. *J. Alloys Compd.* **2009**, *475*, 518–523. [CrossRef]

25. Lee, S.-H.; Cheong, H.M.; Seong, M.J.; Liu, P.; Tracy, C.E.; Mascarenhas, A.; Pitts, J.R.; Deb, S.K. Raman Spectroscopic Studies of Amorphous Vanadium Oxide Thin Films. *Solid State Ion.* **2003**, *165*, 111–116. [CrossRef]

26. Abello, L.; Husson, E.; Repelin, Y.; Lucazeau, G. Vibrational Spectra and Valence Force Field of Crystalline V_2O_5. *Spectrochim. Acta A Mol. Spectrosc.* **1983**, *39*, 641–651. [CrossRef]

27. Julien, C.; Nazri, G.A.; Bergström, O. Raman Scattering Studies of Microcrystalline V_6O_{13}. *Phys. Status Solidi* **1997**, *201*, 319–326. [CrossRef]

28. Jehng, J.M.; Hardcastle, F.D.; Wachs, I.E. The Interaction of V_2O_5 and Nb_2O_5 with Oxide Surface. *Solid State Ion.* **1989**, *32–33*, 904–910.

29. Cvelbar, U.; Mozetic, M.; Sunkara, M.K.; Vaddiraju, S. A Method for the Rapid Synthesis of Large Quantities of Metal Oxide Nanowires at Low Temperatures. *Adv. Mater.* **2005**, *17*, 2138–2142.

30. Ocăna, M.; Garcia, J.V. Low-Temperature Nucleation of Rutile Observed by Raman Spectroscopy during Crystallization of TiO_2. *J. Am. Ceram. Soc.* **1992**, *75*, 2010–2012. [CrossRef]

31. Manning, T.D.; Parkin, I.P.; Clark, R.J.H.; Sheel, D.; Pemble, M.E.; Vernardou, D. Intelligent Window Coatings: Atmospheric Pressure Chemical Vapour Deposition of Vanadium Oxides. *J. Mater. Chem.* **2002**, *12*, 2936–2939. [CrossRef]

32. Awaluddin, A.; Pemble, M.E.; Jones, A.C.; Williams, P.A. Direct Liquid Injection MOCVD Growth of TiO_2 Films Using the Precursor Ti(mpd)(dmae)$_2$. *J. Phys. IV* **2001**, *11*. [CrossRef]

33. Crosbie, M.J.; Lane, P.A.; Wright, P.J.; Williams, D.J.; Jones, A.C.; Leedham, T.J.; Reeves, C.L.; Jones, J. Liquid Injection Metal Organic Chemical Vapour Deposition of Lead-Scandium-Tantalate Thin Films for Infrared Devices. *J. Cryst. Growth* **2000**, *219*, 390–396. [CrossRef]

34. Vernardou, D.; Louloudakis, D.; Katsarakis, N.; Koudoumas, E.; Kazadojev, I.I.; O'Brien, S.; Pemble, M.E.; Povey, I.M. Electrochemical Evaluation of Vanadium Pentoxide Coatings Grown by AACVD. *Sol. Energy Mater. Sol. Cells* **2015**, *43*, 601–605. [CrossRef]

35. Pang, Y.; Chen, Q.; Shen, X.; Tang, X.; Tang, L.; Qian, H. Size-Controlled Ag Nanoparticle Modified WO_3 Composite Films for Adjustment of Electrochromic Properties. *Thin Solid Films* **2010**, *518*, 1920–1924. [CrossRef]

36. Ivanova, T.; Gesheva, K.A.; Popkirov, G.; Ganchev, M.; Tzvetkova, E. Electrochromic Properties of Atmospheric CVD MoO_3 and MoO_3-WO_3 Films and Their Application in Electrochromic Devices. *Mater. Sci. Eng. B* **2005**, *119*, 232–239. [CrossRef]

37. Brezesinski, T.; Fattakhova Rohlfing, D.; Sallard, S.; Antonietti, M.; Smarsly, B.M. Highly Crystalline WO_3 Thin Films with Ordered 3D Mesoporosity and Improved Electrochromic Performance. *Small* **2006**, *2*, 1203–1211. [CrossRef] [PubMed]

38. Psifis, K.; Louloudakis, D.; Vernardou, D.; Spanakis, E.; Papadimitropoulos, G.; Davazoglou, D.; Katsarakis, N.; Koudoumas, E. Effect of O_2 Flow Rate on the Electrochromic Response of WO_3 Grown by LPCVD. *Phys. Status Solidi C* **2015**, *12*, 1011–1015. [CrossRef]

39. Christou, K.; Louloudakis, D.; Vernardou, D.; Katsarakis, N.; Koudoumas, E. One-Pot Synthesis of WO_3 Structures at 95 °C Using HCl. *J. Sol-Gel Sci. Technol.* **2015**, *73*, 520–526. [CrossRef]

40. Jiao, Z.; Wei Sun, X.; Wang, J.; Ke, L.; Volkan Demir, H. Hydrothermally Grown Nanostructured WO_3 Films and Their Electrochromic Characteristics. *J. Phys. D Appl. Phys.* **2010**, *43*, 285501. [CrossRef]

41. Lantelme, F.; Mantoux, A.; Groult, H.; Lincot, D. Electrochemical Study of Phase Transition Processes in Lithium Insertion in V_2O_5 Electrodes. *J. Electrochem. Soc.* **2003**, *150*, A1202–A1208. [CrossRef]

42. Watanabe, T.; Ikeda, Y.; Ono, T.; Hibino, M.; Hododa, M.; Sakai, K.; Kudo, T. Characterization of Vanadium Oxide Sol as a Starting Material for High Rate Intercalation Cathodes. *Solid State Ion.* **2002**, *151*, 313–320. [CrossRef]

43. Sahana, M.B.; Sudakar, C.; Thapa, C.; Lawes, G.; Naik, V.M.; Baird, R.J.; Auner, G.W.; Naik, R.; Padmanabhan, K.R. Electrochemical Properties of V_2O_5 Thin Films Deposited by Spin Coating. *Mater. Sci. Eng. B* **2007**, *143*, 42–50. [CrossRef]

44. Patil, C.E.; Jadhav, P.R.; Tarwal, N.L.; Deshmukh, H.P.; Karanjkard, M.M.; Patil, P.S. Electrochromic Performance of Mixed V_2O_5-MoO_3 Thin Films Synthesized by Pulsed Spray Pyrolysis Technique. *Mater. Chem. Phys.* **2011**, *126*, 711–716. [CrossRef]

45. Bathe, S.R.; Patil, P.S. Electrochromic Characteristics of Fibrous Reticulated WO_3 Thin Films Prepared by Pulsed Spray Pyrolysis Technique. *Sol. Energy Mater. Sol. Cells* **2007**, *91*, 1097–1101. [CrossRef]

Permissions

The contributors of this book come from diverse backgrounds, making this book a truly international effort. This book will bring forth new frontiers with its revolutionizing research information and detailed analysis of the nascent developments around the world.

We would like to thank all the contributing authors for lending their expertise to make the book truly unique. They have played a crucial role in the development of this book. Without their invaluable contributions this book wouldn't have been possible. They have made vital efforts to compile up to date information on the varied aspects of this subject to make this book a valuable addition to the collection of many professionals and students.

This book was conceptualized with the vision of imparting up-to-date information and advanced data in this field. To ensure the same, a matchless editorial board was set up. Every individual on the board went through rigorous rounds of assessment to prove their worth. After which they invested a large part of their time researching and compiling the most relevant data for our readers.

The editorial board has been involved in producing this book since its inception. They have spent rigorous hours researching and exploring the diverse topics which have resulted in the successful publishing of this book. They have passed on their knowledge of decades through this book. To expedite this challenging task, the publisher supported the team at every step. A small team of assistant editors was also appointed to further simplify the editing procedure and attain best results for the readers.

Apart from the editorial board, the designing team has also invested a significant amount of their time in understanding the subject and creating the most relevant covers. They scrutinized every image to scout for the most suitable representation of the subject and create an appropriate cover for the book.

The publishing team has been an ardent support to the editorial, designing and production team. Their endless efforts to recruit the best for this project, has resulted in the accomplishment of this book. They are a veteran in the field of academics and their pool of knowledge is as vast as their experience in printing. Their expertise and guidance has proved useful at every step. Their uncompromising quality standards have made this book an exceptional effort. Their encouragement from time to time has been an inspiration for everyone.

The publisher and the editorial board hope that this book will prove to be a valuable piece of knowledge for researchers, students, practitioners and scholars across the globe.

List of Contributors

Markus Schmid
Fraunhofer-Institute for Process Engineering and Packaging IVV, Giggenhauser Strasse 35, Freising 85354, Germany
Chair for Food Packaging Technology, Technische Universität München, Weihenstephaner Steig 22, Freising 85354, Germany

Martina Lindner
Fraunhofer-Institute for Process Engineering and Packaging IVV, Giggenhauser Strasse 35, Freising 85354, Germany

Arantzazu Valdés, Marina Ramos, Ana Beltrán, Alfonso Jiménez and María Carmen Garrigós
Analytical Chemistry, Nutrition & Food Sciences Department, University of Alicante, 03690 San Vicente del Raspeig (Alicante), Spain

Chunlian Hu
Alloy Powder Co., Ltd., Lanzhou University of Technology, Lanzhou 730050, China

Shanglin Hou
School of Science, Lanzhou University of Technology, Lanzhou 730050, China

Mehran Habibi, Amin Rahimzadeh, Inas Bennouna and Morteza Eslamian
University of Michigan-Shanghai Jiao Tong University Joint Institute, Shanghai 200240, China

Liam Ward
School of Engineering, RMIT University, Melbourne 3001, Australia

Antony Pilkington
School of Science, RMIT University, Melbourne 3001, Australia;
Defence Materials Technology Centre, Melbourne 3122, Australia

Steve Dowey
Defence Materials Technology Centre, Melbourne 3122, Australia
Sutton Tools Pty Ltd., Melbourne 3074, Australia

Chi Zhang, Le Gu, Chongyang Nie, Chuanwei Zhang and Liqin Wang
Research Lab of Space & Aerospace Tribology, Harbin Insititute of Technology, Harbin 150001, China

Qingfan Liu and Susan Tighe
Centre for Pavement and Transportation Technology, University of Waterloo, Waterloo, ON N2L 3G1, Canada

Sina Varamini
McAsphalt Industries Limited, Toronto, ON M1B 5R4, Canada

Kyungmok Kim
School of Aerospace and Mechanical Engineering, Korea Aerospace University, 76 Hanggongdaehang-ro, Deogyang-gu, Goyang-si, Gyeonggi-do 412-791, Korea

Giuseppe Loprencipe and Pablo Zoccali
Department of Civil, Constructional and Environmental Engineering, Sapienza University of Rome, Via Eudossiana 18, 00184 Rome, Italy

María Laura Vera, Mario Roberto Rosenberger, Carlos Enrique Schvezov and Alicia Esther Ares
Instituto de Materiales de Misiones (IMAM), CONICET-UNaM, Posadas 3300, Misiones, Argentina
Facultad de Ciencias Exactas, Químicas y Naturales (FCEQyN), UNaM, Posadas 3300, Misiones, Argentina

Ángeles Colaccio
Facultad de Ciencias Exactas, Químicas y Naturales (FCEQyN), UNaM, Posadas 3300, Misiones, Argentina

Imyhamy M. Dharmadasa, Mohammad L. Madugu and Olajide I. Olusola
Materials & Engineering Research Institute, Faculty of Arts, Computing, Engineering and Sciences, Sheffield Hallam University, Sheffield S1 1WB, UK

Obi K. Echendu
Physics Department, Federal University of Technology Owerri, Ihiagwa PMB 1526, Imo, Nigeria;

Fijay Fauzi
School of Electrical System Engineering, University of Malaysia Perlis, Pauh Putra Campus, 02600 Arau, Perlis, Malaysia

Dahiru G. Diso
Kano University of Science and Technology, Wudil PMB 3244, Kano, Nigeria;

Ajith R. Weerasinghe
California State University, Fresno 2320 E. San Ramon Ave., Fresno, CA 93740, USA;

Thad Druffel, Ruvini Dharmadasa, Brandon Lavery, Jacek B. Jasinski, Tatiana A. Krentsel and Gamini Sumanasekera
Conn Center for Renewable Energy Research, University of Louisville, Louisville, KY 40292, USA

Steffen Drache, Harm Wulff, Christiane A. Helm and Rainer Hippler
Institute of Physics, University of Greifswald, Felix Hausdorff Str. 6, Greifswald 17489, Germany

Arun Kumar Mukhopadhyay
Indian Institute of Engineering Science and Technology, Shibpur, Howrah-3,West Bengal 711103, India

Satyaranjan Bhattacharyya
Surface Physics & Materials Science Division, Saha Institute of Nuclear Physics, 1/AF Bidhan Nagar, Kolkata 700 064, India

Abhijit Majumdar
Institute of Physics, University of Greifswald, Felix Hausdorff Str. 6, Greifswald 17489, Germany

Indian Institute of Engineering Science and Technology, Shibpur, Howrah-3,West Bengal 711103, India

Hsian Sagr Hadi A and Yasutaka Ando
Ashikaga Institute of Technology, 268-1 Omae, Ashikaga, Tochigi, 326-8558, Japan

Dimitra Vernardou
Center of Materials Technology and Photonics, School of Applied Technology, Technological Educational Institute of Crete, 710 04 Heraklion, Crete, Greece

Index

www.ingramcontent.com/pod-product-compliance
Lightning Source LLC
Chambersburg PA
CBHW080641200326
41458CB00013B/4696